The Importance of Fish Phenotype in Aquaculture, Fisheries and Conservation

The Importance of Fish Phenotype in Aquaculture, Fisheries and Conservation

Topic Editors

Zonghang Zhang
Xiumei Zhang

MDPI
Basel • Beijing • Wuhan • Barcelona • Belgrade • Novi Sad • Cluj • Manchester

Topic Editors
Zonghang Zhang
College of Science
Shantou University
Shantou
China

Xiumei Zhang
Fisheries College
Zhejiang Ocean University
Zhoushan
China

Editorial Office
MDPI AG
Grosspeteranlage 5
4052 Basel, Switzerland

This is a reprint of the Topic, published open access by the journals *Animals* (ISSN 2076-2615), *Biology* (ISSN 2079-7737) and *Fishes* (ISSN 2410-3888), freely accessible at: https://www.mdpi.com/topics/A0VO3HK165.

For citation purposes, cite each article independently as indicated on the article page online and as indicated below:

Lastname, A.A.; Lastname, B.B. Article Title. *Journal Name* **Year**, *Volume Number*, Page Range.

ISBN 978-3-7258-5673-2 (Hbk)
ISBN 978-3-7258-5674-9 (PDF)
https://doi.org/10.3390/books978-3-7258-5674-9

Contents

biology

MDPI

Article

Effects of Stocking Density on the Survival, Growth, and Stress Levels of the Juvenile Lined Seahorse (*Hippocampus erectus*) in Recirculating Aquaculture Systems

Tingting Lin [1,2], Siping Li [1,2], Dong Zhang [1,2], Xin Liu [1,2,*] and Yuanhao Ren [1]

1 Key Laboratory of Inland Saline-Alkaline Aquaculture, Ministry of Agriculture and Rural Affairs, East China Sea Fisheries Research Institute, Chinese Academy of Fishery Sciences, Shanghai 200090, China; lintt@ecsf.ac.cn (T.L.); lisiping@ecsf.ac.cn (S.L.); zhangdong@ecsf.ac.cn (D.Z.); renyh@ecsf.ac.cn (Y.R.)
2 Wenchang Innovation Research Center, Fengjiawan Modern Fishery Industry Park, Wenchang 571300, China
* Correspondence: liux@ecsf.ac.cn

Simple Summary: Recirculating aquaculture systems (RASs) have gradually become a highly recommended aquaculture mode due to their environmental friendliness. However, due to their high factory and equipment investments and high electricity consumption, they usually require the farming objects to have high economic value in order to cover their high running costs. Seahorses have high economic value and may be suitable for RASs. Currently, there are almost no reports on the large-scale farming of seahorses using RASs. To provide a new farming object for RASs, in the present study, the lined seahorse (*Hippocampus erectus*) was employed to determine a suitable stocking density in RASs. As is well known, optimizing the stocking density of farmed objects is a prerequisite for their commercial aquaculture. The present study determined suitable stocking densities for lined seahorse juveniles at different growth stages, which will contribute to the development of recirculating aquaculture for the lined seahorse.

Abstract: Seahorses are increasingly regarded as a promising farming object suitable for recirculating aquaculture systems (RASs) due to their high economic value. However, reports on the large-scale farming of seahorses in RASs are rare, and some key parameters, such as stocking densities, are still unclear. In the present study, we employed the lined seahorse (*Hippocampus erectus*), for which large-scale farming has been achieved, to determine the suitable stocking density for three different-sized juveniles in RASs. The three different-sized juveniles had body heights of 4.0, 7.0, and 9.0 cm, and their test density gradients were 1.0, 0.8, 0.6, and 0.4 inds/L; 0.6, 0.5, 0.4, and 0.3 inds/L; and 0.4, 0.3, 0.2, and 0.1 inds/L, respectively. The juveniles were cultivated for one month, and then their survival, growth, and plasma cortisol and brain serotonin contents (two stress-related indicators) were analyzed. The results show that, regardless of the size of the juveniles, a high density can inhibit growth and trigger stress responses. In addition, for small- (4.0 cm) and medium-sized (7.0 cm) juveniles, a high density can also exacerbate size heterogeneity and cause death. Taking into account the welfare and yield of farmed seahorses, the present study suggests that the suitable stocking densities for 4.0, 7.0, and 9.0 cm juveniles in RASs are 0.6, 0.4, and 0.2 inds/L, respectively.

Keywords: stocking density; recirculating aquaculture system; seahorse *Hippocampus erectus*; size heterogeneity

1. Introduction

Aquatic products have become the third-largest source of protein in the human diet after cereals and milk [1]. Due to the rapid decline in wild stock resources caused by overfishing and habitat degradation, aquaculture has replaced wild capture as the main means of supplying aquatic products [2]. There are generally two modes of aquaculture: the traditional extensive mode (e.g., outdoor pond farming) and the modern intensive mode (e.g., indoor factory farming) [3]. The latter can be subdivided into flow-through systems (FTSs) and recirculating aquaculture systems (RASs). FTSs usually pose a high risk of environmental pollution because their wastewater is directly discharged into the environment without treatment, while RASs are environmentally friendly and sustainable because their wastewater is treated and reused with almost zero discharge [4–6]. In the past decade, RASs have become the mainstream mode of aquaculture [3]. Although RASs are widely recommended, they also have limitations, with the most prominent being high factory and equipment investments and high running costs (mainly electricity) [7]. Therefore, this mode is selective in terms of the farming objects, generally requiring them to have high economic value. It is estimated that farming enterprises are likely to make profits only when the market price of the farming objects exceeds USD ten per kilogram [8,9]. Thus far, there are limited objects suitable for RASs, mainly comprising flatfish, salmon, and groupers [7].

The seahorse (*Hippocampus* genus, Syngnathidae family) is a new farming object that has only emerged in the past two decades [10,11]. It has high economic value and is mainly traded as traditional Chinese medicine, ornamental fish, and curios. A brightly colored live specimen is priced at over USD 60, and a dried specimen weighing two grams is sold for over USD 3 [12]. They also have a high market demand: over 37 million specimens are traded every year [13]. Therefore, seahorses may be a promising object suitable for RASs. However, to date, there are almost no reports on the large-scale farming of seahorses using RASs, and some key parameters, such as stocking density, are still unclear.

Stocking density is an important parameter in aquaculture [14]. If the density is too low, it will cause a waste of farming space and result in a low farming yield, while if the density is too high, it may have an impact on the growth and survival of the farmed organisms, which, in turn, may affect the farming yield [15]. In farming production, it is common for the farming yield to decrease rather than increase with an increase in the stocking density. The main reason for this decrease in yield is that high-density farming often produces a crowding effect, which has a series of adverse effects on farmed organisms, such as causing anorexia, stunted growth, inactive behavior, and sluggish reactions and even triggering various diseases [16–18]. Therefore, optimizing the stocking density of different farmed organisms based on their biological characteristics is a prerequisite for their commercial aquaculture.

The lined seahorse (*Hippocampus erectus*) is a seahorse species for which successful commercial aquaculture has been achieved. There is a large lined seahorse farming area in China, and its existing aquaculture model is an FTS [19,20]. In recent years, with the Chinese government's control over the discharge standards and amounts of aquaculture wastewater, as well as farmers' increasing awareness that the economic benefits of seahorse aquaculture are sufficient to cover the high costs of RASs, some farming enterprises have begun to attempt to cultivate lined seahorses in RASs. However, due to the unsuitable stocking density, their farming profits are not ideal. In the present study, we designed various density gradients in RASs and evaluated the effect on the cultivation of lined seahorse juveniles by analyzing their survival, growth, size homogeneity, and stress response. We employed plasma cortisol and brain serotonin (5-hydroxytryptamine, 5-HT) as indicators of stress levels; this is because, in teleosts, plasma cortisol has been proven to be a reliable indicator for indicating the stress response [21,22], and brain 5-HT has been reported to play a crucial role in behavioral and emotional control, as well as in neuroendocrine responses to stress [23,24]. Sustained serotonergic activation usually indicates that the fish is in chronic stress or in a stress-induced pathology (e.g., a depression-like state) [25–27]. The aim of the present study was to determine the optimal stocking density for lined seahorse juveniles at different growth stages in RASs.

2. Materials and Methods

2.1. Ethical Approval

Ethical approval for this study was granted by the Committee on the Ethics of Animal Experiments of the Chinese Academy of Fishery Sciences, with approval reference number CAFS-EC-2023-086. All procedures were conducted in compliance with the recommendations in the Guide for the Management and Use of the Experimental Animals of China Science and Technology Commission.

2.2. Experimental Seahorses

At present, in China, the general operations when using an RAS to cultivate the lined seahorse are as follows: The farming enterprise first purchases one-month-old juveniles with a body height (the straight-line distance from the tip of the head coronet to the tip of the uncurled tail) of about 4.0 cm from a hatchery, and then it cultivates them in an RAS for three months (at this point, the body height of the juveniles reaches about 11.0 cm) before they are put on the market. During these three months, as the juveniles gradually grow, the farming enterprise dilutes the stocking density after each month of cultivation. Therefore, using one month of cultivation as the time point, the present study focused on three different sizes of juveniles, namely, one-month-old juveniles (just entered the system), two-month-old juveniles (cultivated in the system for one month, with a body height of about 7.0 cm), and three-month-old juveniles (cultivated in the system for two months, with a body height of about 9.0 cm), to investigate their optimal stocking densities.

The present study was carried out at Wenchang Innovation Research Center, Fengjiawan Modern Fishery Industry Park, Hainan Province, China, from September 2023 to February 2024. The center houses hundreds of thousands of artificially bred lined seahorses of various sizes and several sets of standardized RASs. The experimental seahorses (i.e., the lined seahorse juveniles with initial body heights of 4.0, 7.0, and 9.0 cm) and experimental systems (i.e., RAS) used in the present study were provided for free by the center.

2.3. Experimental System

The RAS consisted of a water treatment unit and a rearing unit. The water treatment unit included a microscreen filter, two sand filtration tanks filled with quartz sand with a particle diameter of 4–8 mm, a biological fluidized bed reactor, a protein skimmer, a temperature control device, and a UV sterilizer (Figure 1). The rearing unit consisted of 24 rectangular fiberglass tanks. The length, width, and height of each tank were 140 cm, 90 cm, and 90 cm, respectively, with a water capacity of 1000 L. Each tank was equipped with a water inlet and outlet positioned at the water surface. Each tank contained several plastic mesh slices for the seahorses to attach to. The tailwater from each tank flowed into the same large pipe (i.e., the main drainage pipe), then flowed back to the water treatment unit, passed through the water treatment components in sequence, and finally flowed into another large pipe (the main inlet pipe) for distribution to each tank.

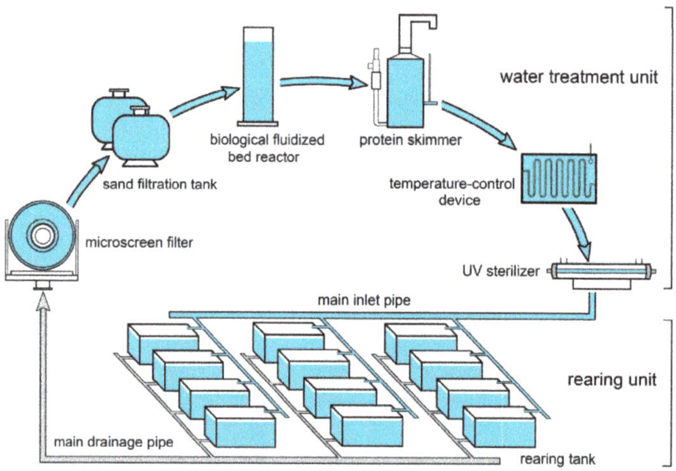

Figure 1. Frame of the recirculating aquaculture system used in the present study.

2.4. Water and Rearing Conditions

The temperature of the rearing water was 25–26 °C, and it was controlled by the temperature control system. The salinity, pH, and dissolved oxygen of the rearing water were 28–29‰, 8.0–8.2, and 6.5–8.0 mg/L, respectively. The ammonia and nitrite were 0.002–0.003 mg/L and 0.01–0.10 mg/L, respectively. The light intensity was 1500–2000 lux, and the photoperiod was 12 h L/12 h D. The inlet water flow rate of each tank was 4.0–5.0 L/min.

2.5. Stocking Density

The test density gradients of the three experimental seahorses, i.e., the 4.0, 7.0, and 9.0 cm juveniles, were 1.0, 0.8, 0.6, and 0.4 inds/L; 0.6, 0.5, 0.4, and 0.3 inds/L; and 0.4, 0.3, 0.2, and 0.1 inds/L, respectively. Each gradient had four replicates (Figure 2).

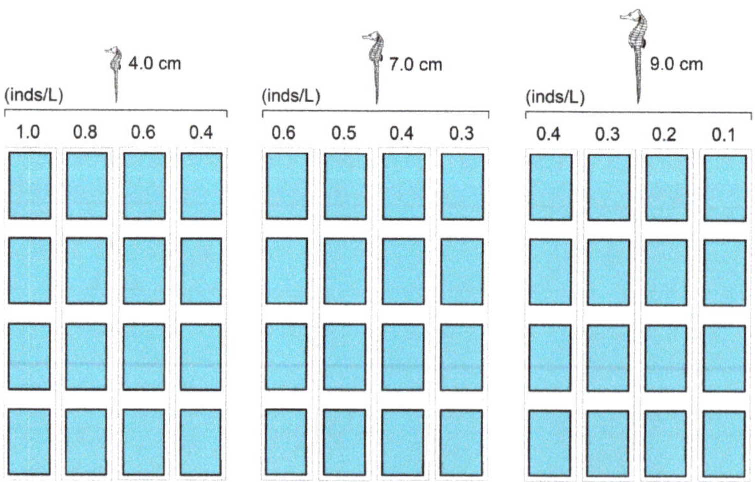

Figure 2. Panoramic view of the experimental design with test stocking densities.

The juveniles with an initial body height of 4.0 cm (i.e., the one-month-old juveniles; the actual measurement values of the body height and wet weight were 4.02 ± 0.14 cm and 0.210 ± 0.031 g) were placed in the RAS rearing tanks according to their corresponding densities. They were fed twice a day (08:00 and 14:00) ad libitum with live copepods for 30 days. The copepods were collected daily in the early morning from outdoor shrimp farming ponds and disinfected with 5 ppm povidone-iodine for one hour before feeding. To prevent the copepods from escaping from the water outlet, a lid made of a plastic fence mesh, which was wrapped in a 120-mesh screen (mesh size of 125 μm), was used to cover the water outlet. Three hours after each feeding, the feces and dead copepods in the tanks were siphoned out, and the lids were brushed clean. Dead seahorses, if any, were picked up and recorded during daily inspections.

The juveniles with initial body heights of 7.0 cm (i.e., the two-month-old juveniles; the actual measurement values of the body height and wet weight were 6.98 ± 0.14 cm and 1.153 ± 0.099 g) and 9.0 cm (i.e., the three-month-old juveniles; the actual measurement values of the body height and wet weight were 8.92 ± 0.14 cm and 3.030 ± 0.175 g) were placed in the RAS rearing tanks according to their corresponding densities. They were fed twice a day (08:00 and 15:00) ad libitum with frozen Mysis for 30 days. The reason why the juveniles' feed was changed from copepods to Mysis is that one-month-old juveniles cannot swallow the entire Mysis due to their small snout, so they prefer copepods during this growth stage; in contrast, two-month-old juveniles are able to swallow the entire Mysis, so from this stage of growth, they prefer Mysis. Frozen Mysis was purchased from Hikari™, Kyorin Industry (Shenzhen) Co., Ltd., Guangzhou, China. It was first thawed, then rinsed with clean water, drained, and finally fed to the juveniles. To allow the shrimp fragments on the water surface to flow out as much as possible from the water outlet, the screen used to wrap the outlet lid was changed to a 10-mesh screen (mesh size of 2 mm). Two hours after each feeding, the feces and uneaten shrimp were siphoned out, and the lids were brushed clean. Dead seahorses, if any, were picked up and recorded during daily inspections.

2.6. Survival and Growth Measurements

After 30 days of feeding, the total number of dead seahorses (N_d) in each replicate of each stocking density was counted. The survival rate (SR) was calculated as

$$SR = 100\ (N_0 - N_d)/N_0, \quad (1)$$

where N_0 is the initial number of live seahorses.

After 30 days of feeding, 40 seahorses were randomly sampled from each replicate of each stocking density to measure their final body height (H_f) and final wet weight (W_f) for the calculation of the specific growth rate (SGR) and Fulton's condition factor (K).

$$SGR = 100\ (\ln W_f - \ln W_i)/\text{time}, \quad (2)$$

$$K = 100\ (W_f/H_f^3), \quad (3)$$

where W_i is the initial wet weight.

The seahorse body height, referring to the straight-line distance from the tip of the head coronet to the tip of the uncurled tail, was measured with a steel ruler with a precision of 0.1 cm (i.e., 1 mm), and the seahorse wet weight was measured using a Mettler Toledo balance with a precision of 0.001 g (i.e., 1 mg).

2.7. Size Heterogeneity Analysis

In production practice, it is often found that juvenile lined seahorses of the same age in the same rearing tank tend to have uneven sizes, with some of them being particularly large and some of them being particularly small. We speculate that this size heterogeneity may be related to the stocking density. To test this speculation, we employed the coefficient of variation (CV) to analyze the size heterogeneity of the juveniles at different densities.

$$CV = 100\ [\text{standard deviation (SD)}/\text{mean}]. \quad (4)$$

2.8. Stress Level Determinations

After 30 days of feeding, another six seahorses were randomly sampled from each replicate of each stocking density and placed one by one in seawater containing 0.02% MS-222 for anesthesia for three minutes. Blood was collected using the tail-cutting method. Specifically, a section of the tail of the anesthetized seahorse was cut off; then, the remaining tail was immediately inserted into a 1.5 mL centrifuge tube containing 0.6 mL of anticoagulant (citric acid 0.48 g, sodium citrate 1.32 g, glucose 1.47 g, and distilled water 100 mL) and dipped into the anticoagulant. Blood spontaneously flowed out of the caudal artery and mixed with the anticoagulant. Due to the small volume of blood in each seahorse, the blood from six seahorses was pooled in the same centrifuge tube. The mixture of blood and anticoagulant was left to stand for 10 min at 4 °C and then centrifuged at 840× *g* for 10 min at 4 °C (Sigma 3K18, Sigma, Kanagawa, Japan) to collect the supernatant (i.e., plasma) [28]. The plasma was stored at −80 °C for a later analysis of cortisol.

After blood collection, the head coronet of the amputated seahorse was cut open, and the brain tissue was quickly collected within 30 s. Due to the small amount of brain tissue in each seahorse, the brain tissues from six seahorses were mixed in the same sampling tube. The brain tissues were homogenized in ice-cold phosphate-buffered saline (PBS, sodium chloride 0.8 g, potassium chloride 0.02 g, disodium hydrogen phosphate 0.29 g, potassium dihydrogen phosphate 0.02 g, and distilled water 100 mL) at pH 7.2 using a Potter–Elvehjem homogenizer and then centrifuged at 840× *g* for 10 min at 4 °C to collect the supernatant [27]. The supernatant was stored at −80 °C for a later analysis of 5-HT.

Cortisol and 5-HT were measured with commercial ELISA kits (MEIMIAN, Jiangsu Meimian Industrial Co., Ltd., Jiangsu, China) suitable for fish following the manufacturer's instructions. The protein concentrations in the plasma and brain tissue supernatant were measured using Bradford's method [29]. The unit of cortisol is expressed as the amount contained in each milligram of plasma protein, and the unit of 5-HT is expressed as the amount contained in each milligram of brain tissue protein.

2.9. Statistical Analysis

All data are expressed as the mean ± standard deviation (SD) and were analyzed using Origin PRO graphing and analysis software (Version 9.2.214). The normality of all data was evaluated using the Shapiro–Wilk test [30], and the homogeneity of the variances was assessed using Levene's test. If the variance was homogeneous, a one-way analysis of variance (ANOVA) was used to analyze the differences among the four stocking densities; if the variance was not homogeneous, the Kruskal–Wallis *H* test (a non-parametric test) [30] was used to analyze the differences among the four stocking densities. If the one-way ANOVA result showed a significant difference ($p < 0.05$) among the four stocking densities, then Tukey's multiple comparison was applied, and if the Kruskal–Wallis *H* test result showed a significant difference, then Dunn's multiple comparison was applied.

3. Results

3.1. Juveniles with a Body Height of 4.0 cm

3.1.1. Survival and Growth

The survival and growth of the juveniles with an initial body height of 4.0 cm at the end of the trial are shown in Table 1. The survival rate results showed significant differences among the four density groups ($F_{3, 15} = 114.21$, $p < 0.001$). Tukey's multiple comparisons showed that there was no significant difference among the 0.8, 0.6, and 0.4 inds/L groups, and their values were 90.88 ± 2.06%, 92.79 ± 1.14%, and 94.00 ± 1.32%, respectively, while the value of the 1.0 inds/L group was only 71.18 ± 2.99%, which was significantly lower than that of the other three groups.

Table 1. The survival rate and growth parameters (mean ± SD) of the juveniles with an initial body height of 4.0 cm after one month of rearing.

	Stocking Density (inds/L)				*p*-Value
	1.0	0.8	0.6	0.4	
Survival rate (%)	71.18 ± 2.99 [b]	90.88 ± 2.06 [a]	92.79 ± 1.14 [a]	94.00 ± 1.32 [a]	<0.001
Final BH (cm)	6.03 ± 0.89 [b]	6.16 ± 0.79 [b]	6.87 ± 0.70 [a]	6.95 ± 0.65 [a]	<0.001
Final WW (g)	0.773 ± 0.325 [b]	0.812 ± 0.289 [b]	1.107 ± 0.304 [a]	1.166 ± 0.286 [a]	<0.001
SGR	3.99 ± 1.61 [b]	4.24 ± 1.44 [b]	5.36 ± 1.24 [a]	5.56 ± 1.11 [a]	<0.001
K	0.329 ± 0.025 [b]	0.330 ± 0.022 [b]	0.331 ± 0.024 [b]	0.338 ± 0.022 [a]	0.002
CV of final BH	14.84 ± 0.70 [a]	12.94 ± 0.91 [b]	9.75 ± 0.99 [c]	9.38 ± 0.85 [c]	<0.001
CV of final WW	42.24 ± 3.21 [a]	35.83 ± 2.72 [b]	26.25 ± 3.13 [c]	24.57 ± 2.91 [c]	<0.001

BH: body height; WW: wet weight; SGR: specific growth rate; *K*: Fulton's condition factor; CV: coefficient of variation. The superscript lowercase letters (a, b, and c) after each value in the same row indicate the results of multiple comparisons, and the absence of shared letters between two values indicates significant differences between them.

The results of the final body height ($H_3 = 197.04$, $p < 0.001$), final wet weight ($H_3 = 190.00$, $p < 0.001$), and SGR ($H_3 = 190.00$, $p < 0.001$) also showed significant differences among the four density groups. Dunn's multiple comparisons showed that the low-density groups (i.e., 0.6 and 0.4 inds/L) had better growth performance, and their final body height, final wet weight, and SGR were 1.13-, 1.43-, and 1.33-fold higher than those in the high-density groups (i.e., 1.0 inds/L and 0.8 inds/L), respectively. The *K* (Fulton's condition factor) results also showed significant differences among the four density groups ($F_{3, 639} = 4.86$, $p = 0.002$). Tukey's multiple comparisons showed that the 0.4 inds/L group had a significantly higher value, and the other three groups showed no significant difference.

3.1.2. Size Heterogeneity

The CVs of the final body height ($F_{3, 15} = 36.46$, $p < 0.001$) and final wet weight ($F_{3, 15} = 30.82$, $p < 0.001$) showed significant differences among the four density groups. Tukey's multiple comparisons showed that the variation was the highest in the 1.0 inds/L group, followed by the 0.8 inds/L group, and it was the lowest in the 0.6 and 0.4 inds/L groups. The final body height and final wet weight CVs of the seahorses reared at 1.0 inds/L were 1.55- and 1.66-fold higher than those of the seahorses reared at 0.4 and 0.6 inds/L, respectively (Table 1). The final body height and final wet weight of each seahorse were examined in scatter plots. The scatter plots of the high-density groups (i.e., 1.0 and 0.8 inds/L) were more scattered and could be roughly divided into two size populations (green and red circles), while the scatter plots of the low-density groups (i.e., 0.6 and 0.4 inds/L) were more concentrated and contained one size population (green circle) (Figure 3).

3.1.3. Stress Level

The cortisol results showed significant differences among the four density groups ($F_{3, 15} = 24.20$, $p < 0.001$). Tukey's multiple comparisons showed that the 1.0 inds/L group had the highest level of cortisol, which was significantly higher than that of the other three groups. The lowest level of cortisol was present in the low-density groups (i.e., 0.6 and 0.4 inds/L), whose value was 36.09% lower than that of the 1.0 inds/L group (Figure 4A). Similarly, the 5-HT results also showed significant differences among the four density groups ($F_{3, 15} = 8.73$, $p = 0.002$). Tukey's multiple comparisons showed that the 1.0 inds/L group had the highest level, while the 0.6 and 0.4 inds/L groups had the lowest level, and the highest level was 1.10 times higher than the lowest level (Figure 4B).

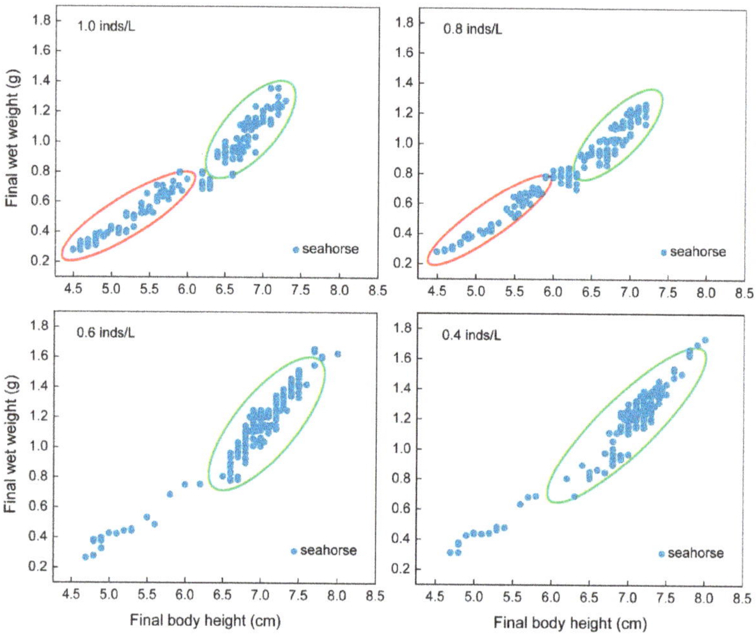

Figure 3. The final body height and wet weight of the juveniles with an initial body height of 4.0 cm (real body height: 4.02 ± 0.14 cm; real wet weight: 0.210 ± 0.031 g) after one month of rearing. Each solid blue dot represents a seahorse juvenile. The red and green circles represent populations of different body sizes.

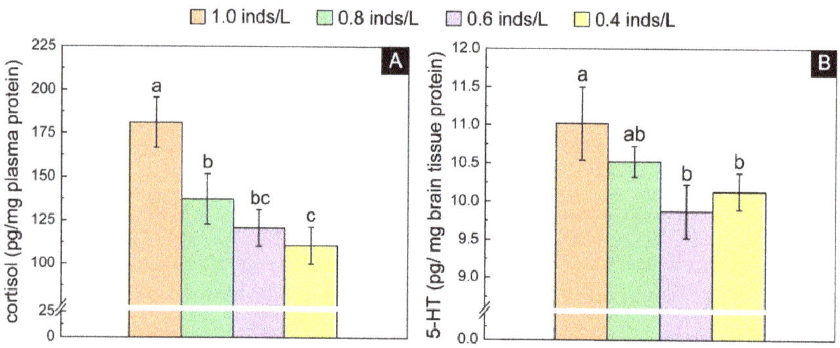

Figure 4. The plasma cortisol (**A**) and brain 5-HT (**B**) contents in the juveniles with an initial body height of 4.0 cm (real body height: 4.02 ± 0.14 cm; real wet weight: 0.210 ± 0.031 g) after one month of rearing. The lowercase letters (a, b, and c) on each density bar chart indicate the results of multiple comparisons, and the absence of shared letters between two densities indicates significant differences between them.

3.2. *Juveniles with a Body Height of 7.0 cm*

3.2.1. Survival and Growth

The survival and growth of the juveniles with an initial body height of 7.0 cm at the end of the trial are shown in Table 2. The survival rate results showed significant differences among the four density groups ($F_{3, 15}$ = 16.46, p < 0.001). Tukey's multiple comparisons showed that there was no significant difference among the 0.5, 0.4, and 0.3 inds/L groups, and their values were 83.40 ± 2.57%, 84.63 ± 2.10%, and 87.83 ± 2.53%, respectively, while the value of the 0.6 inds/L group was only 76.38 ± 2.30%, which was significantly lower than that of the other three groups.

Table 2. The survival rate and growth parameters (mean ± SD) of the juveniles with an initial body height of 7.0 cm after one month of rearing.

	Stocking Density (inds/L)				*p*-Value
	0.6	0.5	0.4	0.3	
Survival rate (%)	76.38 ± 2.30 [b]	83.40 ± 2.57 [a]	84.63 ± 2.10 [a]	87.83 ± 2.53 [a]	<0.001
Final BH (cm)	8.49 ± 0.60 [b]	8.69 ± 0.48 [b]	8.98 ± 0.50 [a]	9.04 ± 0.55 [a]	<0.001
Final WW (g)	2.409 ± 0.679 [c]	2.681 ± 0.569 [b]	3.041 ± 0.575 [a]	3.089 ± 0.624 [a]	<0.001
SGR	2.31 ± 1.00 [c]	2.73 ± 0.78 [b]	3.16 ± 0.75 [a]	3.20 ± 0.81 [a]	<0.001
K	0.382 ± 0.038 [c]	0.401 ± 0.032 [b]	0.414 ± 0.028 [a]	0.410 ± 0.027 [a]	<0.001
CV of final BH	7.08 ± 0.30 [a]	5.50 ± 0.50 [b]	5.52 ± 0.48 [b]	6.00 ± 0.30 [b]	<0.001
CV of final WW	28.36 ± 1.36 [a]	21.22 ± 1.88 [b]	18.81 ± 1.13 [b]	20.11 ± 1.10 [b]	<0.001

BH: body height; WW: wet weight; SGR: specific growth rate; *K*: Fulton's condition factor; CV: coefficient of variation. The superscript lowercase letters (a, b, and c) after each value in the same row indicate the results of multiple comparisons, and the absence of shared letters between two values indicates significant differences between them.

The final body height results showed significant differences among the four density groups ($H_3 = 104.38$, $p < 0.001$). Dunn's multiple comparisons showed that the 0.4 and 0.3 inds/L groups had significantly higher values, and there was no significant difference between the values of the 0.6 inds/L and 0.5 inds/L groups. The value in the low-density groups (i.e., 0.4 and 0.3 inds/L) was 1.05-fold higher than that in the high-density groups (i.e., 0.6 and 0.5 inds/L). The results of the final wet weight ($H_3 = 132.89$, $p < 0.001$), SGR ($H_3 = 132.89$, $p < 0.001$), and *K* ($H_3 = 78.72$, $p < 0.001$) all showed significant differences among the four density groups. Dunn's multiple comparisons showed that the low-density groups (i.e., 0.4 and 0.3 inds/L) had the best growth performance, followed by the 0.5 inds/L group, and the 0.6 inds/L group had the worst. The final wet weight, SGR, and *K* in the best groups were 1.27-, 1.38-, and 1.08-fold higher than those in the worst group, respectively.

3.2.2. Size Heterogeneity

The CVs of the final body height ($F_{3, 15} = 13.31$, $p < 0.001$) and final wet weight ($F_{3, 15} = 37.00$, $p < 0.001$) showed significant differences among the four density groups (Table 2). Tukey's multiple comparisons showed that the 0.6 inds/L group had the highest variations, which were significantly higher than those of the other three groups, while there was no significant difference among the other three groups. The final body height and final wet weight CVs of the 0.6 inds/L group were 1.25- and 1.41-fold higher than those of the other three groups, respectively. The final body height and final wet weight of each seahorse were examined in scatter plots. The scatter plot of the 0.6 inds/L group was more scattered and could be roughly divided into two size populations (green and red circles), while the scatter plots of the other three groups were more concentrated and contained one size population (green circle) (Figure 5).

3.2.3. Stress Level

The cortisol results showed significant differences among the four density groups ($F_{3, 15} = 9.89$, $p = 0.001$). Tukey's multiple comparisons showed that the 0.6 inds/L group had the highest value, which was significantly higher than that of the other three groups, whereas there was no significant difference among the other three groups (Figure 6A). The value of the 0.6 inds/L group was 1.30 times higher than the mean of the other three groups. Similarly, the 5-HT results also showed significant differences among the four density groups ($F_{3, 15} = 10.26$, $p = 0.001$). High values were found in the high-density groups (i.e.,

0.6 and 0.5 inds/L), and low values were found in the low-density groups (i.e., 0.4 and 0.3 inds/L), and the high values were 1.09 times higher than the low values (Figure 6B).

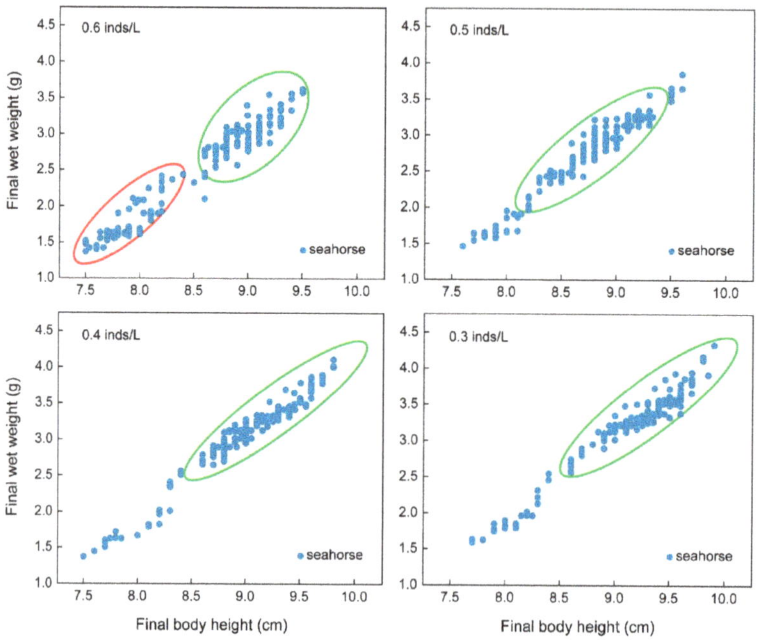

Figure 5. The final body height and wet weight of the juveniles with an initial body height of 7.0 cm (real body height: 6.98 ± 0.14 cm; real wet weight: 1.153 ± 0.099 g) after one month of rearing. Each solid blue dot represents a seahorse juvenile. The red and green circles represent populations of different body sizes.

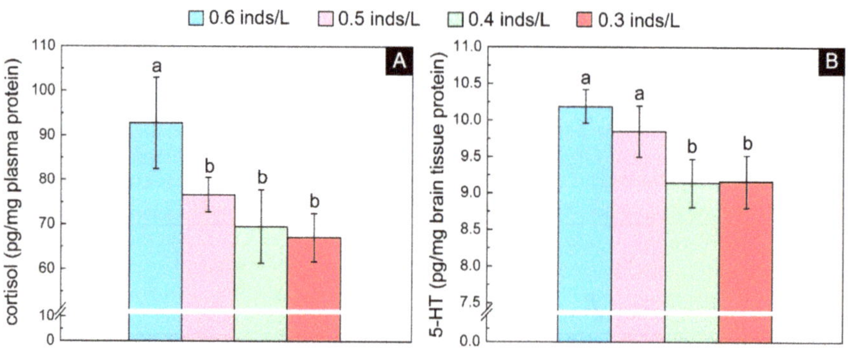

Figure 6. The plasma cortisol (**A**) and brain 5-HT (**B**) contents in the juveniles with an initial body height of 7.0 cm (real body height: 4.02 ± 0.14 cm; real wet weight: 0.210 ± 0.031 g) after one month of rearing. The lowercase letters (a and b) on each density bar chart indicate the results of multiple comparisons, and the absence of shared letters between two densities indicates significant differences between them.

3.3. *Juveniles with a Body Height of 9.0 cm*

3.3.1. Survival and Growth

The survival and growth of the juveniles with an initial body height of 9.0 cm at the end of the trial are shown in Table 3. The survival rate results showed no significant difference among the four density groups ($F_{3,15} = 1.26$, $p = 0.331$), and their mean value was 96.61 ± 1.13%.

Table 3. The survival rate and growth parameters (mean ± SD) of the juveniles with an initial body height of 9.0 cm after one month of rearing.

	Stocking Density (inds/L)				*p*-Value
	0.4	0.3	0.2	0.1	
Survival rate (%)	95.94 ± 0.85	96.25 ± 0.83	97.25 ± 1.19	97.00 ± 1.41	0.331
Final BH (cm)	10.53 ± 0.47 ^{c}	10.71 ± 0.50 ^{b}	10.87 ± 0.49 ^{a}	10.93 ± 0.49 ^{a}	<0.001
Final WW (g)	5.352 ± 0.792 ^{c}	5.713 ± 0.941 ^{b}	6.089 ± 1.014 ^{a}	6.207 ± 1.035 ^{a}	<0.001
SGR	1.86 ± 0.49 ^{c}	2.07 ± 0.56 ^{b}	2.27 ± 0.61 ^{a}	2.34 ± 0.62 ^{a}	<0.001
K	0.455 ± 0.021 ^{b}	0.460 ± 0.019 ^{b}	0.468 ± 0.023 ^{a}	0.470 ± 0.024 ^{a}	<0.001
CV of final BH	4.45 ± 0.42	4.66 ± 0.35	4.57 ± 0.31	4.53 ± 0.46	0.888
CV of final WW	14.80 ± 0.62	16.53 ± 1.42	16.66 ± 1.43	16.72 ± 2.02	0.243

BH: body height; WW: wet weight; SGR: specific growth rate; *K*: Fulton's condition factor; CV: coefficient of variation. The superscript lowercase letters (a, b, and c) after each value in the same row indicate the results of multiple comparisons, and the absence of shared letters between two values indicates significant differences between them.

The results of the final body height ($F_{3,639} = 22.08$, $p < 0.001$), final wet weight ($H_3 = 81.16$, $p < 0.001$), and SGR ($F_{3,639} = 22.76$, $p < 0.001$) showed significant differences among the four density groups. Multiple comparisons showed that the low-density groups (i.e., 0.1 and 0.2 inds/L) had the best growth performance, followed by the 0.3 inds/L group, and the 0.4 inds/L group had the worst. The final body height, final wet weight, and SGR in the best groups were 1.04-, 1.15-, and 1.24-fold higher than those in the worst group, respectively. The *K* results also showed significant differences among the four density groups ($F_{3,639} = 15.20$, $p < 0.001$). Tukey's multiple comparisons showed that the 0.1 and 0.2 inds/L groups had significantly higher values, and there was no significant difference between the values of the 0.4 inds/L and 0.3 inds/L groups.

3.3.2. Size Heterogeneity

The CVs of the final body height ($F_{3,15} = 0.21$, $p = 0.888$) and final wet weight ($F_{3,15} = 1.59$, $p = 0.243$) showed no significant differences among the four density groups (Table 3). The final body height and final wet weight of each seahorse were examined in scatter plots. The scatter plots of the four density groups all roughly contained one size population (green circle) (Figure 7).

3.3.3. Stress Level

The cortisol results showed significant differences among the four density groups ($F_{3,15} = 17.92$, $p < 0.001$). Tukey's multiple comparisons showed that the 0.4 inds/L group had a significantly higher value, and the other three groups showed no significant difference. The value of the 0.4 inds/L group was 1.62 times higher than the mean of the other three groups (Figure 8A). Similarly, the 5-HT results also showed significant differences among the four density groups ($F_{3,15} = 8.19$, $p = 0.003$). Tukey's multiple comparisons showed that the 0.4 inds/L group had the highest value, followed by the 0.3 inds/L group, and the 0.2 and 0.1 inds/L groups had the lowest value; furthermore, the highest value was 1.18 times higher than the lowest value (Figure 8B).

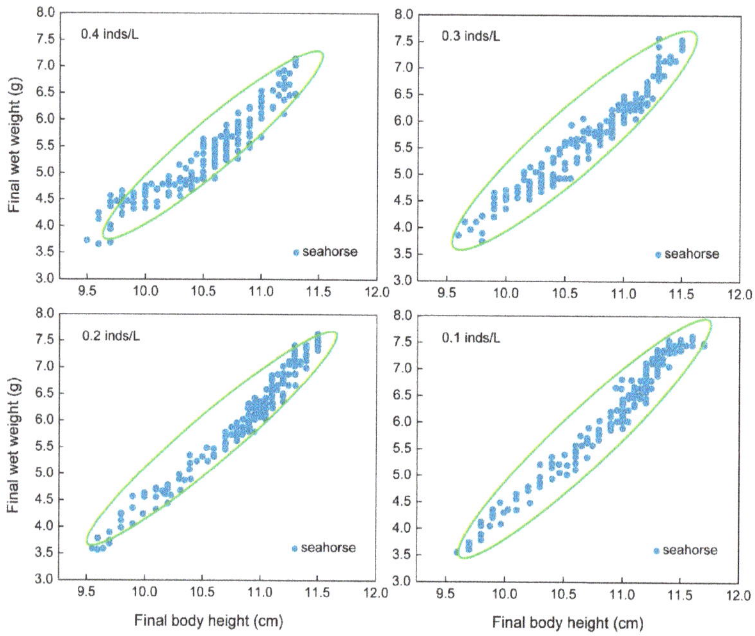

Figure 7. The final body height and wet weight of the juveniles with an initial body height of 9.0 cm (real body height: 8.92 ± 0.14 cm; real wet weight: 3.030 ± 0.175 g) after one month of rearing. Each solid blue dot represents a seahorse juvenile. The green circle represents a population of one body size.

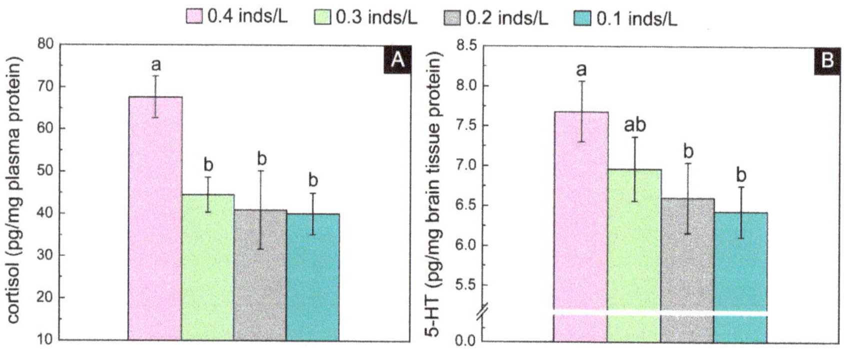

Figure 8. The plasma cortisol (**A**) and brain 5-HT (**B**) contents in the juveniles with an initial body height of 9.0 cm (real body height: 8.92 ± 0.14 cm; real wet weight: 3.030 ± 0.175 g) after one month of rearing. The lowercase letters (a and b) on each density bar chart indicate the results of multiple comparisons, and the absence of shared letters between two densities indicates significant differences between them.

4. Discussion

The present study demonstrated that the stocking density has a significant impact on the survival, growth, size homogeneity, and physiological stress of the juvenile lined seahorse (*H. erectus*). Specifically, a high density typically leads to a reduced survival rate, slowed growth, and an increased stress response. A high density also exacerbates the size heterogeneity of small- and medium-sized juveniles (e.g., juveniles with initial body heights of 4.0 cm and 7.0 cm). Taking into account the welfare and yield per cubic meter of water of farmed seahorses, the present study suggests that the suitable stocking densities for lined seahorses with initial body heights of 4.0 cm, 7.0 cm, and 9.0 cm in RASs are 0.6 inds/L, 0.4 inds/L, and 0.2 inds/L, respectively.

A high stocking density has been demonstrated to inhibit growth and even cause death in many farmed organisms [15,17,18,31]. The results of the present study show that this adverse effect also occurs in the lined seahorse, and, regardless of the size of the juveniles, their growth rates decreased with an increasing stocking density. In the case of sufficient food and almost identical rearing facilities and water (the present study used recirculating water), the growth rate in high-density cultivation was lower than that in low-density cultivation, which was likely related to the crowding stress caused by a high density. In teleosts, plasma cortisol is a reliable indicator of stress, and brain 5-HT is also gradually being recognized as a neuroendocrine indicator related to emotions and stress; high levels of these usually indicate that the organism is in a stressed state, even in a depression-like state [21,22,27]. In the present study, the stress response data showed that, regardless of the size of the juveniles, if they were under a high density, their plasma cortisol and brain 5-HT were at a high level, proving that the juveniles under a high density were in a state of crowding stress. The most direct effect of crowding stress is that it drives the farmed organisms to allocate more energy to compete for the space that they need for survival, resulting in a decrease in the energy used for growth or immunity [18,32]. In addition, crowding stress can also trigger a series of indirect effects, such as decreased appetite, reduced food conversion efficiency, inflammatory reactions, and weakened antioxidant capacity [18,33–35]. These direct and indirect effects are largely the main causes of slow growth.

In the present study, the stress response data also showed that the cortisol and 5-HT contents decreased with the growth of the seahorses. A similar result was also observed in Atlantic salmon (*Salmo salar*): the plasma cortisol and brain 5-HT contents in fish sampled in the second month (length: 24.7 ± 0.6 cm, weight: 176 ± 13.1 g) were higher than those in fish sampled in the fifth month (length: 38.2 ± 1.2 cm, weight: 721 ± 81.4 g) [27]. The decreases in the cortisol (pg/mg plasma protein) and 5-HT contents (pg/mg brain tissue protein) may be due to a decrease in their secretion (i.e., numerator decrease) or an increase in other protein components in plasma and brain tissue (denominator increase) with the growth of farmed organisms. The lower levels of cortisol and 5-HT in larger fish may suggest that they have a greater tolerance or adaptability to stress than smaller fish, as inferred by Stevens et al. [36], who stated that small fish are likely to be more vulnerable and sensitive than large fish when exposed to stress. In support of this, the present study found that, when suffering from crowding stress, small-sized seahorses (an initial body height of 4.0 cm at a 1.0 inds/L density) showed a number of deaths, with a survival rate of about 71%, while the large-sized seahorses (an initial body height of 9.0 cm at a 0.4 inds/L density) showed almost no deaths, with a survival rate of about 96%, which proves that large fish have a stronger resistance to stress than small fish. In addition, we also found that some deaths occurred in juveniles with an initial body height of 7.0 cm, regardless of the density, and their survival rates were all below 90%. These deaths may be related to the change in food, as the seahorse juveniles with an initial body height of 7.0 cm began to switch from feeding on live copepods to frozen Mysis. During this process, a small portion of juveniles died due to their inability to adapt to frozen food [19,37]. When the seahorses were fully adapted to the frozen food, there were no more deaths.

The results of the present study also show that the final body height and final wet weight CVs for the small- and medium-sized juveniles in high-density cultivation (e.g., 4.0 cm juveniles at 1.0 and 0.8 inds/L and 7.0 cm juveniles at 0.6 inds/L) were higher than those in low-density cultivation, indicating that high-density cultivation exacerbates the size heterogeneity of small- and medium-sized juveniles. This size heterogeneity may be related to the social hierarchy of the lined seahorse. It is generally believed that the higher the population density, the more likely it is that a social hierarchy forms [38]. In high-density cultivation, there is an intense intraspecific competition for food and space between fish.

Due to differences in body size, personality traits, and social experience, their competitive abilities also vary, which determines their different social ranks in competition [39–41]. Dominant fish usually have a larger body size and food intake, stronger food and space acquisition abilities, and lower predation risks and stress responses, which contribute to their rapid growth. On the contrary, subordinate fish usually have a smaller body size, a lower food intake, and weaker food and space acquisition abilities, and they are subjected to stresses such as attack and intimidation from dominant fish, which lead to their growth retardation [41–43]. Therefore, the social hierarchy is likely to be the reason for the large deviation of fish in high-density cultivation. Unlike the small- and medium-sized juveniles, the size heterogeneity of the large-sized juveniles (initial body height of 9.0 cm) did not exacerbate in high-density cultivation. A possible reason for this is that, among the large-sized juveniles, dominant fish began to allocate a portion of energy to gonadal and brood pouch development, leading to a slowdown in their growth [44,45]. The growth slowdown of dominant fish led to a weakening of their body size advantage, thereby improving the social rank of subordinate fish. The abilities of subordinate fish to acquire food and space were strengthened, and their growth also accelerated accordingly [38,46]. Thus, the growth slowdown of dominant fish and the growth acceleration of subordinate fish may be the reason why the size heterogeneity of the large-sized juveniles was not exacerbated in high-density cultivation.

5. Conclusions

In the present study, we demonstrated that the stocking density has a significant impact on the survival, growth, size homogeneity, and physiological stress of the juvenile lined seahorse (*H. erectus*). A high density can lead to reduced survivorship, slowed growth, an increased stress response, and an exacerbated size heterogeneity. Therefore, the recommended stocking densities for 4.0 cm, 7.0 cm, and 9.0 cm juvenile lined seahorses in RASs are 0.6 inds/L, 0.4 inds/L, and 0.2 inds/L, respectively.

Author Contributions: Conceptualization, T.L. and X.L.; methodology, T.L.; software, S.L.; validation, T.L., S.L. and X.L.; formal analysis, T.L.; investigation, Y.R.; resources, D.Z.; data curation, T.L.; writing—original draft preparation, T.L.; writing—review and editing, X.L. and D.Z.; visualization, Y.R.; supervision, D.Z.; project administration, T.L.; funding acquisition, X.L. All authors have read and agreed to the published version of the manuscript.

Funding: This research was funded by the National Natural Science Foundation of China (Grant No. 32172950), the Key Research and Development Projects of Hainan (Grant No. ZDYF2022XDNY-350), and the Central Public-Interest Scientific Institution Basal Research Funds of CAFS (Grant No. 2023TD56) and of ECSF (Grant No. 2024QT02).

Institutional Review Board Statement: All procedures were conducted in compliance with the recommendations in the Guide for the Management and Use of the Experimental Animals of China Science and Technology Commission. The animal study protocol was authorized by the Committee on the Ethics of Animal Experiments of Chinese Academy of Fishery Sciences, with approval reference number: CAFS-EC-2023-086.

Informed Consent Statement: Not applicable.

Data Availability Statement: The original data presented in the study are included in the article; further inquiries can be directed to the corresponding author.

Acknowledgments: We are very grateful to the Wenchang Innovation Research Center for the free provision of the experimental seahorses and recirculating aquaculture system, as well as their staff's assistance in seahorse rearing and management.

Conflicts of Interest: The authors declare no conflicts of interest.

References

1. Tacon, A.G.J.; Metian, M. Fish matters: Importance of aquatic foods in human nutrition and global food supply. *Rev. Fish. Sci.* **2013**, *21*, 22–38. [CrossRef]

2. Tacon, A.G.J.; Metian, M. Food matters: Fish, income, and food supply—A comparative analysis. *Rev. Fish. Sci. Aquac.* **2017**, *26*, 15–28. [CrossRef]
3. Li, H.; Cui, Z.; Cui, H.; Bai, Y.; Yin, Z.; Qu, K. Hazardous substances and their removal in recirculating aquaculture systems: A review. *Aquaculture* **2023**, *569*, 739399. [CrossRef]
4. Nie, X.; Mubashar, M.; Zhang, S.; Qin, Y.; Zhang, X. Current progress, challenges and perspectives in microalgae-based nutrient removal for aquaculture waste: A comprehensive review. *J. Clean. Prod.* **2020**, *277*, 124209. [CrossRef]
5. Jasmin, M.Y.; Syukri, F.; Kamarudin, M.S.; Karim, M. Potential of bioremediation in treating aquaculture sludge: Review article. *Aquaculture* **2020**, *519*, 734905. [CrossRef]
6. Calderini, M.L.; Stevčić, C.; Taipale, S.; Pulkkinen, K. Filtration of Nordic recirculating aquaculture system wastewater: Effects on microalgal growth, nutrient removal, and nutritional value. *Algal Res.* **2021**, *60*, 102486. [CrossRef]
7. Liu, Y.; Liu, B.; Shi, C.; Sun, G. Recirculating aquaculture systems in China-Current application and prospects. *Fish. Aquac. J.* **2015**, *6*, 134. [CrossRef]
8. Xin, N.; Zhang, S.; Yang, Y.; Peng, L.; Yu, X.; Wang, Z. Analysis of the production cost of *Epinephelus* spp. in recirculating aquaculture system in northern China. *Fish. Mod.* **2018**, *45*, 50–54, (In Chinese with English abstract).
9. Zhao, C. Turbot Factory Circulating Water Aquaculture and Water Aquaculture Economic Benefit Comparison Research. Master's Thesis, Shanghai Ocean University, Shanghai, China, 2015.
10. Koldewey, H.J.; Martin-Smith, K.M. A global review of seahorse aquaculture. *Aquaculture* **2010**, *302*, 131–152. [CrossRef]
11. Vincent, A.C.J.; Foster, S.J.; Koldewey, H.J. Conservation and management of seahorses and other Syngnathidae. *J. Fish Biol.* **2011**, *78*, 1681–1724. [CrossRef]
12. Cohen, F.P.A.; Valenti, W.C. Opportunities and constraints for developing low-cost aquaculture of seahorses in mangrove estuaries. *Aquaculture* **2019**, *502*, 121–127. [CrossRef]
13. Kuo, T.C.; Vincent, A. Assessing the changes in international trade of marine fishes under CITES regulations—A case study of seahorses. *Mar. Policy* **2018**, *88*, 48–57. [CrossRef]
14. Abd El-Hack, M.E.; El-Saadony, M.T.; Nader, M.M.; Salem, H.M.; El-Tahan, A.M.; Soliman, S.M.; Khafaga, A.F. Effect of environmental factors on growth performance of Nile tilapia (*Oreochromis niloticus*). *Int. J. Biometeorol.* **2022**, *66*, 2183–2194. [CrossRef]
15. Zhang, G.; Jiang, X.; Zhou, W.; Chen, W.; Levy, T.; Wu, X. Stocking density affects culture performance and economic profit of adult all-female Chinese mitten crabs (*Eriocheir sinensis*) reared in earthen ponds. *Aquaculture* **2024**, *581*, 740352. [CrossRef]
16. Andrade, T.; Afonso, A.; Pérez-Jiménez, A.; Oliva-Teles, A.; de las Heras, V.; Mancera, J.M.; Serradeiro, R.; Costas, B. Evaluation of different stocking densities in a Senegalese sole (*Solea senegalensis*) farm: Implications for growth, humoral immune parameters and oxidative status. *Aquaculture* **2015**, *438*, 6–11. [CrossRef]
17. Long, L.; Zhang, H.; Ni, Q.; Liu, H.; Wu, F.; Wang, X. Effects of stocking density on growth, stress, and immune responses of juvenile Chinese sturgeon (*Acipenser sinensis*) in a recirculating aquaculture system. *Comp. Biochem. Phys. C* **2019**, *219*, 25–34. [CrossRef]
18. Zheng, J.; Qian, Y.; Zheng, X. Effects of stocking density on juvenile *Amphioctopus fangsiao* (Mollusca: Cephalopodasca: Cephalopoda): Survival, growth, behavior, stress tolerance and biochemical response. *Aquaculture* **2023**, *567*, 739243. [CrossRef]
19. Lin, Q.; Lin, J.; Zhang, D. Breeding and juvenile culture of the lined seahorse, *Hippocampus erectus* Perry, 1810. *Aquaculture* **2008**, *277*, 287–292. [CrossRef]
20. Huang, J.; Qin, G.; Zhang, B.; Tan, S.; Sun, J.; Lin, Q. Effects of food, salinity, and ammonia-nitrogen on the physiology of juvenile seahorse (*Hippocampus erectus*) in two typical culture models in China. *Aquaculture* **2020**, *520*, 734965. [CrossRef]
21. Ramsay, J.M.; Feist, G.W.; Varga, Z.M.; Westerfield, M.; Kent, M.L.; Schreck, C.B. Whole-body cortisol is an indicator of crowding stress in adult zebrafish, *Danio rerio*. *Aquaculture* **2006**, *258*, 565–574. [CrossRef]
22. Sadoul, B.; Geffroy, B. Measuring cortisol, the major stress hormone in fishes. *J. Fish Biol.* **2019**, *94*, 540–555. [CrossRef] [PubMed]
23. Lanfumey, L.; Mongeau, R.; Cohen-Salmon, C.; Hamon, M. Corticosteroid-serotonin interactions in the neurobiological mechanisms of stress-related disorders. *Neurosci. Biobehav. Rev.* **2008**, *32*, 1174–1184. [CrossRef] [PubMed]
24. Andrews, P.W.; Bharwani, A.; Lee, K.R.; Fox, M.; Thomson, J.A., Jr. Is serotonin an upper or a downer? The evolution of the serotonergic system and its role in depression and the antidepressant response. *Neurosci. Biobehav. Rev.* **2015**, *51*, 164–188. [CrossRef] [PubMed]
25. Lillesaar, C. The serotonergic system in fish. *J. Chem. Neuroanat.* **2011**, *41*, 294–308. [CrossRef] [PubMed]
26. Braithwaite, V.A.; Ebbesson, L.O.E. Pain and stress responses in farmed fish. *Rev. Sci. Tech. Off. int. Epiz.* **2014**, *33*, 245–253. [CrossRef]
27. Vindas, M.A.; Johansen, I.B.; Folkedal, O.; Höglund, E.; Gorissen, M.; Flik, G.; Kristiansen, T.S.; Øverli, Ø. Brain serotonergic activation in growth-stunted farmed salmon: Adaption versus pathology. *R. Soc. Open Sci.* **2016**, *3*, 160030. [CrossRef]
28. Lin, T.; Liu, X.; Xiao, D.; Zhang, D. Plasma levels of immune factors and sex steroids in the male seahorse *Hippocampus erectus* during a breeding cycle. *Fish Physiol. Biochem.* **2017**, *43*, 889–899. [CrossRef]
29. Bradford, M.M. A rapid and sensitive method for quantification of microgram quantities of protein utilizing the principle of protein-dye-binding. *Anal. Biochem.* **1976**, *72*, 248–254. [CrossRef]
30. NCSS. *One-Way Analysis of Variance*; NCSS: Silver Spring, MD, USA, 2012. Available online: https://www.ncss.com/wp-content/themes/ncss/pdf/Procedures/NCSS/One-Way_Analysis_of_Variance.pdf (accessed on 5 April 2024).

31. Braun, N.; de Lima, R.L.; Baldisserotto, B.; Dafre, A.L.; de Oliveira Nuñer, A.P. Growth, biochemical and physiological responses of *Salminus brasiliensis* with different stocking densities and handling. *Aquaculture* **2010**, *301*, 22–30. [CrossRef]
32. North, B.P.; Turnbull, J.F.; Ellis, T.; Porter, M.J.; Migaud, H.; Bron, J.; Bromage, N.R. The impact of stocking density on the welfare of rainbow trout (*Oncorhynchus mykiss*). *Aquaculture* **2006**, *255*, 466–479. [CrossRef]
33. Santos, G.A.; Schrama, J.W.; Mamauag, R.E.P.; Rombout, J.H.W.M.; Verreth, J.A.J. Chronic stress impairs performance, energy metabolism and welfare indicators in European seabass (*Dicentrarchus labrax*): The combined effects of fish crowding and water quality deterioration. *Aquaculture* **2010**, *299*, 73–80. [CrossRef]
34. Li, W.; Li, D.; Yang, Q.; Liu, L.; Liu, J.; Lu, J.; Wang, Y.; Tang, R.; Li, L.; Zhang, X. Long-term crowding stress induces chronic inflammatory response and declines the immunity of grass carp (*Ctenopharyngodon idella*). *Aquaculture* **2023**, *577*, 739976. [CrossRef]
35. Bi, B.; Yuan, Y.; Zhao, Y.; He, M.; Song, H.; Kong, L.; Gao, Y. Effect of crowding stress on growth performance, the antioxidant system and humoral immunity in hybrid sturgeon. *Aquacult. Rep.* **2023**, *28*, 101468. [CrossRef]
36. Stevens, C.H.; Croft, D.P.; Paull, G.C.; Tyler, C.R. Stress and welfare in ornamental fishes: What can be learned from aquaculture? *J. Fish Biol.* **2017**, *91*, 409–428. [CrossRef] [PubMed]
37. Lin, Q.; Lin, J.; Zhang, D.; Wang, Y. Weaning of juvenile seahorses *Hippocampus erectus* perry, 1810 from live to frozen food. *Aquaculture* **2009**, *291*, 224–229. [CrossRef]
38. Zhang, Z.H.; Guo, H.Y.; Zhang, X.M.; Zhang, X.M. Fish social hierarchy and its application in aquaculture and stock enhancement. *Chin. J. Ecol.* **2018**, *37*, 1257–1264. [CrossRef]
39. Colléter, M.; Brown, C. Personality traits predict hierarchy rank in male rainbowfish social groups. *Anim. Behav.* **2011**, *81*, 1231–1237. [CrossRef]
40. Akbaripasand, A.; Krkosek, M.; Lokman, P.M.; Closs, G.P. Does social status within a dominance hierarchy mediate individual growth, residency and relocation? *Oecologia* **2014**, *176*, 771–779. [CrossRef]
41. Bessa, E.; Sadoul, B.; Mckenzie, D.J.; Geffroy, B. Group size, temperature and body size modulate the effects of social hierarchy on basal cortisol levels in fishes. *Horm. Behav.* **2021**, *136*, 105077. [CrossRef]
42. Dou, S.Z.; Masuda, R.; Tanaka, M.; Tsukamoto, K. Size hierarchies affecting the social interactions and growth of juvenile Japanese flounder, *Paralichthys olivaceus*. *Aquaculture* **2004**, *233*, 237–249. [CrossRef]
43. Wilson, A.J.; Grimmer, A.; Rosenthal, G.G. Causes and consequences of contest outcome: Aggressiveness, dominance and growth in the sheepshead swordtail, *Xiphophorus birchmanni*. *Behav. Ecol. Sociobiol.* **2013**, *67*, 1151–1161. [CrossRef]
44. Lin, Q.; Zhang, D.; Lin, J. Effects of light intensity, stocking density, feeding frequency and salinity on the growth of sub-adult seahorses *Hippocampus erectus* Perry, 1810. *Aquaculture* **2009**, *292*, 111–116. [CrossRef]
45. Planas, M.; Quintas, P.; Chamorro, A.; Silva, C. Female maturation, egg characteristics and fatty acids profile in the seahorse *Hippocampus guttulatus*. *Anim. Reprod. Sci.* **2010**, *122*, 66–73. [CrossRef] [PubMed]
46. Pettersson, J.; Johnsson, J.I.; Bohlin, T. The competitive advantage of large body size declines with increasing group size in rainbow trout. *J. Fish Biol.* **1996**, *49*, 370–372. [CrossRef]

animals

MDPI

Article

Mate Choice in Western Mosquitofish (*Gambusia affinis*) in Response to Virtual Mates: A Method for the Investigation of Fish Mate Choice Using Maya 3D Simulation Technology

Bowen Feng [1,†], Liming Chen [1,†], Liangmin Huang [1], Jun Li [1] and Kai Liu [1,2,3,*]

1 Fisheries College, Jimei University, Xiamen 361021, China; fengbowen0323@163.com (B.F.); chen1015085915@163.com (L.C.); lmhuang@jmu.edu.cn (L.H.); lijun1982@jmu.edu.cn (J.L.)
2 Fujian Provincial Key Laboratory of Marine Fishery Resources and Eco-Environment, Fisheries College, Jimei University, Xiamen 361021, China
3 State Key Laboratory of Mariculture Breeding, Fisheries College, Jimei University, Xiamen 361021, China
* Correspondence: liukai1218@jmu.edu.cn
† These authors contributed equally to this work.

Simple Summary: This study explored how visual signals affect mate choice in animals, focusing on the western mosquitofish (*Gambusia affinis*). We created 3D simulation animations using the Maya 2018 software to accurately replicate key movements and behaviors essential for our research. Through preference tests, we validated these animations and found that the fish could effectively identify 3D simulated mates. Both male and female mosquitofish exhibited a strong preference for larger animations. Additionally, the fish displayed significantly greater attraction to 3D simulations compared to 2D ones. These findings underscore the potential of 3D simulation technology in studying fish behavior, offering an efficient, precise, and non-invasive method for future research on mate choice. This work not only enhances our understanding of animal behavior but also contributes to conservation strategies aimed at preserving fish populations and their ecosystems.

Abstract: Visual signals are crucial for animals to obtain information about their environment, and they play a significant role in mate choice. However, individual variability and factors such as movement patterns can hinder research flexibility. A key challenge in this field is the accurate simulation of specific movements and behaviors. In this study, we investigated the western mosquitofish (*Gambusia affinis*) by creating 3D simulation animations using the Maya 2018 software. Meanwhile, we validated the effectiveness of this animation through dichotomous association preference tests. The results showed that our subjects could successfully identify 3D simulated mates, and both males and females demonstrated pronounced preferences for larger simulated animations. Moreover, our findings revealed that this species exhibited a notably stronger preference for 3D simulations compared to 2D animations. These findings suggest that 3D simulation technology holds significant potential for the investigation of fish mate choice, offering an efficient, precise, standardized, and easily manageable non-invasive method for future research in fish behavior.

Keywords: mate choice; Maya 3D simulation animation; visual signals; *Gambusia affinis*

1. Introduction

Visual signals play a crucial role in many species within the animal kingdom, particularly in aspects such as mate selection, foraging, and predator avoidance [1–4]. Studies have shown that fish can detect visual signals from predators and adjust their behavior in territory defense, the evasion of threats, and courtship activities [5,6]. For example, juvenile coho salmon (*Oncorhynchus kisutch*) exhibited a significant reduction in aggressive behavior towards mirror images when exposed to visual signals from avian predators [7]. Mate choice is vital to sexual selection and biological evolution [8–10], influencing many species' reproductive success and population dynamics. It involves a range of complex and variable signals, including size, coloration, distinctive ornamental features, and movement patterns, which are often dynamic and change over time and space [11–14]. In fish, mate choice is typically driven by visual, chemical, and auditory cues, with visual signals playing a critical role in many species. Currently, many methods in fish mate choice studies are based on visual signals [15]. Utilizing these visual characteristics, researchers have developed various approaches for investigation, including models, images, and video playback [16–19].

With the advancement of computer imaging technology, an increasing number of studies have employed computer-simulated animations in the research of fish behavior, successfully applying this technique to investigate mate choice in various social animals, such as three-spined sticklebacks (*Gasterosteus aculeatus*) [20], pipefish (*Syngnathus typhle*) [21], zebrafish and their allies (genus *Danio*) [22,23], and pygmy swordtails (*Xiphophorus nigrensis*) [24]. Two-dimensional (2D) simulation animation technology has been widely utilized to explore fish mate choice preferences. For instance, in studies on mate selection in the striped cichlid (*Pelvicachromis taeniatus*), it was found that the stimulatory effects of 2D simulation animations outperformed those of static images [25]. Two-dimensional simulation animations offer several advantages in studying fish mate choice, including preventing habituation due to the stimuli's movement characteristics, maintaining subjects' attention, consistently eliciting social responses, and ease of production. However, this technology still has limitations in terms of animation fidelity and realism. For example, fish's nuanced and flexible behaviors during swimming and courtship, such as turning, fin waving, and reproductive fin display, are challenging to accurately represent using traditional 2D simulation animations. Therefore, adopting a method with higher simulation fidelity and more precise behavioral characteristics would greatly enhance the study of fish mate choice behaviors.

As three-dimensional (3D) simulation animation technology advances, the use of 3D simulation animations in the study of animal behavior has matured, offering an innovative visual presentation method to simulate, analyze, and showcase the natural behaviors of animals and their habitats [26]. This technology enables the creation of complex animal behaviors and interaction scenarios without disturbing the animals or their living environments [27]. Compared to 2D simulation animations, 3D simulation animations possess unique advantages, enhancing their applicability in specific contexts. Firstly, 3D simulation animations provide a realistic sense of depth and space, offering experimental subjects a more immersive experience. Secondly, 3D simulation animations can produce highly realistic textures, lighting effects, and surface details through advanced rendering techniques, improving the visual quality. Recently, 3D simulation animation technology has been employed in animal behavior research by establishing 3D simulation animated models, yielding positive results. For example, in studies examining how the jumping spider (*Portia fimbriata*) responds to visual cues in its prey's eyes, researchers used computer-rendered virtual 3D lures to manipulate these lures and investigate the spider's hunting behavior and decision-making processes based on visual information from the prey's eyes [28]. Additionally, to study the key role of visual signals during lizard movement, the Maya 3D simulation animation software was used to simulate the movement of individuals in a three-dimensional space, representing an innovative approach to understanding dynamic environments that merges evolutionary biology with digital art, demonstrating the method's practicality through the examination of lizard movements in a plant motion noise environment [29,30]. However, the full potential of 3D simulation animations in fish behavior research remains underutilized, with only a limited number of studies reported. For instance, researchers created 3D simulation animations of the swimming behavior of *Oryzias latipes*, establishing a solid technical foundation for the application of 3D simulation animations in fish behavior studies [31]. Furthermore, in research on mate choice in *Poecilia latipinna*, animations constructed using the Blender 3D software demonstrated that the simulation intensity could match that of live animals [32,33]. In a study examining the mate choice of male *Etheostoma zonale* for conspecific versus heterospecific females of

E. barrenense, the researchers found no significant difference in the mate preference intensity between live heterospecific individuals and animated simulations [34]. Overall, 3D simulation animations offer high manipulability and standardized approaches for fish behavior research, with the potential to reduce and replace live animal stimuli while optimizing the experimental protocols. Therefore, effectively utilizing 3D simulation animations to study fish behavior is a worthwhile area for exploration.

This study focuses on the western mosquitofish (*Gambusia affinis*), investigating its mate choice behavior when exposed to 3D simulation animations. Previous studies on this species have highlighted several factors influencing its mate choice, including its body size, coloration, courtship behavior, and male ornamentation [35,36]. Prior research has demonstrated high consistency between the mate choice behaviors of live *G. affinis* and those in computer-animated experiments, with both males and females showing apparent mating behaviors towards animated partners, thereby validating the effectiveness of 2D simulation animations in mate choice studies for this species [37]. Hence, we aim to create an animated model of *G. affinis* using the Maya 3D technology to comprehensively examine both sexes' morphological and behavioral characteristics and explore the potential of 3D simulation animations in mate choice research. We hypothesize that (1) *G. affinis* can effectively recognize 3D simulation animated mates, and (2) compared to 2D simulation animations, 3D simulation animations offer a greater advantage in studying mate choice in *G. affinis*.

2. Materials and Methods

2.1. Origin and Maintenance of Test Fish

All test fish used in this study were wild-caught individuals collected from the field area of Longyan City, Fujian Province, China (N 25°05.33′, E 117°00.32′; Figure S1). We placed the collected *G. affinis* in 200 L transparent glass tanks equipped with an automatic oxygen supply and a circulating filtration system. The bottom of each tank was layered with gravel, aquatic plants, and stones to simulate a natural environment. A 12-h light/12-h dark cycle was maintained, and each tank housed 40 experimental fish with a 1:1 male-to-female ratio for an acclimation period of 21 days. The fish were fed twice daily (at 09:00 and 17:00) with commercially available flake food and frozen bloodworms. The water temperature in both the holding and experimental tanks was maintained at 25 ± 1 °C. We maintained the water quality by exchanging half of the water in each holding tank every two weeks, using filtered tap water throughout the experiment [37].

2.2. Three-Dimensional Simulation Animation Modeling

2.2.1. Images Collection

We collected high-resolution images of *G. affinis*, following Chen et al. [37]. To create 3D simulation animations that accurately reflected the natural variations in the body size and swimming speed of the subjects, we anesthetized the fish using a 10% clove oil solution after photographing (Sony a7R III Mirrorless, Sony, Tokyo, Japan) them, and we measured the standard length (SL) of 30 wild *G. affinis* individuals (15 females and 15 males). Additionally, the swimming speed of another 30 individuals (15 females and 15 males) was assessed using an activity measurement method [37]. Finally, image data of 15 females and 15 males *G. affinis* were collected (mean ± SD, females SL: 28.49 ± 3.3 mm, n = 15; male SL: 21.83 ± 2.66 mm, n = 15; Figure S2), along with video data from 10 individuals (5 females and 5 males; female swimming speed: 2.96 ± 0.64 cm/s, n = 5; male swimming speed: 2.81 ± 0.88 cm/s, n = 5). These data were used to create the 3D simulation animations.

2.2.2. Modeling

The 3D simulation animation models were created using Autodesk Maya 2018 (Autodesk, San Francisco, CA, USA) for 3D graphics, Adobe Premiere Pro 2020 (Adobe, San Francisco, CA, USA) for video processing, Adobe Substance 3D Painter 7.4.1.1418 (Adobe, San Francisco, CA, USA) for 3D model texturing, and Adobe Photoshop 2020 for image editing. First, an initial model was established in Maya, using a high-resolution image of *G. affinis* as a reference image (Figure 1a). A cube (polycube) was added as the basic structure of the model (Figure 1b). The model was then switched to Edit Mode, where tools such as Rotate, Extrude, and Cut were used to create and adjust new vertices. These vertices were moved to fit the fish's external contours, as shown in the reference image (Figure 1c). Next, mirroring was performed along the X, Y, or Z axis of the scene to create a symmetrical half of the head, and a mirror command was used to generate the mirrored geometry, which was then merged with the other side to complete the body model. Finally, the Smooth command was used to smooth the body, completing the model's construction (Figure 1d) [38].

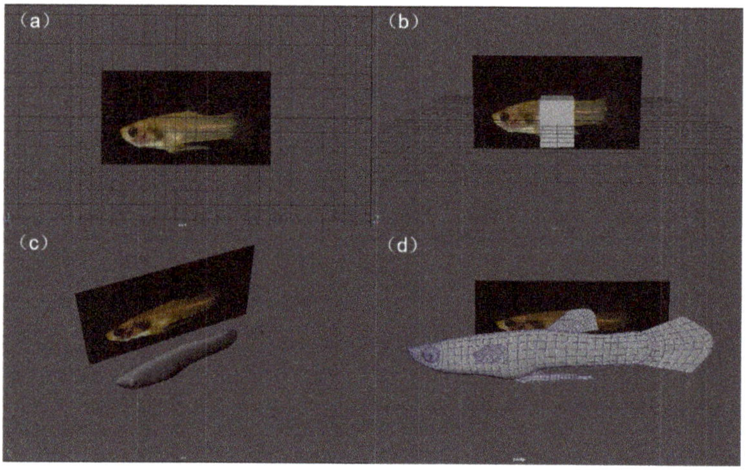

Figure 1. The modeling procedure. (**a**) Import reference image. (**b**) Add cube. (**c**) Fit fish body contour. (**d**) *G. affinis* model.

2.2.3. UV Unfolding and Mapping

To create the UV mapping for the fish model, we first selected the geometric components of the fish body using the model's component mode. An appropriate axis, typically the Z-axis, was then chosen to establish an initial planar mapping, ensuring that the entire fish body was encompassed within the UV (U-VEEZ) space. Next, we utilized the Cut UV Edges tool in the UV editor to sever the UV shell along the edges of the fish body, thereby minimizing UV stretching and distortion. Subsequently, we selected the Automatic option from the UV menu to enable Maya to automatically unfold the UV shell. Afterward, the fish body UV shell was selected, ensuring that it remained within the 0 to 1 UV space without overlapping. Manual adjustments to the position and scale of the UV shell were made to guarantee even texture distribution. The UV unfolding process for the fish's eyes and fins followed the same methodology. Then, the image editing software (e.g., Substance 3D Painter 7.4.1.1418) was employed to paint the textures for the fish body, eyes, and fins based on the UV layout. The painted texture files were then imported into Maya and applied to the model, verifying the texture fitting. An Arnold material was subsequently created and applied to the model within the material properties. These steps were repeated to generate materials and apply textures for the fish's eyes and fins. Finally, Maya's Arnold rendering engine was utilized for a preview render (Figure 2) [38].

2.2.4. Skeletal Binding

We initiated the process by accessing the tool panel and selecting the "Fish Structure" option to create a basic skeletal chain using the Advanced Skeleton menu in Maya (Figure 3a). The generated skeletal chain was subsequently adjusted to align with the fish's anatomical structure. The Move tool was utilized to manually position each bone, ensuring alignment with the respective components of the fish model, including the head, spine, and fins. The Orient Joint tool was then employed to guarantee the correct orientation of each joint, ensuring that the joint axes were properly aligned (Figure 3b). Next, the fish body model was selected, and, while holding the Shift key, the root joint of the Advanced Skeleton chain was chosen to bind the model to the skeleton. Within the weight adjustment tool, various joints were selected, and the brush tool was employed to modify their influence areas, ensuring that each joint controlled the model's deformation within a reasonable range by adjusting the weights, thereby preventing unnatural stretching or distortion. Subsequently, the auto-generated controllers feature in Advanced Skeleton was utilized to create controllers for the skeletal chain (Figure 3c). Finally, the positions of the controllers were adjusted to ensure that each major joint had a corresponding controller for animation and deformation purposes (Figure 3d) [38].

Figure 2. The model mapping. (**a**) Model (side view). (**b**) Model (top view).

2.2.5. Animation Creation

In Maya, the animation frame range was established on the timeline. This study required a 10 s animation at a frame rate of 24 frames per second (fps), leading to a timeline spanning from 0 to 240 frames. Keyframes were then positioned on the timeline to denote the controller's position, rotation, and scale. The timeline slider was moved to the midpoint of the animation (frame 120), where the controller's position and rotation were adjusted to create the fish's intermediate pose, followed by the placement of another keyframe. Subsequently, the slider was moved to the end of the animation (frame 240), and the controller's position and rotation were again modified to finalize the fish's pose, after which another keyframe was set. The animation was previewed in Maya, and the keyframes and animation curves were further refined to ensure that all actions and transitions met the intended expectations (Figure 4a). Arnold was selected as the renderer, and a spherical light source was incorporated into the scene to ensure the adequate lighting of the fish (Figure 4b). The render resolution, frame range (0–240 frames), and output format were configured. After confirming that all materials and textures were correctly applied and displayed in the render, the rendering of the entire animation sequence was initiated. Upon the completion of rendering, the output frame sequence was imported into the video editing software (Adobe Premiere Pro 2020) for compositing and post-production [38].

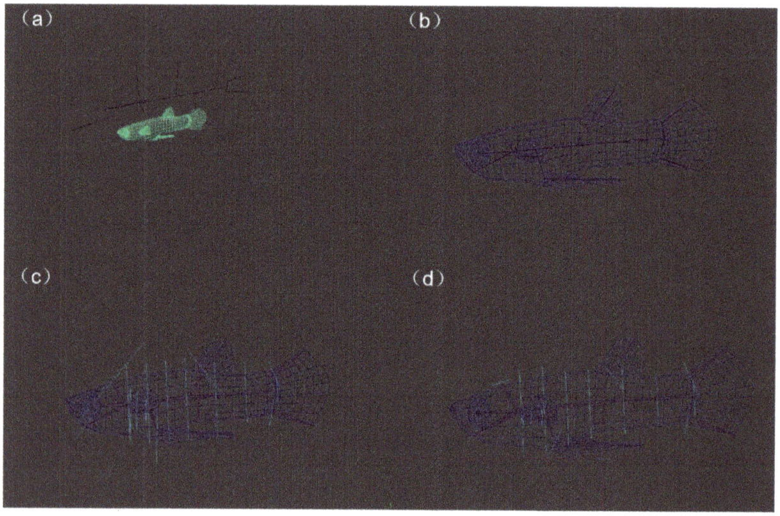

Figure 3. Skeletal binding. (**a**) Create bones. (**b**) Position bones. (**c**) Create controllers. (**d**) Adjust controllers.

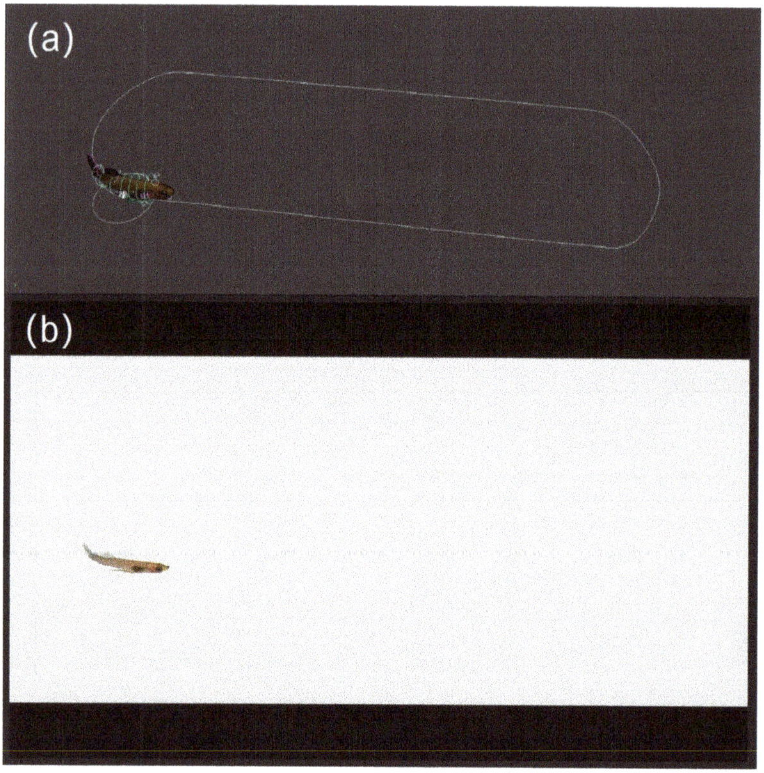

Figure 4. Animation creation. (**a**) Insert keyframes. (**b**) Lighting.

2.3. Mate Choice Test

At 24 h before the mate choice experiment, we randomly selected adult test fish from the holding tanks and segregated them by sex into 96 L tanks for isolation. To mitigate aggressive interactions, each fish was housed individually in a 1.5 L clear perforated plastic bottle, allowing for water exchange with the environment [39]. The mate choice test employed a dichotomous association preference test, with the setup and procedure adapted from Chen et al. [37]. The experimental apparatus comprised a glass tank measuring 60 × 35 × 40 cm and two 22-inch computer screens (E2216H, Dell, Guangzhou, China) positioned on the short sides of the tank. To ensure uniformity in the display, both screens were calibrated to identical brightness and color settings before the experiment. The choice areas within the tank were delineated with a marker, creating two 10 cm preference zones adjacent to the screens and a central 40 cm neutral zone. The long sides of the tank were covered with gray plastic sheets to minimize external disturbances. A foam board was placed at the bottom to ensure the visibility of the screens above the water surface, with the water level maintained at 25 cm to align with the screens' height. A high-definition camera (Logitech C920, Beijing, China) was mounted approximately 70 cm above the tank to facilitate the remote monitoring of the experiment via a computer, thereby preventing human interference. Illumination was provided by a 35 W LED light positioned 40 cm above the tank, and an air stone connected to an air pump was placed within the tank for aeration. The water was changed after every four trials to maintain its quality. At the beginning of the test, the focal individual was placed into a transparent plexiglass cylinder (10 cm diameter), positioned at the center of the neutral zone within the tank, while the first pair of animations was played. Following a 3 min habituation period, during which the fish were able to view both animations, the cylinder was gently removed. Subsequently, a 5 min observation period was conducted, during which we recorded the association times, defined as the duration spent in each preference zone. To minimize potential side biases, we relocated the focal fish to the central cylinder, swapped the side assignments of the 3D stimulus animations, and repeated the measurement of the association preferences after an additional 3 min habituation period [37]. Additionally, the standard length (SL) of each test fish was measured. The mate choice preference was assessed by calculating the strength of preference (SOP), based on the total time spent by each fish in the preference zones; more time spent in the zone with the preferred mate indicated a stronger preference. We tested $n = 15$ females and $n = 15$ males for their mate choice preferences in each trial, i.e., a total of 45 females (mean ± SD, SL: 28.49 ± 3.30 mm) and 45 males (SL: 21.83 ± 2.90 mm). The experiment consisted of three components: (1) the validation of the effectiveness of 3D simulation animations as mate choices, utilizing both 3D animations and cylindrical objects of equivalent size and color as test stimuli; (2) the exploration of individual preferences for virtual mates based on body size differences using 3D animations, where only the body length varied between the left and right animations; (3) the investigation of individual preferences for 2D versus 3D simulation animations, employing animations of the same body length for both 2D and 3D mate choice tests.

2.4. Statistical Analyses

All statistical analyses were conducted in SPSS 19.0 (IBM, Armonk, NY, USA). All descriptive statistics are presented as mean ± SD values, and $p < 0.05$ indicated a statistically significant difference in the results.

We estimated each focal individual's strength of preference (SOP) for cylindrical objects versus 3D animations, large 3D animations versus small 3D animations, and 2D animations versus 3D animations as mating partners as follows: SOP = time spent associating with preferred stimulus fish/time spent with both stimulus fish. Therefore, it can be concluded that the SOP values indicate the intensity of the experimental fish's preferences for various animations. By comparing the magnitudes of the SOP values, we can determine the degree of individual preference for different virtual mate objects. To test for overall preferences within each group and test situation, we compared the SOP values against a null assumption (i.e., SOP = 0) by using one-sample *t*-tests. Subsequently, a generalized linear model (GLM) was used to analyze the effect of the standard length (SL) of the experimental fish on the mate preference, with the SOP value as the dependent variable, the animation number as a random factor, and the standard length (SL) of the experimental fish as a covariate [40].

3. Results

3.1. Three-Dimensional Simulation Animations

A total of twenty 3D simulation animations of the *G. affinis* were created, employing the body length as a variable. These animations were organized into five groups for each sex, with each group consisting of one large and one small specimen (Figure 5a,b). The animation videos can be found in the online Supplementary Materials.

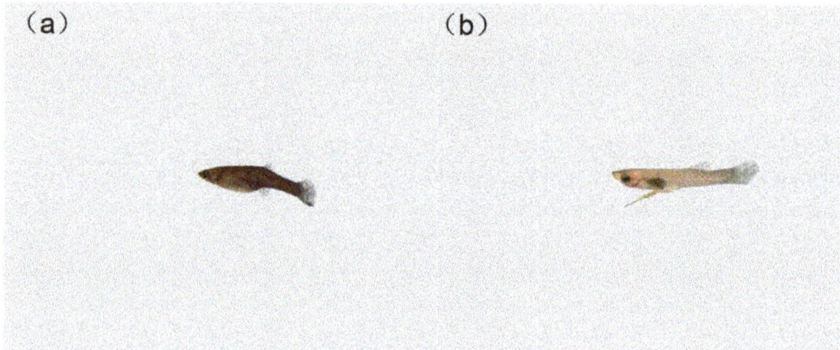

Figure 5. Three-dimensional simulation animation model of *G. affinis*. (**a**) Female. (**b**) Male.

3.2. Validity of 3D Simulation Animations

The results of the GLM test indicated that the "Animation ID" was not significantly different from the "SL" when considered as a random variable and a covariate, respectively (Table 1). We initially predicted that both male and female *G. affinis* would exhibit a significant preference for the 3D virtual fish, and the results indeed confirmed our hypothesis. The results of the mate choice tests showed that, compared to the time spent in the preference zone for the cylindrical object animations (75.07 ± 35.05 s), females spent significantly more time near the 3D simulation animations (207.20 ± 38.10 s; $p < 0.001$; Figure 6a). Similarly, males also spent significantly more time with the 3D simulated animations (182.27 ± 58.30 s; $p < 0.001$) compared to males with the cylindrical object animations (137.47 ± 66.63 s; Figure 6b).

Table 1. Results of generalized linear model (GLM) testing for the effects of the chosen animation type (3D/cylindrical object animations) of the individual on the strength of female and male preferences.

Sex	Factor	*df*	*F*	*p*	Wilks' Partial np^2
Female	Animation ID	10	0.40	0.88	0.02
	SL	1	1.42	0.29	0.06
	Error	15			
Male	Animation ID	10	0.93	0.51	0.03
	SL	1	0.03	0.87	0.001
	Error	15			

3.3. Overall Preferences for Mate Choice Based on 3D Simulation Animations

The results of the GLM test revealed no significant differences between the random variable "Animation ID" and the covariate "SL" (Table 2). A statistically significant effect was detected in both males and females, and pronounced variation in individual preferences (i.e., SOP values) was observed. Females spent, on average, significantly more time in association with larger male 3D simulation animations (190.93 ± 63.81 s; $p < 0.001$) than with small-bodied male animations (148.20 ± 66.41 s; Figure 7a). Likewise, males also exhibited significantly more time near large-bodied female 3D animations (200.40 ± 39.07 s; $p < 0.001$) compared to males with smaller females (107.67 ± 34.89 s; Figure 7b).

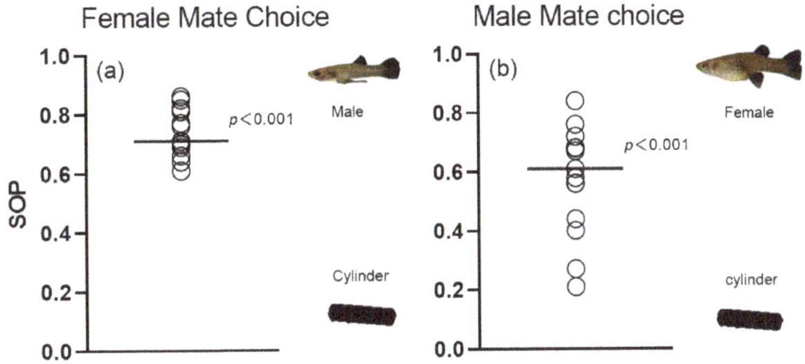

Figure 6. Distribution of individual strength of preference (SOP) values derived from dichotomous association preference tests. (**a**) Female choice for 3D versus cylindrical object animations. (**b**) Male choice for 3D versus cylindrical object animations. Solid lines represent the mean SOP across individuals. Results from one-sample *t*-test testing against the null assumption (SOP = 0) are presented.

Table 2. Results of generalized linear model (GLM) testing for the effects of the choosing individual's body size (SL) on the strength of female and male preferences.

Sex	Factor	*df*	*F*	*p*	Wilks' Partial np^2
Female	Animation ID	10	1.22	0.40	0.04
	SL	1	2.31	0.17	0.08
	Error	15			
Male	Animation ID	10	1.10	0.48	0.03
	SL	1	0.001	0.97	2.83
	Error	15			

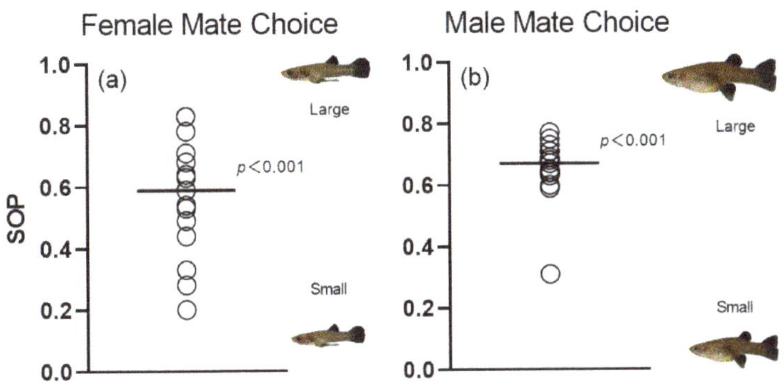

Figure 7. Distribution of individual strength of preference (SOP) values derived from dichotomous association preference tests. (**a**) Female choice for large versus small male 3D animations. (**b**) Male choice for large versus small female 3D animations. Solid lines represent the mean SOP across individuals. Results from one-sample *t*-test testing against the null assumption (SOP = 0) are presented.

3.4. Comparison Between 3D and 2D Simulation Animations

The results of the GLM test revealed no significant differences between the random variable "Animation ID" and the covariate "SL" (Table 3). The results indicate that, compared to the time spent near 2D animation males (151.13 ± 68.53 s), females exhibited significantly more time with 3D simulation animations (211.00 ± 77.68 s; $p < 0.001$; Figure 8a). Meanwhile, males also showed significantly more time in association with 3D simulation animations (186.67 ± 87.03 s; $p < 0.001$) than with 2D animations (137.6 ± 66.04 s; Figure 8b).

Table 3. Results of generalized linear model (GLM) testing for the effects of the chosen animation type (3D/2D animations) of the individual on the strength of female and male preferences.

Sex	Factor	*df*	*F*	*p*	Wilks' Partial np^2
Female	Animation ID	10	1.11	0.48	0.04
	SL	1	0.02	0.90	0.001
	Error	15			
Male	Animation ID	10	1.82	0.24	0.05
	SL	1	4.12	0.09	0.11
	Error	15			

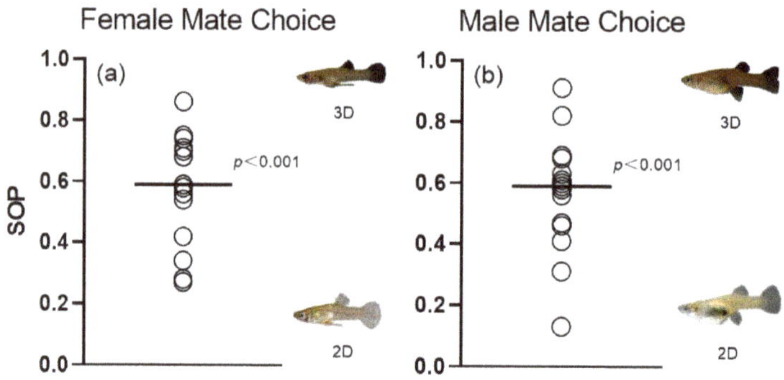

Figure 8. Distribution of individual strength of preference (SOP) values derived from dichotomous association preference tests. (**a**) Female choice for male 3D versus 2D animations. (**b**) Male choice for female 3D versus 2D animations. Solid lines represent the mean SOP across individuals. Results from one-sample *t*-test testing against the null assumption (SOP = 0) are presented.

4. Discussion

4.1. Quality of 3D Simulation Animations

We used the Maya 3D technology to produce a total of twenty 3D simulation animations of both male and female *G. affinis* in this study, with five groups for each sex. We are dedicated to investigating mate choice preferences based on 3D animation models, aiming to explore the application of 3D animation technology in the field of fish mate choice. Recently, 3D simulated animation technology has been reported in behavioral studies of reptiles due to its outstanding image processing and precise behavioral feature simulation capabilities. For instance, by employing the Maya 3D technology to examine the movement of the Jacky dragon (*Amphibolurus muricatus*) surrounded by plant noise, researchers have demonstrated the practicality of this technology in studying the visual perceptions of this species [29]. However, in the actual production process of 3D simulation animations, species differences can lead to variations in the application of skeletal binding techniques and methods [41]. This study utilized the Maya software to establish 3D animated models for *G. affinis*, and, by comparing the animation models of our study object and *A. muricatus*, we identified several key differences in the skeletal binding techniques. These differences arise from the anatomical structures and locomotion mechanisms of the two types of species. Generally, the body structures of reptiles are more complex, featuring prominent limbs, long necks, tails, and other characteristics. This skeletal structure includes multiple joints and intricate skeletal connections, particularly in the limbs and head. The accurate simulation of the movement of these parts is required during 3D modeling, as well as considering how muscles propel limb movement. In contrast, the bodies of fish are typically streamlined, primarily relying on the tail fin and other fins for swimming [42,43]. The streamlined shape of their bodies is more pronounced to facilitate rapid movement in water. Due to the simpler skeletal systems of fish, primarily consisting of vertebrae and fewer joints, this implies that 3D modeling must focus on simulating body flexion and tail fin oscillation. Therefore, this movement pattern necessitates dynamic spinal activity and tail oscillation in the 3D model, which requires the model to not only have smooth skeletal animations but also simulate the effects of water flow on the body.

The application of 3D animation technology in fish behavior research has currently received only limited reports. For example, researchers created a 3D simulation animation of the swimming Medaka (*Oryzias latipes*) using the Blender software to explore the potential application of 3D animation technology in fish behavior studies [31]. In comparison to the Medaka model created in Blender, the model produced in this study using the Maya software demonstrated superior smoothness and realism in the body and head, closely resembling the actual *G. affinis*. First, the Maya software is equipped with advanced modeling tools, making it particularly suitable for the creation of smooth biological surfaces, such as the streamlined shape of fish [44]. By precisely controlling the surface details, highly realistic 3D animation models can be produced. Secondly, the Arnold rendering engine used in this study provides exceptional light and shadow effects, optimizing the treatment of light transmission and material properties, including transparency, detailed skin textures, and color variations. Therefore, the skin texture constructed based on Maya's 3D technology appears more natural and realistic, which are key factors in achieving lifelike fish models. Although Blender is a powerful open-source tool capable of producing high-quality animations, Maya excels in terms of its advanced features and professional support. The advantages of Maya mainly lie in its robust modeling capabilities, advanced dynamic systems, high-end material and rendering options, and extensive industry plugin support. These factors collectively ensure that Maya can produce more realistic animation effects, making it particularly prominent in creating highly lifelike fish animations and providing new potential solutions for fish behavior research.

4.2. *Validity of 3D Simulation Animations in Fish Mate Choice*

This study validated the effective recognition of 3D simulation animations by *G. affinis* based on binary mate choice tests. Additionally, the motivation for individual sexual behavior in male mate choice, represented by the continuous flaring of the reproductive fins and mating attempts, also demonstrated a valid response to the 3D simulation animations. This is consistent with the results obtained from previous observations and experiments in our laboratory [45]. Furthermore, we explored the choice preferences of individuals faced with virtual mates of different body sizes, finding consistency with our earlier experiments on mate choice in *G. affinis* using 2D simulation animations [37]. Finally, to investigate the potential advantages of 3D simulation animations in the study of fish mate choice, we compared individual mate preferences based on 2D and 3D simulation animations. The results revealed a significant preference for 3D simulation animations among individuals. This also demonstrates the strong advantages of 3D animations in studying the impact of complex visual signals on mate choice, as it can more accurately manipulate and present these signals, proving to be more effective than 2D animations. While 2D animations have advantages in terms of the production costs and the control of single features, 3D animations exhibit greater realism, multidimensional manipulation, and dynamic interaction capabilities in mate choice research [46]. These advantages enable 3D animations to accurately simulate complex behaviors in natural environments, enhancing the ecological validity and standardization levels of experiments [26]. Therefore, the application of 3D animations in mate choice research offers significant advantages, particularly in studies requiring high precision and complex behavior demonstration.

Recently, research on fish mate choice behavior based on 3D simulation technology has been reported in a study of the sailfin molly (*Poecilia latipinna*). The researchers created animations of *P. latipinna* using the FishSim animation Toolchain (FishSim Animation Toolchain, https://bitbucket.org/EZLS/fish_animation_toolchain/, accessed on 15 November 2024) and Blender v.2.70a software (Blender Foundation, Amsterdam, the Netherlands), conducting mating choice experiments by controlling the swimming behavior of animated fish with a game controller, thereby introducing new perspectives and techniques into the field of animal behavior research [33]. The advantage of this method lies in the ability to investigate individual mate selection behavior through the artificial control of fish behavior. Using animation technology to study fish mate choice through visual signals raises a key question: can individuals distinguish between live fish and 3D animations? If they can reliably identify animations and show consistent preferences for both live fish and animations, then animations could effectively replace live fish in future studies. However, comparative studies on this topic are limited, as many have only tested stimuli individually, without direct comparisons. For instance, a study on *P. latipinna* found that females responded equally to both animations and videos, and individuals showed a stronger response to 3D animations than to 3D boxes. Additionally, males could also differentiate between animations of both sexes [32]. This study demonstrated the advantages of 3D animation in mate choice research on *P. latipinna* and its potential to replace live fish. However, it did not directly compare the preferences for live fish versus animations, which is a gap in this field. While this may not significantly affect the use of 3D animations in fish mate choice studies, future research that includes direct comparisons between live fish and animation stimuli would enrich the findings and deepen our understanding of how fish phenotypic cues influence courtship behavior. Compared to animations created with FishSim and the Blender software, Maya's 3D animations demonstrate superior precision, realistically representing the true coloration and behavioral characteristics of fish in natural environments. Therefore, it is recommended that future studies combine external controllers with the Maya 3D technology, which will not only enhance the control over fish behavior but also improve the quality of fish animation models. This technology is expected to be widely applicable in various fish behavioral experiments, not limited to mate choice studies.

We found that significant technical differences existed between 2D and 3D animations when creating animation models of *G. affinis*. The first is the display quality of the animation: 2D animations are created through hand drawing or digital drawing techniques, presenting flat images with a relatively simple visual style, limited in depth and three-dimensionality [47]. In contrast, 3D animations use 3D modeling software to create three-dimensional images, generating realistic visuals through rendering techniques. This can portray complex light and shadow effects, authentic textures, and intricate details. Secondly, regarding the fineness of behavior, the precision of 2D animations depends on the number of drawn frames and details. Although the smoothness of the animation is high, the representation of movements is constrained to the two-dimensional plane. Meanwhile, 3D animation, through skeletal rigging and motion capture technology, can more intricately simulate complex behaviors and actions, providing a stronger sense of realism [48,49]. Meanwhile, in terms of skin texture, 2D animation primarily conveys texture through color, shading, and texture drawings. Although texture effects can be enhanced through technical means, they remain inherently flat and lack realism. Meanwhile, 3D animation utilizes texture mapping and rendering techniques to simulate authentic skin textures and light effects, making the appearance of the fish more lifelike [50,51]. Overall, both 2D and 3D animations have their respective advantages and characteristics when creating animation models of *G. affinis*. Specifically, 2D animation is more advantageous in terms of the production cost and efficiency, while 3D animation exhibits significant advantages in the display quality, behavioral precision, and skin texture. In the future, it is essential to select the most appropriate research method based on the specific situational needs in fish behavioral research, and the Maya 3D simulation technology is undoubtedly an excellent choice.

5. Conclusions

This study established an animation model of *G. affinis* using the Maya 3D technology, validating the effectiveness of 3D animation in researching mate choice in this species and exploring their mate selection patterns. Maya 3D animation holds great promise for the study of fish mate choice, as its powerful animation and simulation tools enable researchers to accurately simulate visual signals such as the body posture, coloration, and movement patterns of fish, creating highly realistic virtual fish models. This technology aids in controlling variables, allowing the testing of different visual stimuli on mate choice behavior within a virtual environment, thus providing new means to understand the mechanisms of sexual selection in fish. Furthermore, Maya can simulate complex underwater environments and lighting conditions, making experiments more closely aligned with natural scenarios, thereby enhancing the validity and ecological significance of the experimental results. In future research, we will focus on more efficiently producing 3D simulation animations of fish using deep learning technologies and will conduct further exploration in scene rendering.

Supplementary Materials: The following supporting information can be downloaded at: https://www.mdpi.com/article/10.3390/ani14233369/s1, Figure S1: A map showing the distribution of the sampling sites in Longyan City, Fujian Province, China; Figure S2: Images of (a) males (M1–M10) and (b) females (F1–F10) used to generate 3D simulation animations; Video S1: Male 3D simulation animation; Video S2: Female 3D simulation animation.

Author Contributions: Conceptualization, experimental design and project administration, K.L.; Methodology, K.L., B.F. and L.C.; Software, L.C.; Validation, L.C. and B.F.; Formal analysis: L.C.; Investigation: L.C.; Data curation: B.F. and J.L.; Writing—original draft preparation: B.F. and L.C.; Writing—review and editing: K.L. and L.H.; Visualization: L.C.; Supervision: K.L.; Funding acquisition: K.L. All authors have read and agreed to the published version of the manuscript.

Funding: This research was funded by the Natural Science Foundation of Fujian Province, China, Grant No. 2024J01705.

Institutional Review Board Statement: This study met all relevant ethical standards for experimental research, approved by the Animal Care and Use Committee of Fisheries College of Jimei University (Animal Ethics No. 1067).

Informed Consent Statement: Not applicable.

Data Availability Statement: The original contributions presented in the study are included in the article; further inquiries can be directed to the corresponding author.

Acknowledgments: We sincerely thank the reviewers for their insightful comments that helped to improve the manuscript.

Conflicts of Interest: The authors declare no conflicts of interest.

References

1. Andersson, M. *Sexual Selection*; Princeton University Press: Princeton, NJ, USA, 1994.
2. Darwin, C. *The Descent of Man, and Selection in Relation to Sex*; D. Appleton: New York, NY, USA, 1872; Volume 2.
3. Ladich, F. Recent advances in hearing and sound communication by fishes. *J. Acoust. Soc. Am.* **2006**, *120*, 3102–3103. [CrossRef]
4. Ladich, F. Ecology of sound communication in fishes. *Fish Fish.* **2019**, *20*, 552–563. [CrossRef] [PubMed]
5. Ferrari, M.C.; Wisenden, B.D.; Chivers, D.P. Chemical ecology of predator–prey interactions in aquatic ecosystems: A review and prospectus. *Can. J. Zool.* **2010**, *88*, 698–724. [CrossRef]
6. Weiss, L.; Laforsch, C.; Tollrian, R. The taste of predation and the defences of prey. In *Chemical Ecology in Aquatic Systems*; OUP Oxford: Oxford, UK, 2012; pp. 111–126.
7. Martel, G.; Dill, L.M. Feeding and aggressive behaviours in juvenile coho salmon (*Oncorhynchus kisutch*) under chemically-mediated risk of predation. *Behav. Ecol. Sociobiol.* **1993**, *32*, 365–370. [CrossRef]

8. Andersson, M.; Simmons, L.W. Sexual selection and mate choice. *Trends Ecol. Evol.* **2006**, *21*, 296–302. [CrossRef]
9. Ryan, M.J. Sexual selection and mate choice. *Behav. Ecol. Evol. Approach* **1997**, *4*, 179–202.
10. Jones, A.G.; Ratterman, N.L. Mate choice and sexual selection: What have we learned since Darwin? *Proc. Natl. Acad. Sci. USA* **2009**, *106*, 10001–10008. [CrossRef]
11. Cornuau, J.H.; Rat, M.; Schmeller, D.S.; Loyau, A. Multiple signals in the palmate newt: Ornaments help when courting. *Behav. Ecol. Sociobiol.* **2012**, *66*, 1045–1055. [CrossRef]
12. Herdman, E.J.; Kelly, C.D.; Godin, J.G.J. Male mate choice in the guppy (*Poecilia reticulata*): Do males prefer larger females as mates? *Ethology* **2004**, *110*, 97–111. [CrossRef]
13. Dosen, L.D.; Montgomerie, R. Female size influences mate preferences of male guppies. *Ethology* **2004**, *110*, 245–255. [CrossRef]
14. Matthews, I.; Evans, J.; Magurran, A.E. Male display rate reveals ejaculate characteristics in the Trinidadian guppy, *Poecilia reticulata*. *Proc. R. Soc. Lond. Ser. B Biol. Sci.* **1997**, *264*, 695–700. [CrossRef]
15. Rosenthal, G.G. *Mate Choice: The Evolution of Sexual Decision Making from Microbes to Humans*; Princeton University Press: Princeton, NJ, USA, 2017.
16. Rowland, W.J. Studying visual cues in fish behavior: A review of ethological techniques. *Environ. Biol. Fishes* **1999**, *56*, 285–305. [CrossRef]
17. Kane, A.S.; Salierno, J.D.; Gipson, G.T.; Molteno, T.C.; Hunter, C. A video-based movement analysis system to quantify behavioral stress responses of fish. *Water Res.* **2004**, *38*, 3993–4001. [CrossRef] [PubMed]
18. Fleishman, L.; Endler, J. Some comments on visual perception and the use of video playback in animal behavior studies. *Acta Ethologica* **2000**, *3*, 15–27. [CrossRef]
19. Boom, B.J.; He, J.; Palazzo, S.; Huang, P.X.; Beyan, C.; Chou, H.-M.; Lin, F.-P.; Spampinato, C.; Fisher, R.B. A research tool for long-term and continuous analysis of fish assemblage in coral-reefs using underwater camera footage. *Ecol. Inform.* **2014**, *23*, 83–97. [CrossRef]
20. Mazzi, D.; Künzler, R.; Bakker, T.C. Female preference for symmetry in computer-animated three-spined sticklebacks, *Gasterosteus aculeatus*. *Behav. Ecol. Sociobiol.* **2003**, *54*, 156–161. [CrossRef]
21. Robinson-Wolrath, S.I. Video playback versus live stimuli for assessing mate choice in a pipefish. *Environ. Biol. Fishes* **2006**, *75*, 409–414. [CrossRef]
22. Rosenthal, G.G.; Ryan, M.J. Assortative preferences for stripes in danios. *Anim. Behav.* **2005**, *70*, 1063–1066. [CrossRef]
23. Turnell, E.; Mann, K.; Rosenthal, G.G.; Gerlach, G. Mate choice in zebrafish (*Danio rerio*) analyzed with video-stimulus techniques. *Biol. Bull.* **2003**, *205*, 225–226. [CrossRef]
24. Rosenthal, G.G.; Wagner, W.E., Jr.; Ryan, M.J. Secondary reduction of preference for the sword ornament in the pygmy swordtail *Xiphophorus nigrensis* (Pisces: Poeciliidae). *Anim. Behav.* **2002**, *63*, 37–45. [CrossRef]
25. Baldauf, S.; Kullmann, H.; Thünken, T.; Winter, S.; Bakker, T. Computer animation as a tool to study preferences in the cichlid *Pelvicachromis taeniatus*. *J. Fish Biol.* **2009**, *75*, 738–746. [CrossRef] [PubMed]
26. Chouinard-Thuly, L.; Gierszewski, S.; Rosenthal, G.G.; Reader, S.M.; Rieucau, G.; Woo, K.L.; Gerlai, R.; Tedore, C.; Ingley, S.J.; Stowers, J.R.; et al. Technical and conceptual considerations for using animated stimuli in studies of animal behavior. *Curr. Zool.* **2016**, *63*, 5–19. [CrossRef]
27. Naik, H.; Bastien, R.; Navab, N.; Couzin, I.D. Animals in virtual environments. *IEEE Trans. Vis. Comput. Graph.* **2020**, *26*, 2073–2083. [CrossRef] [PubMed]
28. Harland, D.P.; Jackson, R.R. Influence of cues from the anterior medial eyes of virtual prey on *Portia fimbriata*, an araneophagic jumping spider. *J. Exp. Biol.* **2002**, *205*, 1861–1868. [CrossRef]
29. Bian, X.; Chandler, T.; Laird, W.; Pinilla, A.; Peters, R. Integrating evolutionary biology with digital arts to quantify ecological constraints on vision-based behaviour. *Methods Ecol. Evol.* **2018**, *9*, 544–559. [CrossRef]
30. Bian, X.; Chandler, T.; Pinilla, A.; Peters, R.A. Now you see me, now you don't: Environmental conditions, signaler behavior, and receiver response thresholds interact to determine the efficacy of a movement-based animal signal. *Front. Ecol. Evol.* **2019**, *7*, 130. [CrossRef]
31. Nakayasua, T.; Watanabea, E. Analysis of movements of medaka fish and construction of 3D animation. In Proceedings of the 6th International Symposium on Aero Aqua Bio-mechanisms, Honolulu, HI, USA, 13–16 November 2014.
32. Gierszewski, S.; Müller, K.; Smielik, I.; Hütwohl, J.-M.; Kuhnert, K.-D.; Witte, K. The virtual lover: Variable and easily guided 3D fish animations as an innovative tool in mate-choice experiments with sailfin mollies-II. Validation. *Curr. Zool.* **2017**, *63*, 65–74. [CrossRef]
33. Gierszewski, S.; Baker, D.; Müller, K.; Hütwohl, J.-M.; Kuhnert, K.-D.; Witte, K. Using the FishSim animation toolchain to investigate fish behavior: A case study on mate-choice copying in Sailfin Mollies. *J. Vis. Exp.* **2018**, *141*, e58435. [CrossRef]
34. Roberts, N.S.; McCaulley, C.; Mendelson, T.C. Validating the use of computer animations in male *Etheostoma zonale*: A comparison of individual response to live and artificial stimuli. *Curr. Zool.* **2019**, *65*, 725–727. [CrossRef]
35. Langerhans, R.B.; Makowicz, A.M. Sexual selection paves the road to sexual isolation during ecological speciation. *Evol. Ecol. Res.* **2013**, *15*, 633–651.
36. Endler, J.A. Natural and sexual selection on color patterns in poeciliid fishes. In *Evolutionary Ecology of Neotropical Freshwater Fishes: PROCEEDINGS of the 1st International Symposium on Systematics and Evolutionary Ecology of Neotropical Freshwater Fishes, Held at DeKalb, IL, USA, 14–18 June 1982*; Springer: Dordrecht, The Netherlands, 1984; pp. 95–111.

37. Chen, B.-j.; Liu, K.; Zhou, L.-j.; Gomes-Silva, G.; Sommer-Trembo, C.; Plath, M. Personality differentially affects individual mate choice decisions in female and male Western mosquitofish (*Gambusia affinis*). *PLoS ONE* **2018**, *13*, e0197197. [CrossRef] [PubMed]
38. Tickoo, S. *Autodesk Maya 2018: A Comprehensive Guid*; Cadcim Technologies: Schererville, IN, USA, 2018.
39. Polverino, G.; Cigliano, C.; Nakayama, S.; Mehner, T. Emergence and development of personality over the ontogeny of fish in absence of environmental stress factors. *Behav. Ecol. Sociobiol.* **2016**, *70*, 2027–2037. [CrossRef]
40. Sommer-Trembo, C.; Bierbach, D.; Arias-Rodriguez, L.; Verel, Y.; Jourdan, J.; Zimmer, C.; Riesch, R.; Streit, B.; Plath, M. Does personality affect premating isolation between locally-adapted populations? *BMC Evol. Biol.* **2016**, *16*, 138. [CrossRef] [PubMed]
41. Gatesy, S.M.; Baier, D.B.; Jenkins, F.A.; Dial, K.P. Scientific rotoscoping: A morphology-based method of 3-D motion analysis and visualization. *J. Exp. Zool. Part A Ecol. Genet. Physiol.* **2010**, *313*, 244–261. [CrossRef]
42. Sfakiotakis, M.; Lane, D.M.; Davies, J.B.C. Review of fish swimming modes for aquatic locomotion. *IEEE J. Ocean. Eng.* **1999**, *24*, 237–252. [CrossRef]
43. Blake, R.W. Fish functional design and swimming performance. *J. Fish Biol.* **2004**, *65*, 1193–1222. [CrossRef]
44. Khan, A.H.; Hussmann, K.R.; Powalla, D.; Hoerner, S.; Kruusmaa, M.; Tuhtan, J.A. An open 3D CFD model for the investigation of flow environments experienced by freshwater fish. *Ecol. Inform.* **2022**, *69*, 101652. [CrossRef]
45. Gomes-Silva, G.; Liu, K.; Chen, B.; Plath, M.; Sommer-Trembo, C. Does individual variation in male mate choice copying reflect differences in social responsiveness? *Choice* **2017**, *14*, 31–34. [CrossRef]
46. Zhao, Y.; Jiang, J.; Chen, Y.; Liu, R.; Yang, Y.; Xue, X.; Chen, S. Metaverse: Perspectives from graphics, interactions and visualization. *Vis. Inform.* **2022**, *6*, 56–67. [CrossRef]
47. Rall, H. *Animation: From Concepts and Production*; CRC Press: Boca Raton, FL, USA, 2017.
48. Maestri, G. *Digital Character Animation 3*; New Riders: Indianapolis, IN, USA, 2006.
49. Ratner, P. *3-D Human Modeling and Animation*; John Wiley & Sons: Hoboken, NJ, USA, 2012.
50. Agusanto, K.; Li, L.; Chuangui, Z.; Sing, N.W. Photorealistic rendering for augmented reality using environment illumination. In Proceedings of the Second IEEE and ACM International Symposium on Mixed and Augmented Reality, Tokyo, Japan, 10 October 2003; pp. 208–216.
51. Alexander, O.; Rogers, M.; Lambeth, W.; Chiang, M.; Debevec, P. The digital emily project: Photoreal facial modeling and animation. In *Acm Siggraph 2009 Courses*; ACM: New York, NY, USA, 2009; pp. 1–15.

animals

MDPI

Article

Boat Noise Increases the Oxygen Consumption Rate of the Captive Juvenile Large Yellow Croaker, *Larimichthys crocea*

Ruijie Xu [1,†], Shouguo Yang [2,†], Yiyu Li [1], Xuguang Zhang [1,*] and Xianming Tang [2,*]

1 Engineering Technology Research Center of Marine Ranching, College of Oceanography and Ecological Science, Shanghai Ocean University, Shanghai 201306, China; m230501259@st.shou.edu.cn (R.X.); eyufd31@163.com (Y.L.)
2 Hainan Provincial Key Laboratory of Tropical Maricultural Technology, Hainan Academy of Ocean and Fisheries Sciences, Haikou 571126, China; yangshouguo82@163.com
* Correspondence: zhang_xuguang@163.com (X.Z.); hn.tangxm@aliyun.com (X.T.)
† These authors contributed equally to this work and share first authorship.

Simple Summary: Coastal noise pollution threatens wild large yellow croaker (*Larimichthys crocea*) survival. Despite artificial restocking efforts, populations of this ecologically and commercially vital fish continue to decline. Our lab experiments reveal that boat noise elevates oxygen demand in juvenile croakers at high temperatures, pushing them beyond metabolic limits. These findings highlight a hidden threat: noise disrupts energy balance, particularly when fish struggle to cope with warming seas. We urge revised noise regulations in climate-vulnerable habitats to prevent croaker population collapse.

Abstract: Anthropogenic noise pollution is increasingly acknowledged as a major threat to marine ecosystems, especially for sound-sensitive species, such as the large yellow croaker (*Larimichthys crocea*). While the effects of underwater noise on fish behavior and physiology have been well-documented, its influence on oxygen metabolism across varying temperatures remains poorly understood. This study examines the impact of boat noise on the oxygen consumption rate (OCR) of juvenile large yellow croakers at different temperatures, a key factor in their metabolic activity. The underwater noise generated by a fishing boat spans a broad frequency range, with a peak spectrum level of 130 dB re 1 μPa at low frequencies between 100 and 200 Hz. Our findings reveal that boat noise significantly elevates the OCR of juvenile fish, with mass-specific OCR increasing by 65.0%, 35.3%, and 28.9% at 18 °C, 25 °C, and 30 °C, respectively. Similarly, individual OCR rose by 60.7%, 35.3%, and 17.1% at these temperatures. These results demonstrate that boat noise triggers a stress response in fish, resulting in heightened metabolic demands across different seasonal conditions. Notably, the impact of boat noise on respiratory metabolism is most significant at lower temperatures. In aquatic environments with stable oxygen levels, the noise-induced rise in oxygen consumption could lead to hypoxia and provoke maladaptive behavioral changes in fish.

Keywords: noise pollution; metabolic stress; optimal temperature

1. Introduction

Recently, there has been growing concern about the detrimental influence of anthropogenic underwater noise on marine life, including fish and other aquatic animals [1,2]. In fish, human-made underwater noise can mask acoustic signals, change behavior, induce physiological changes, such as high stress responses, and result in physical harm or mortality [3–6]. The extent of the impact depends on several factors, including noise levels, frequency, temporal patterns, and the distance from the source. Noise levels determine the intensity of the sound, with higher levels causing more pronounced physiological and behavioral responses in fish [7–9]. Frequency is also critical, as fish are most sensitive to sounds within specific frequency ranges, typically between 100 and 1000 Hz [10,11]. Temporal patterns, such as the duration, intermittency, and predictability of noise, play a significant role in determining its impact. For example, continuous noise can lead to chronic stress, while intermittent noise may cause acute stress responses [12–14]. Additionally, the distance from the noise source influences the sound pressure level (SPL) experienced by the fish, with closer proximity resulting in higher SPL and greater physiological impacts [7–9]. Temporal patterns of noise exposure, including duration, intermittency, and predictability, are critical factors influencing fish responses. For instance, prolonged exposure to continuous noise can lead to chronic stress, resulting in suppressed immune function, reduced growth, and altered behavior [12–14]. In contrast, intermittent noise, such as that produced by passing boats, can cause acute stress responses, including increased swimming activity and elevated cortisol levels [10,15]. Predictable noise patterns, such as regular boat traffic, may allow fish to habituate over time, whereas unpredictable noise can lead to sustained stress responses [7,8]. Recent studies have shown that temporal variability in noise exposure can significantly affect fish physiology and behavior, highlighting the need for further research on these dynamics [10,12–14].

The large yellow croaker, native to East Asia, is one of the most commercially important marine fish species, with extensive aquaculture development along the southeastern coast of China [16]. Nearshore floating net cages are commonly used for the rearing of large yellow croaker from juveniles to adults. At the same time, juvenile fish will also be released into the nearshore to replenish fishery stocks in the ocean. Anthropogenic underwater noise, especially originating from boat traffic, is concentrated not only in coastal waters but also extends to the open ocean. Juvenile large yellow croakers are inevitably disturbed by boat noise during their development in both artificial aquaculture cages and natural marine environments. The large yellow croaker is characterized by its large sagittae otolith, rendering it highly sensitive to sound [17]. Fish experience physiological and biochemical changes when exposed to environmental stressors, like high temperature and noise. These stress responses affect their metabolism, immunity, and behavior including oxygen demand [18,19]. As water temperature rises, fish metabolic rate increases, requiring more oxygen for critical organs, like the heart and gill [20,21]. If the temperature exceeds a certain threshold, fish may not receive enough oxygen to meet the increased demand, resulting in cellular dysfunction and tissue instability [21].

Environmental noise is a significant stressor for fish, imposing a metabolic burden. Noise pollution, especially from certain sources, like vessel noise, can increase stress hormone levels in fish. Stress-induced physiological disruptions, including immune suppression and impaired growth, are well-documented in fish under environmental stressors [22–24]. However, the present study focuses on oxygen consumption as a direct metabolic response to noise exposure. Similar to other organisms, fish utilize oxygen during respiration to sustain cellular respiration and provide energy for physiological processes and behaviors. Oxygen consumption rate is a crucial indicator of the fish's metabolic activity and energy expenditure, as well as its ability to respond to environmental fluctuations [25]. Conversely, for fish, direct measurements of respiratory oxygen consumption are employed to gauge their metabolic status [26].

Temperature is a critical environmental factor that significantly influences the metabolic activity, growth, and survival of the large yellow croaker. To assess the combined effects of noise and temperature on oxygen consumption, we selected three temperature gradients (18 °C, 25 °C, and 30 °C) that reflect both natural seasonal variations and the species' thermal tolerance range [27,28].

This study employed intermittent respiration measurement technology to investigate the effect of boat noise exposure on the oxygen consumption of juvenile yellow croaker at varying temperatures. As a coastal fish species sensitive to sound, understanding the potential impact of noise on the metabolic function of the large yellow croaker is crucial for sustainable fisheries management of this species.

2. Materials and Methods

2.1. Fish Collection and Husbandry

The juvenile large yellow croakers were all healthy and disease-free individuals and were obtained from the Zhoushan Marine Fish Hatchery (Zhejiang, China) in April 2024. After arrival at the lab of Shanghai Ocean University, to minimize the influence of housing conditions on oxygen consumption rates, fish were housed in a large indoor tank (300 × 300 × 60 cm) with a recirculating water system that maintained stable temperature (18 °C, 25 °C, 30 °C), salinity (25‰), and dissolved oxygen levels (7.97 ± 0.47 mg/L). Fish were acclimated to these conditions for at least 10 days before the experiments to reduce stress and ensure they were in a stable physiological state. A 12 h light–dark cycle was maintained to simulate natural conditions and minimize disruptions to the fish's circadian rhythm. The fish were maintained in water bodies with sufficient space and oxygen levels, and no significant mortality occurred throughout the experiments.

2.2. Intermittent (Stop-Flow) Respirometry

The impact of boat noise on the oxygen consumption rates of *L. crocea* was assessed by measuring the oxygen uptake of a group of four fish using an intermittent flow respirometer [26,29,30] equipped with a four-channel resting respirometry system. The four-channel resting respirometry system comprises four respirometry chambers (Loligo® Systems, Viborg, Denmark), with each chamber measuring 20 cm in length and 10 cm in diameter, submerged parallel in a rectangular water tank filled with seawater. The rectangular tank has dimensions of 120 × 90 × 30 cm (length × width × height), with a water depth of 25 cm. During the experiment, the water temperature (18 °C, 25 °C, 30 °C) and salinity (25‰) were maintained to match the conditions of the fish captivity environment. Each respirometry chamber is circulated by two water pumps interconnected with rubber tubes (Figure 1). One pump (flush pump) operates within an external circuit to exchange of water between the respirometry chamber and surrounding water body. The second pump (recirculation pump) controls the water flow within the respirometry chamber. Intermittent respirometry allows the measurement of aquatic oxygen consumption rate through the alternating activation and deactivation of water flow via two circuits. The oxygen content within the chamber was measured using a fiber optic oxygen sensor positioned inside a recirculation pump tube.

During the experiment, oxygen levels in the chamber were recorded at a frequency of once every 5 s to ensure continuous mixing of the water within the system. Due to the potential risk of fish suffocation and mortality with extended exposure, each experiment was limited to a duration of 30 min.

2.3. Experiment on Oxygen Consumption

The oxygen consumption rate of juvenile fish was assessed by comparing a control group left undisturbed (no noise exposure) with a treatment group exposed to boat noise. To minimize stress during the transfer of juvenile croakers into the respirometry chambers, the fish were gently netted and allowed to acclimate in the chambers for 30 min prior to measurements. During this period, the chambers remained submerged in the main tank to maintain consistent environmental conditions, such as temperature and salinity. This acclimation period helped reduce handling stress and ensured that the fish were in a stable physiological state before oxygen consumption measurements commenced. To ensure that the observed effects on oxygen consumption rates reflected the acute impact of noise exposure, measurements were conducted immediately after sound exposure. However, the fish in the control group were undisturbed by the noise. These fish were subjected to the same experimental conditions as the treatment group (e.g., placed in a respirometry chambers, temperature gradients, and oxygen consumption measurements).

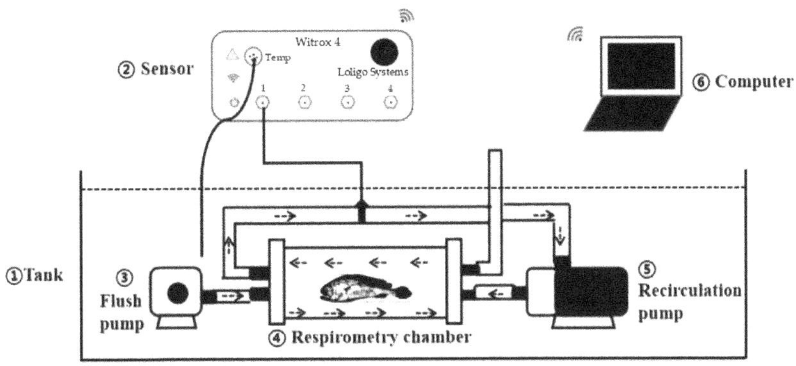

Figure 1. Schematic diagram of the respirometry system. ① Tank: the water reservoir that holds the experimental medium. ② Sensor (Witrox 4, Loligo® Systems, Viborg, Denmark): measures the oxygen concentration and temperature in the system. It includes an integrated temperature sensor (Temp) and other monitoring components. ③ Flush pump: circulates water to ensure a steady flow through the respirometry chamber for flushing. ④ Respirometry chamber: houses the experimental organism (e.g., fish) and is the primary site for measuring oxygen consumption. ⑤ Recirculation pump: maintains water flow within the system during closed respirometry periods. ⑥ Computer: records and analyzes data collected by the sensor system, including oxygen levels and temperature. The water current flows in the direction shown by the arrows, with water being flushed through the respirometry chamber by the flush pump (③) and recirculated during measurement by the recirculation pump (⑤). The recirculation pump (⑤) was always open. During the measuring phases (30 min), the flush pump (③) was closed and opened during refresh phases (5 min), so that oxygen-saturated water from the surrounding tank was pumped into the respirometry chamber.

Based on the varying water temperature during the life cycle of the large yellow croaker, three water temperature gradients at 18 °C, 25 °C, and 30 °C were set to evaluate the influence of boat noise on its respiratory metabolism across different water temperature. To minimize stress on the fish, we did not simultaneously expose them to multiple temperature gradients. Instead, we took advantage of the natural increase in water temperature from April to June, raising the temperature for each experimental group. The respirometry experiments were conducted on April 29th, May 20th, and June 28th at temperatures spanning 14–23 °C, 22–27 °C, and 27–31 °C, respectively. The water temperature in the fish holding environment was maintained at 18 °C, 25 °C, and 30 °C using an aquarium heater (ARH, 500W; SUNSUN, Zhoushan, China) on the corresponding dates.

During the experiments, each temperature group consisted of 20 fish for both the control and treatment groups. For each temperature condition, we conducted two trials (one for the control group and one for the treatment group), resulting in a total of six trials across all temperature conditions. Each trial involved four fish tested simultaneously in the respirometry chambers, meaning that there were five replicates in each condition. Throughout the entire experimental process, the water quality conditions in the holding tanks for the experimental fish remained the same (as described in Section 2.2). The fish had an average total length of (10.94 ± 0.82) cm, an average standard length of (8.46 ± 0.57) cm, and an average body weight of (10.26 ± 1.97) g. No significant difference in fish body weight was observed across groups (F = 2.094, $p = 0.128$), ensuring the reliability of the experimental data.

2.4. Noise Exposure

The boat noise emitted by a wooden fishing boat powered by a 22-horsepower outboard engine was recorded in a mussel farming area abundant with croaker populations, including large yellow croaker. The recording was performed using a Reson hydrophone (TC4032 with an inbuilt preamplifier, sensitivity −170 dB re 1V/μPa, calibrated by the manufacturers; Teledyne Marine, Slangerup, Denmark) and an acquisition module (BK3050-A-040, 48 kHz sampling rate; Brüel & Kjær, Skodsborgvej, Denmark), approximately 20 m away from the fishing boat as it passed by, and the recording was saved as a Wav sound file. The file was played back through a portable computer with the Audacity 2.4 software. After passing through the power amplifier (Sanshui S2-350, Guangzhou, China), the signal was fed to an underwater speaker (UW30 frequency response of 0.1–10 kHz, Lubell Labs Inc., Columbus, OH, USA). An annular plate was utilized to mount the underwater speaker facing upward at the center of the bottom of a cylindrical tank with a diameter of 60 cm; the tank was filled with seawater to a depth of 70 cm.

Figure 2 shows the mean sound pressure levels of tank boat noise obtained through spectral analysis using BK Connect software (version 2009.1) with a 1 Hz resolution and a Hanning window (66.7% overlap). Considering that most fish species are sensitive to frequencies between 100 and 1000 Hz, the analysis was limited to the 1–1600 Hz range in the spectrum, covering the auditory range of fish. The mean SPLrms from 1 to 1600 Hz was measured as (106 ± 9.1) dB re 1 μPa (mean ± S.D.), with a peak spectrum level of 130 dB re 1 μPa at low frequencies between 100 and 200 Hz at a distance of 10 cm above the speaker. The juvenile large yellow croaker was randomly exposed to boat noise for durations of 20–40 s, repeated six times within a 10 min period in the tank. This aimed to simulate the time it takes for fish to escape when encountering boat and to avoid excessive exposure that could cause harm to the fish.

2.5. Calculations and Analysis

The oxygen consumption rate was determined by analyzing the fluctuations in oxygen levels within the chamber during measurement intervals. The oxygen consumption rate of juvenile large yellow croaker was calculated using the following formula [31,32]:

$$Q = (C_0 - C_t)\,V/W/t \quad (1)$$

$$Q_1 = (C_0 - C_t)\,V/t \quad (2)$$

where Q is the mass-specific oxygen consumption rate ($mg \cdot g^{-1} \cdot h^{-1}$), Q_1 is the individual oxygen consumption rate ($mg \cdot ind^{-1} \cdot h^{-1}$), C_0 and C_t are the dissolved oxygen content (mg/L) at the beginning and end of the experiment, respectively, V is the water volume of the respirometry chamber (L), W is the body weight of the fish (g), and t is the closed respirometry period (h).

2.6. Statistical Analysis

The statistical analysis of the data was performed using SPSS 26.0 software. Data were first tested for normality with the D'Agostino–Pearson omnibus normality test. Datasets that passed the normality test were analyzed with a one-way ANOVA and t-test, with the least significant difference (LSD) method used for the multiple comparisons test. The significance level was set at 0.05, and when differences were significant ($p < 0.05$), multiple comparisons between group means were conducted using the Duncan's multiple comparisons test. Datasets that did not pass the normality test were analyzed with a non-parametric ANOVA on ranks (Kruskal–Wallis's test) with a Duncan's multiple comparisons test. Graphs and regression analysis were carried out using Origin 2024, with regression curves established ($p < 0.05$), and the curve fitting was based on maximizing the R^2 value to obtain the optimal theoretical value. Experimental data were expressed as mean ± standard deviation (mean ± S.D.).

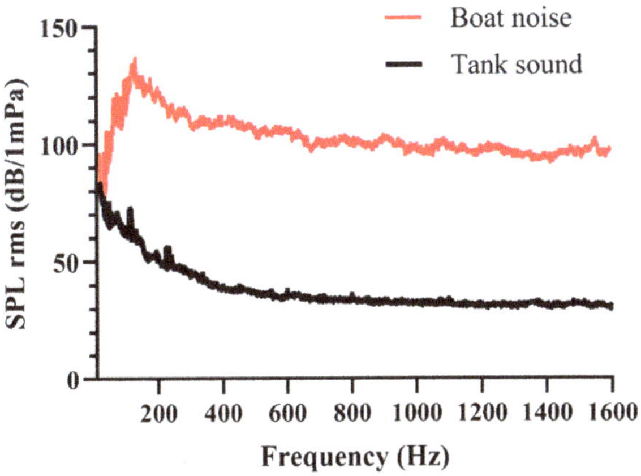

Figure 2. The sound pressure level of boat noise playback and ambient sound.

3. Results

3.1. Effects of Temperature on Oxygen Consumption Rate

The mass-specific oxygen consumption rate and individual oxygen consumption rate of fish exhibit a significant increase as temperature rises across three temperature levels (18 °C, 25 °C, and 30 °C; $p < 0.05$). At 18 °C, the mass-specific oxygen consumption rate was 0.20 ± 0.04 $mg \cdot g^{-1} \cdot h^{-1}$, and the individual oxygen consumption rate was 2.01 ± 0.46 $mg \cdot ind^{-1} \cdot h^{-1}$. At 25 °C and 30 °C, the mass-specific oxygen consumption rates were 0.34 ± 0.08 and 0.45 ± 0.10 $mg \cdot g^{-1} \cdot h^{-1}$, and the individual oxygen consumption rates were 3.57 ± 1.62 and 4.84 ± 1.55 $mg \cdot ind^{-1} \cdot h^{-1}$, respectively (Table 1).

Table 1. The mass-specific oxygen consumption rate and individual oxygen consumption rate exposure to boat noise.

Temperature (°C)	Weight (g)		Mass-Specific Oxygen Consumption Rate ($mg \cdot g^{-1} \cdot h^{-1}$)		Individual Oxygen Consumption Rate ($mg \cdot ind^{-1} \cdot h^{-1}$)	
	Control	Noise	Control	Noise	Control	Noise
18.58 ± 0.64	10.25 ± 1.25 ^{Aa}	9.99 ± 1.29 ^{Aa}	0.20 ± 0.04 ^{Aa}	0.33 ± 0.08 ^{Ba}	2.01 ± 0.46 ^{Aa}	3.23 ± 0.91 ^{Ba}
25.26 ± 0.02	10.25 ± 3.16 ^{Aa}	10.35 ± 1.53 ^{Aa}	0.34 ± 0.08 ^{Ab}	0.46 ± 0.11 ^{Bb}	3.57 ± 1.62 ^{Ab}	4.83 ± 1.40 ^{Bb}
30.07 ± 0.09	10.75 ± 2.02 ^{Aa}	10.03 ± 2.05 ^{Aa}	0.45 ± 0.10 ^{Ac}	0.58 ± 0.16 ^{Bc}	4.84 ± 1.55 ^{Ac}	5.67 ± 1.59 ^{Bc}

Uppercase letter indicate significant differences ($p < 0.05$) in body weight and oxygen consumption rate of juvenile large yellow croaker between the control and noise-exposed groups under the same temperature conditions. Lowercase letters indicate significant differences ($p < 0.05$) in body weight and oxygen consumption rate of juvenile large yellow croaker across different temperatures under the same treatment conditions.

3.2. Changes in Oxygen Consumption Rate After Boat Noise Exposure

The presence of boat noise had a significant impact on the oxygen consumption rates of juvenile large yellow croaker. This effect was especially evident at elevated temperatures. At 18 °C, 25 °C, and 30 °C, both the mass-specific oxygen consumption rates and individual oxygen consumption rates were significantly higher in the noise treatment group compared to the control group (Table 1, Figure 3). As the temperature rose, both groups showed an increase in oxygen consumption rates, with a more pronounced rise observed in the noise-exposed group. In the control group, the mass-specific oxygen consumption rates were 0.20 ± 0.04 mg·g^{-1}·h^{-1}, 0.34 ± 0.08 mg·g^{-1}·h^{-1}, and 0.45 ± 0.10 mg·g^{-1}·h^{-1} at 18 °C, 25 °C, and 30 °C, respectively. When exposed to boat noise, the mass-specific oxygen consumption rate increases by 65.0%, 35.3%, and 28.9% at the respective temperatures (Table 1, Figure 3A). In the control group, the individual oxygen consumption rates were 2.01 ± 0.46 mg·ind^{-1}·h^{-1}, 3.57 ± 1.62 mg·ind^{-1}·h^{-1}, and 4.84 ± 1.55 mg·ind^{-1}·h^{-1} at 18 °C, 25 °C, and 30 °C, respectively. After exposure to boat noise, the individual oxygen consumption rates increased to 3.23 ± 0.91 mg·ind^{-1}·h^{-1}, 4.83 ± 1.40 mg·ind^{-1}·h^{-1}, and 5.67 ± 1.59 mg·ind^{-1}·h^{-1}, representing an increase of 60.7%, 35.3%, and 17.1%, respectively (Table 1, Figure 3B).

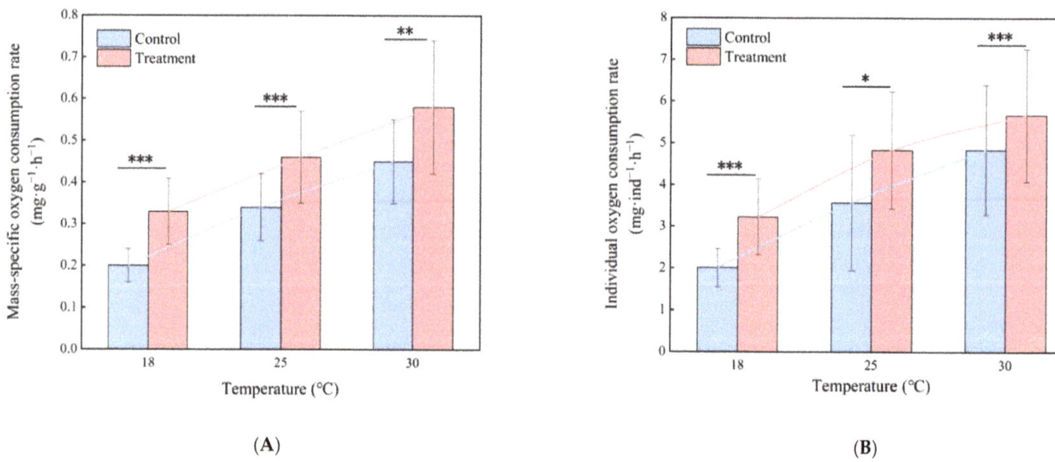

Figure 3. The mass-specific oxygen consumption rate (**A**) and individual oxygen consumption rate (**B**) in control fish (blue bar) and treatment fish (pink bar). The significance levels for each comparison are shown for the corresponding pairs through *t*-tests.* $p < 0.05$; ** $p < 0.01$; *** $p < 0.001$.

It is evident that as body weight increases, individual oxygen consumption rate also increases (Supplementary Table S1). In order to account for weight variations, this study focused solely on analyzing changes in mass-specific oxygen consumption rate over time.

At all temperature conditions, the mass-specific oxygen consumption rate of the treatment group was significantly higher than that of the control group, with a statistically significant difference ($p < 0.05$, Figure 4). However, at all temperatures, the oxygen consumption rate of both the control and treatment groups decreased over time (Figure 4). At 18 °C, the initial specific oxygen consumption rate of the treatment group was approximately 0.39 mg·g^{-1}·h^{-1}, while the control group had a rate of 0.29 mg·g^{-1}·h^{-1} at 5 min. By 30 min, both groups had stabilized at around 0.13 mg·g^{-1}·h^{-1}. At 25 °C, the treatment group had an initial specific oxygen consumption rate of approximately 0.64 mg·g^{-1}·h^{-1} at 5 min, higher than the control group's rate of 0.45 mg·g^{-1}·h^{-1}. By 30 min, the control group's rate had decreased to approximately 0.31 mg·g^{-1}·h^{-1}, while the treatment group's rate had decreased to approximately 0.26 mg·g^{-1}·h^{-1}. At 30 °C, the treatment group had the highest initial specific oxygen consumption rate at 5 min, at approximately 0.72 mg·g^{-1}·h^{-1}, while the control group had a rate of 0.56 mg·g^{-1}·h^{-1}. By 30 min, the control group's rate had decreased to approximately 0.35 mg·g^{-1}·h^{-1}, while the treatment group's rate had decreased to approximately 0.43 mg·g^{-1}·h^{-1}.

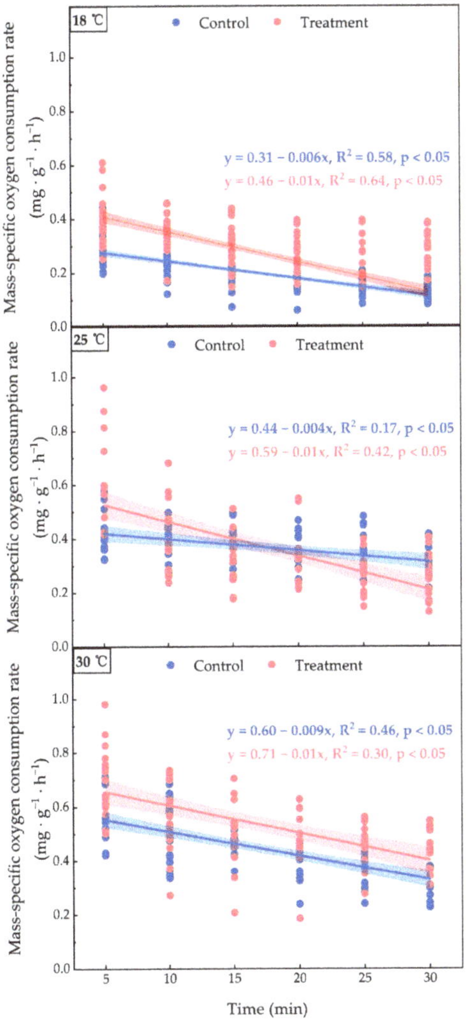

Figure 4. Changes in mass-specific oxygen consumption rate over time before and after exposure to boat noise. The figure illustrates the changes in mass-specific oxygen consumption rate (mg·g^{-1}·h^{-1}) over time (minutes) for the control and treatment groups under three temperature conditions: 18 °C, 25 °C, and 30 °C. The *y*-axis represents the mass-specific oxygen consumption rate (mg·g^{-1}·h^{-1}), and the *x*-axis represents time (minutes). The data are presented as box plots, with medians indicated by horizontal lines and means represented by white dots. The significance levels for each comparison are shown for the corresponding pairs through *t*-tests. * $p < 0.05$; ** $p < 0.01$; *** $p < 0.001$; ns indicates no significant difference.

4. Discussion

The oxygen metabolism of fish is predominantly controlled by internal physiological mechanisms, representing a consistent intrinsic trait that remains relatively unaffected by minor fluctuations in environmental conditions [33,34]. In similar environmental conditions, variations in measurement times, such as day or night, do not significantly affect the results (Supplementary Information). This physiological regulatory process is essential for ensuring the sustained and stable life processes of fish and for maintaining the reliability of the data collected in this experiment. While the transfer process, though its handling stress was minimized, may have influenced absolute oxygen consumption rates, the comparative validity between the noise-exposed and noise-free conditions remained unaffected.

Boat noise exposure, on the other hand, significantly elevated oxygen consumption rates in juvenile large yellow croaker, with temperature-dependent amplification (Figures 3 and 4). The temperature-driven escalation in oxygen consumption was markedly amplified under noise exposure, consistent with established noise-induced metabolic demand mechanisms in fish [35].

Temperature is a crucial factor in the oxygen consumption of organisms, particularly for fish, as changes in temperature can impact metabolic activities and adaptive behavioral responses. With the ongoing increase in global temperatures, fish and other aquatic creatures must adjust to new temperature ranges in order to sustain their normal physiological functions [20,36]. Specifically, 18 °C represents the lower end of their optimal temperature range, 25 °C falls within their optimal growth range (20 °C–28 °C), and 30 °C approaches the upper limit of their thermal tolerance, mimicking conditions that may occur during warmer seasons or due to climate change [37,38]. The experimental findings show a notable increase in the oxygen consumption rate of juvenile large yellow croaker with rising temperatures, followed by a gradual decrease over time across different temperature levels. The mass-specific oxygen consumption rate increased significantly by 70.0% when the temperature rose from 18 °C to 25 °C ($p < 0.05$), compared to a 32.4% increase from 25 °C to 30 °C ($p < 0.05$). These temperatures were tested over a short period, and they reflect the range of thermal conditions that could influence fish metabolism and growth [35,39]. The highest mass-specific and individual oxygen consumption rates were seen at 30 °C (Table 1). This suggests that metabolic activity in juvenile large yellow croaker peaks at higher temperatures, approaching the upper limit of their thermal tolerance.

Exposure to noise resulted in a significant rise in oxygen consumption rates, even under extreme temperatures critical for survival. It has been demonstrated that when juvenile large yellow croakers are exposed to underwater noise, they experience a stress response that results in increased cortisol and catecholamine levels in their bodies [15,22]. Cortisol and catecholamine help the fish mobilize energy reserves by promoting the production of glucose and breaking down fats, while also inhibiting protein synthesis [15,23,24]. These metabolic changes lead to higher energy expenditure, causing an increase in the oxygen consumption rate. Initially, the elevated oxygen consumption helps the fish cope with the stress caused by noise.

When juvenile large yellow croakers were transferred to the respirometry chamber, they experienced stress responses that had not fully subsided by the time oxygen consumption measurements were taken. Over time, the fish adapt to the noisy environment by reducing unnecessary activities to conserve energy, improving aerobic metabolic processes to minimize energy loss and optimizing energy production [23,24]. They also suppress energy-intensive processes, like protein synthesis, while activating energy-generating pathways, like glycolysis and the tricarboxylic acid cycle (TCA), to generate energy [23,24]. These adaptive strategies ultimately result in a decrease in mass-specific oxygen consumption rate in juvenile large yellow croaker. As such, this resulted in a decrease in mass-specific oxygen consumption rates for both the control and treatment groups across all three temperature conditions (Figure 4). The observed metabolic downregulation—likely a compensatory mechanism to reduce oxygen demand under thermal stress—aligns with established thermoregulatory strategies in fish [40,41]. Specifically, the rate gradually decreased as time progressed, with a more significant reduction in oxygen consumption at higher temperatures. This aligns with previous research indicating that fish will make behavioral and physiological adjustments to conserve energy and prevent excessive oxygen consumption in consistent high-temperature environments [20,42].

When the environment changes, fish need to consume extra energy to sustain metabolic homeostasis and adjust to the new environmental conditions [43]. The temperature of 25 °C is generally considered optimal for the growth and metabolic activities of large yellow croaker [39,44,45]. Temperatures of 18 °C and 30 °C are below and above the optimal growth temperature of large yellow croaker (approximately 25 °C), respectively. At low temperatures, metabolic processes slow down, resulting in a relative increase in energy allocation to stress responses. In contrast, at high temperature, elevated temperatures accelerate metabolic activities, thereby increasing energy demand [43,46]. The presence of noise, as an additional environmental stressor, further exacerbates energy consumption in fish [15,47]. Consequently, the mass-specific oxygen consumption rate in the treatment group was consistently higher than that in the control group across all time points.

The mass-specific oxygen consumption rate in the noise group initially exceeded that of the control group at 25 °C but later decreased below baseline levels. This biphasic response suggests that noise exposure disrupted metabolic regulation in large yellow croaker, despite their capacity for short-term stress adaptation at optimal temperatures [22]. Our experimental design minimized pseudoreplication risks through the individual isolation of fish and randomized allocation to noise/control groups, following established protocols for aquatic bioacoustics research [48]. These findings highlight the need to evaluate chronic noise effects under climate-driven thermal shifts, as acute experimental conditions may underestimate ecological impacts.

Based on standard experimental protocols in bioacoustic research, as well as the prevalence of local fishing vessels and their representative acoustic characteristics, we used the noise generated by a 22-horsepower wooden fishing boat as the sole noise source in our study [49–51]. In this study, the use of a single noise source helped control noise-related confounding factors, allowing us to investigate the impact of boat noise on fish oxygen consumption. This approach is supported by relevant studies in the literature [52–54]. However, underwater noise in marine environments is the cumulative effect of multiple sources. Therefore, the results of this study may only be applicable to the specific noise profile used and may not fully represent the effects of other types of boat noise on fish physiology. Future studies should incorporate recordings from multiple boats with varying engine types, sizes, and operational conditions. This would provide a more comprehensive understanding of how different noise profiles affect fish metabolisms and behavior.

5. Conclusions

From an ecological perspective, the increased oxygen consumption rates observed in this study suggest that noise pollution could disrupt the energy balance of wild fish populations, potentially affecting their survival, reproduction, and behavior [11,55,56]. In coastal habitats, where anthropogenic noise is concentrated, these effects could exacerbate the challenges faced by already vulnerable species, such as the large yellow croaker. Conservation efforts should, therefore, prioritize the reduction of noise pollution in critical habitats, such as breeding and nursery grounds, to support the recovery and sustainability of wild fish populations.

Future research should focus on the long-term effects of noise pollution on fish physiology and behavior, particularly in aquaculture and natural environments. Additionally, studies should investigate the effectiveness of noise mitigation strategies, such as acoustic barriers or habitat restoration, in reducing the impacts of anthropogenic noise on fish populations. These efforts will provide valuable insights for the sustainable management of aquatic resources and the conservation of marine biodiversity.

Supplementary Materials: The following supporting information can be downloaded at: https://www.mdpi.com/article/10.3390/ani15050714/s1, Figure S1: Changes in individual oxygen consumption rate with body weight; Table S1: Two-factor ANOVA for different times of the day and different experimental dates.

Author Contributions: X.Z. and X.T. designed the study and provided technical guidance; R.X., S.Y., and Y.L. carried out the experiments and the formal analysis. R.X. and X.Z. wrote the manuscript; X.T. reviewed and edited the manuscript. All authors have read and agreed to the published version of the manuscript.

Funding: This work was supported by the National Key R&D Program of China (2023YFD2401902); the National Natural Science Foundation of China (32373100); the Marine Genetic Resource Center and the fund of the Hainan Provincial Key Laboratory of Tropical Maricultural Technologies (HNTMTOF202305).

Institutional Review Board Statement: The animal study protocol was approved by the Animal Ethics Committee of Shanghai Ocean University (series number: SHOU-DW-2023-001). The experimental protocol for fish was performed in strict accordance with the requirements of Animal Ethics Procedures and Guidelines of the People's Republic of China.

Informed Consent Statement: Not applicable.

Data Availability Statement: The data included in this study can be provided on request from the corresponding author.

Acknowledgments: The authors express their gratitude to the editors and anonymous reviewers for their valuable comments and suggestions on this manuscript.

Conflicts of Interest: The authors declare no conflicts of interest.

References

1. Casper, B.M.; Halvorsen, M.B.; Matthews, F.; Carlson, T.J.; Popper, A.N. Recovery of barotrauma injuries resulting from exposure to pile driving sound in two sizes of hybrid striped bass. *PLoS ONE* **2013**, *8*, e73844. [CrossRef] [PubMed]
2. Kight, C.R.; Swaddle, J.P. How and why environmental noise impacts animals: An integrative, mechanistic review. *Ecol. Lett.* **2011**, *14*, 1052–1061. [CrossRef] [PubMed]
3. Wallenius, M.A. The interaction of noise stress and personal project stress on subjective health. *J. Environ. Psychol.* **2004**, *24*, 167–177. [CrossRef]
4. Hildebrand, J.A. Anthropogenic and natural sources of ambient noise in the ocean. *Mar. Ecol. Prog. Ser.* **2009**, *395*, 5–20. [CrossRef]
5. Neo, Y.Y.; Ufkes, E.; Kastelein, R.A.; Winter, H.V.; Ten, C.C.; Slabbekoorn, H. Impulsive sounds change European seabass swimming patterns: Influence of pulse repetition interval. *Mar. Pollut. Bull.* **2015**, *97*, 111–117. [CrossRef]
6. Purser, J.; Radford, A.N. Acoustic noise induces attention shifts and reduces foraging performance in three-spined sticklebacks (*Gasterosteus aculeatus*). *PLoS ONE* **2011**, *6*, e17478. [CrossRef] [PubMed]
7. Jong, K.D.; Forland, T.N.; Amorim, M.C.P.; Rieucau, G.; Slabbekoorn, H.; Sivle, L.D. Predicting the effects of anthropogenic noise on fish reproduction. *Rev. Fish Biol. Fish.* **2020**, *30*, 245–268. [CrossRef]
8. Francis, J.; Kieran, C.; Lawrence, B.; Sarah, D. The effect of anthropogenic and biological noise on fish behavior and physiology: A meta-analysis. *J. Acoust. Soc. Am.* **2017**, *141*, 3862. [CrossRef]
9. Cox, K.; Brennan, L.P.; Gerwing, T.G.; Dudas, S.E.; Juanes, F. Sound the alarm: A meta-analysis on the effect of aquatic noise on fish behavior and physiology. *Glob. Change Biol.* **2018**, *24*, 3105–3116. [CrossRef]

10. Zhang, X.G.; Tang, X.M.; Zou, L.H. Transcriptome analysis reveals dysfunction of the endoplasmic reticulum protein processing in the sonic muscle of small yellow croaker (*Larimichthys polyactis*) following noise exposure. *Mar. Environ. Res.* **2024**, *194*, 106299. [CrossRef]
11. Popper, A.N.; Hawkins, A.D. An overview of fish bioacoustics and the impacts of anthropogenic sounds on fishes. *J. Fish Biol.* **2019**, *94*, 692–713. [CrossRef] [PubMed]
12. Wong, M.L.; Lau, L.H.; Gordillo-Martinez, F.; Vasconcelos, R.O. The effect of time regime in noise exposure on the auditory system and behavioural stress in the zebrafish. *Sci. Rep.* **2022**, *12*, 15353. [CrossRef] [PubMed]
13. Neo, Y.Y.; Seitz, J.; Kastelein, R.A.; Winter, H.V.; Cate, C.T.; Slabbekoorn, H. Temporal structure of sound affects behavioural recovery from noise impact in European seabass. *Biol. Conserv.* **2014**, *178*, 65–73. [CrossRef]
14. Hubert, J.; Neo, Y.Y.; Winter, H.V.; Slabbekoorn, H. The role of ambient sound levels, signal-to-noise ratio, and stimulus pulse rate on behavioural disturbance of seabass in a net pen. *Behav. Processes* **2020**, *170*, 103992. [CrossRef]
15. Lara, R.A.; Vasconcelos, R.O. Impact of noise on development, physiological stress and behavioural patterns in larval zebrafish. *Sci. Rep.* **2021**, *11*, 6615. [CrossRef]
16. Cai, H.W.; Ross, L.G.; Telfer, T.C.; Wu, C.W.; Zhu, A.Y.; Zhao, S.; Xu, M.Y. Modelling the nitrogen loadings from large yellow croaker (*Larimichthys crocea*) cage aquaculture. *Environ. Sci. Pollut. Res. Int.* **2016**, *23*, 7529–7542. [CrossRef]
17. Ramcharitar, J.; Gannon, D.P.; Popper, A.N. Bioacoustics of fishes of the family sciaenidae (croakers and drums). *Trans. Am. Fish. Soc.* **2006**, *135*, 1409–1431. [CrossRef]
18. Guillen, A.C.; Borges, M.E.; Herrerias, T.; Kandalski, P.K.; Souza, M.R.D.P.D.; Donatti, L. Gradual increase of temperature trigger metabolic and oxidative responses in plasma and body tissues in the Antarctic fish *Notothenia rossii*. *Fish Physiol. Biochem.* **2022**, *48*, 337–354. [CrossRef]
19. Egorova, V.I.; Sveshnikova, E.V.; Naumova, V.V.; Kiryanov, D.A.; Smirnova, A.N. Influence of water temperature on metabolic energy structure in fish. *Vestn. Astrakhan State Tech. Univ. Ser. Fish. Ind.* **2019**, *2019*, 110–116. [CrossRef]
20. Ern, R.; Andreassen, A.H.; Jutfelt, F. Physiological mechanisms of acute upper thermal tolerance in fish. *Physiology* **2023**, *38*, 141–158. [CrossRef]
21. Schreck, C.B. Stress and fish reproduction: The roles of allostasis and hormesis. *Gen. Comp. Endocrinol.* **2010**, *165*, 549–556. [CrossRef]
22. Petitjean, Q.; Jean, S.; Gandar, A.; Cte, J.; Jacquin, L. Stress responses in fish: From molecular to evolutionary processes. *Sci. Total Environ.* **2019**, *20*, 371–380. [CrossRef] [PubMed]
23. Ao, J.Q.; Mu, Y.N.; Xiang, L.X.; Fan, D.D.; Feng, M.J.; Zhang, S.C.; Shi, Q.; Zhu, L.Y.; Li, T.; Ding, Y.; et al. Genome sequencing of the perciform fish *Larimichthys crocea* provides insights into molecular and genetic mechanisms of stress adaptation. *PLoS Genet.* **2015**, *11*, e1005118. [CrossRef] [PubMed]
24. Van der Meer, D.L.; van den Thillart, G.E.; Witte, F.; de Bakker, M.A.; Besser, J.; Richardson, M.K.; Spaink, H.P.; Leito, J.T.; Bagowski, C.P. Gene expression profiling of the long-term adaptive response to hypoxia in the gills of adult zebrafish. *Am. J. Physiol. Regul. Integr. Comp. Physiol.* **2005**, *289*, 1512–1519. [CrossRef] [PubMed]
25. Khater, E.S.; Bahnasawy, A.; El-Ghobashy, H.; Shaban, Y.; Elsheikh, F.; El-Reheem, S.A.; Aboegela, M. Mathematical model for predicting oxygen concentration in tilapia fish farms. *Sci. Rep.* **2021**, *11*, 24130. [CrossRef]
26. Green, L.; Jutfelt, F. Elevated carbon dioxide alters the plasma composition and behaviour of a shark. *Biol. Lett.* **2014**, *10*, 1127–1132. [CrossRef]
27. Bickler, P.E.; Buck, L.T. Hypoxia tolerance in reptiles, amphibians, and fishes: Life with variable oxygen availability. *Annu. Rev. Physiol.* **2007**, *69*, 145–170. [CrossRef]
28. Nour, E.; Wang, H.P. Transcriptional stress responses to environmental and husbandry stressors in aquaculture species. *Rev. Aquac.* **2016**, *8*, 61–88. [CrossRef]
29. Meskendahl, L.; Fontes, R.P.; Herrmann, J.P.; Temming, A. Metabolic costs of spontaneous swimming in *Sprattus sprattus* L., at different water temperatures. *PLoS ONE* **2019**, *14*, e0225568. [CrossRef]
30. Borowiec, B.G.; Firth, B.L.; Craig, P.M. Oxygen consumption rate during recovery from loss of equilibrium induced by warming, hypoxia, or exhaustive exercise in rainbow darter (*Etheostoma caeruleum*). *J. Fish Biol.* **2024**, *105*, 23–33. [CrossRef]
31. Yan, Y.L.; Yuan, X.C.; Shi, Y.H.; Liu, Y.H.; Xie, Y.D. Effect of salinity on growth, oxygen consumption rate and ammonia excretion rate of false kelpfish (*Sebastiscus marmoratus*). *J. Dalian Ocean Univ.* **2019**, *34*, 545–551. [CrossRef]
32. Jing, R.; Song, X.; Liu, R.; Guo, X. Effects of Temperature on Growth and Oxygen Consumption Rate in Juvenile Goldfish (*Carassius auratus*). *Hebei Fish.* **2015**, 1–3.
33. Damsgaard, C.; Baliga, V.B.; Bates, E.; Burggren, W.; McKenzie, D.J.; Taylor, E.; Wright, P.A. Evolutionary and cardio-respiratory physiology of air-breathing and amphibious fishes. *Acta Physiol.* **2020**, *228*, e13406. [CrossRef] [PubMed]
34. Reid, S.G.; Sundin, L.; Milsom, W.K. The cardiorespiratory system in tropical fishes: Structure, function, and control. *Fish Physiol.* **2005**, *21*, 225–275. [CrossRef]
35. Wang, Q.F.; Shen, W.L.; Liu, C.; Mu, D.L.; Wu, X.F.; Guo, N.G.; Zhu, J.Q. Effects of multi-environmental factors on physiological and biochemical responses of large yellow croaker, *Larimichthys crocea*. *Chemosphere* **2017**, *184*, 907–915. [CrossRef]

36. Rubalcaba, J.G. Metabolic responses to cold and warm extremes in the ocean. *PLoS Biol.* **2024**, *22*, e3002479. [CrossRef]
37. Cai, L.; Yin, J.; Yan, X.; Zhou, Y.; Tang, R.; Yu, M. The environmental analysis and site selection of mussel and large yellow croaker aquaculture areas based on high resolution remote sensing. *Acta Oceanol. Sin.* **2024**, *43*, 66–86. [CrossRef]
38. Liu, J.F. *Large Yellow Croaker Aquaculture and Biology*; Xiamen University Press: Xiamen, China, 2013.
39. Li, M.Y.; Miao, L.; Chen, J.; Chen, Q.J.; Yu, C. Establishment and discussion on the new model of simulated ecological and grading aquaculture for large yellow croaker. *J. Ningbo Univ. Nat. Sci. Eng. Ed.* **2017**, *30*, 1–5. [CrossRef]
40. Khan, F.U.; Younas, W.; Shang, Y.; Tu, Z.; Khan, M.I.; Zuberi, A.; Wang, Y. Effects of acclimation temperature on growth, physiology and thermal tolerance of the juvenile grass carp *Ctenopharyngodon idella*. *Aquaculture* **2024**, *581*, 740421. [CrossRef]
41. Jayasundara, N.; Somero, G.N. Physiological plasticity of cardiorespiratory function in a eurythermal marine teleost, the longjaw mudsucker, *Gillichthys mirabilis*. *J. Exp. Biol.* **2013**, *216*, 2111–2121. [CrossRef]
42. Carter, M.J.; Cortes, P.A.; Rezende, E.L. Temperature variability and metabolic adaptation in terrestrial and aquatic ectotherms. *J. Therm. Biol.* **2023**, *115*, 103565. [CrossRef] [PubMed]
43. Sandblom, E.; Gräns, A.; Axelsson, M.; Seth, H. Temperature acclimation rate of aerobic scope and feeding metabolism in fishes: Implications in a thermally extreme future. *Proc. Biol. Sci.* **2014**, *281*, 4482–4500. [CrossRef] [PubMed]
44. Liu, Y.T.; Yao, C.W.; Cui, K.; Hao, T.T.; Yin, Z.Y.; Xu, W.X.; Huang, W.X.; Mai, K.S.; Ai, Q.H. Nutritional programming of large yellow croaker (*Larimichthys crocea*) larvae by dietary vegetable oil: Effects on growth performance, lipid metabolism and antioxidant capacity. *Br. J. Nutr.* **2022**, *129*, 967–980. [CrossRef]
45. Chen, Y.L.; Huang, W.Q.; Shan, X.J.; Chen, J.; Wang, H.B. Growth characteristics of cage-cultured large yellow croaker *Larimichthys crocea*. *Aquac. Rep.* **2020**, *16*, 100242. [CrossRef]
46. Kullgren, A.; Jutfelt, F.; Fontanillas, R.; Sundell, K.; Samuelsson, L.M.; Wiklander, K.; Kling, P.; Koppe, W.; Larsson, D.G.; Björnsson, B.T.; et al. The impact of temperature on the metabolome and endocrine metabolic signals in atlantic salmon (*Salmo salar*). *Comp. Biochem. Physiol. Part A Mol. Integr. Physiol.* **2013**, *164*, 44–53. [CrossRef]
47. Kathleen, M.G.; Brittany, B. Social buffering of the stress response: Insights from fishes. *Biol. Lett.* **2022**, *18*, 20220332. [CrossRef]
48. Zotin, A.A.; Klemenov, S.Y. Endogenous biorhythms of the oxygen consumption rate in individual development of *Lymnaea stagnalis* (Lymnaeidae, Gastropoda). *Biol. Bull. Russ. Acad. Sci.* **2013**, *40*, 500–507. [CrossRef]
49. Slabbekoorn, N.; Bouton, H. Soundscape orientation: A new field in need of sound investigation. *Anim. Behav.* **2008**, *76*, e5–e8. [CrossRef]
50. Melvin, G.D.; Kloser, R.J.; Honkalehto, T. The adaptation of acoustic data from commercial fishing vessels in resource assessment and ecosystem monitoring. *Fish. Res.* **2016**, *178*, 13–25. [CrossRef]
51. Parsons, M.; Meekan, M. Acoustic characteristics of small research vessels. *J. Mar. Sci. Eng.* **2020**, *8*, 970. [CrossRef]
52. Jain-Schlaepfer, S.; Fakan, E.; Rummer, J.L.; Simpson, S.D.; Mccormick, M.I. Impact of motorboats on fish embryos depends on engine type. *Conserv. Physiol.* **2018**, *6*, coy014. [CrossRef] [PubMed]
53. McCormick, M.I.; Allan, B.J.M.; Harding, H.; Simpson, S.D. Boat noise impacts risk assessment in a coral reef fish but effects depend on engine type. *Sci. Rep.* **2018**, *8*, 3847. [CrossRef] [PubMed]
54. McCormick, M.I.; Fakan, E.P.; Nedelec, S.L.; Allan, B.J.M. Effects of boat noise on fish fast-start escape response depend on engine type. *Sci. Rep.* **2019**, *9*, 6554. [CrossRef] [PubMed]
55. Erbe, C.; Reichmuth, C.; Cunningham, K.; Lucke, K.; Dooling, R. Communication masking in marine mammals: A review and research strategy. *Mar. Pollut. Bull.* **2016**, *103*, 15–38. [CrossRef]
56. Filiciotto, F.; Cecchini, S.; Buscaino, G.; Maccarrone, V.; Piccione, G.; Fazio, F. Impact of aquatic acoustic noise on oxidative status and some immune parameters in gilthead sea bream *Sparus aurata* (Linnaeus, 1758) juveniles. *Aquac. Res.* **2017**, *48*, 1895–1903. [CrossRef]

fishes

MDPI

Article

Study on the Flow Velocity Preference of the Four Major Chinese Carps Using Convolutional Neural Networks

Ning Qiu [1], Wenjing Li [2], Yi Yu [3], Jianna Jia [1], Guoqiang Ma [1] and Shitao Peng [1,*]

1 Key Laboratory of Environmental Protection Technology on Water Transport, Ministry of Transport, Tianjin Research Institute for Water Transport Engineering, Tianjin 300456, China; qiuning@tiwte.ac.cn (N.Q.); jiajn@tiwte.ac.cn (J.J.); magq@tiwte.ac.cn (G.M.)

2 National Engineering Research Center of Eco-Environment in the Yangtze River Economic Belt, China Three Gorges Corporation, Wuhan 430014, China; li_wenjing@ctg.com.cn

3 Fisheries Institute, Sichuan Academy of Agricultural Sciences, Chengdu 611731, China; 17784411664@163.com

* Correspondence: pengshitaotj@163.com

Abstract: Flow velocity is a critical factor in determining the suitability of fish habitats. Understanding the preference patterns of the four major Chinese carps (FMCCs) for different flow velocities is crucial for their habitat conservation and restoration. In this study, the preference of individual fish species, approximately 15 cm in length, for flow velocity was investigated at flow velocity gradients of 0.0, 0.4, 0.8, 1.2, 1.6, and 2.0 times their body length. Additionally, a deep learning algorithm based on convolutional neural networks (CNNs) was employed for fish target detection. The results showed that, at this length, black carp (*Mylopharyngodon piceus*) preferred fast currents when the inlet flow velocity was between 0.4 and 1.6 times their body length, while grass carp (*Ctenopharyngodon idella*), silver carp (*Hypophthalmichthys molitrix*), and bighead carp (*Hypophthalmichthys nobilis*) preferred fast currents when the inlet flow velocity of the test flume was between 0.4 and 2.0 times their body length. However, this preference for fast currents decreased as the overall flow velocity increased to a specific threshold, eventually leading to their avoidance. The highest preference for fast currents among the four species was observed at inlet flow velocities of 1.2, 0.4, 0.8, and 0.8 times their body length, respectively. The findings of this study provide important insights into habitat conservation and restoration for the FMCCs in projects focused on the construction of navigation channels and water conservancy.

Keywords: Chinese carps; deep learning; flow velocity preference; flow velocity gradient

Key Contribution: The results of this study not only provide a reference for the improvement and optimization of artificial breeding environments for the four major Chinese carp species but also offer insights into the protection, restoration, or reconstruction of their habitats in navigation channels and water conservancy engineering. Furthermore, this research introduces an innovative approach using a deep learning model based on convolutional neural networks for the position detection of experimental fish targets in fish behavior studies. This study provides novel ideas for methodological innovations in fish ethology research.

1. Introduction

Flow velocity and flow patterns within the aquatic living environment are decisive for the survival and growth of fish. Variations in fish morphology are significantly influenced by the flow velocity in their habitats, with fish body shapes similarly responding to flow velocity across diverse watersheds [1]. Fluctuations in water flow velocity significantly affect not only fish behavioral patterns, such as tail flicking and schooling [2], but also their feeding behavior [3], growth [4], material metabolism [5], muscle quality [6], and even survival and reproduction [7,8]. Consequently, flow velocity and flow patterns are considered crucial determinants of the suitability of fish habitats.

The four major Chinese carps (FMCCs)—black carp (*Mylopharyngodon piceus*), grass carp (*Ctenopharyngodon idella*), silver carp (*Hypophthalmichthys molitrix*), and bighead carp (*Hypophthalmichthys nobilis*)—are significant economic fish species in China [9]. The ecological habits of these species and the impacts of water engineering projects on them [9,10], as well as methods for the conservation and restoration of their habitats [11], have been popular topics in fish ecology research in China. Determining the optimal and preferred flow velocities of the FMCCs can offer crucial data and theoretical support for the conservation and restoration of their wild habitats. Previous studies have explored the flow velocity preferences of FMCCs at the larval stage [7]. However, in-depth comparative studies involving laboratory-controlled experiments at different life history stages can contribute to a better understanding of their preference patterns and adaptation strategies regarding flow velocity during these distinct periods.

With the development of numerous deep learning models such as Image Net [12], VGGNet (Visual Geometry Group Net) [13], Residual Net [14], Dense Net [15], and Mobile Net [16], deep learning-based object detection methods have been widely applied for image classification. Some researchers have employed convolutional neural network models for the object detection of aquatic organisms such as fish and have achieved some results [17,18]. However, most of these studies are limited to the automatic identification of fish species in natural or aquaculture waters, and few attempts have involved the application of these models for fish location identification in behavioral experiments. Deep learning models detect objects by learning their features using a large volume of training data and extracting them from images with noisy and complex backgrounds. Moreover, their detection accuracy is less affected by environmental changes than commonly used image threshold segmentation methods in fish behavior research, and they have relatively high efficiency and accuracy [19]. Therefore, in fish ethology research, the application of deep learning-based image classification and object detection technologies to the detection and location identification of experimental fish could significantly improve the efficiency and accuracy of behavioral data analysis.

In this study, we introduced a deep learning object detection algorithm based on convolutional neural networks (CNNs) to investigate the preference behavior of FMCCs (with a body length of approximately 15 cm) across different flow velocities. This research not only provides a reference for optimal flow velocities in the artificial breeding of the four major Chinese carps but also offers theoretical insights for the restoration and reconstruction of their habitats in water engineering projects.

2. Materials and Methods

2.1. Animals and Housing

For this study, black carp (body length: 14.90 ± 0.36 cm), grass carp (body length: 15.13 ± 0.40 cm), silver carp (body length: 15.02 ± 0.39 cm), and bighead carp (body length: 15.09 ± 0.40 cm) were sourced from, Yuzhenduo Aquaculture Co., Ltd., Meishan, China. The appearance of four kinds of fish was shown in Figure 1. The experimental fish were reared in a recirculating aquaculture system similar to the experimental environment. During the experiment, from 08:00 to 22:00 every day, this aquaculture system was illuminated using fluorescent tubes placed at the top, with 80–100 lux of light intensity. Water temperature, pH, and dissolved oxygen were monitored daily (with an HQ30d device; Hach,

Loveland, CO, USA) and maintained at 25 to 27 °C, 7.32 to 8.75, and 7.22 to 8.54 mg/L, respectively. All fish were fed to satiation twice daily with ozone-disinfected frozen red worms (*Chironomidae flaviplumus* larva; Yuerle, Tianjin, China).

Figure 1. The external appearance of the four major Chinese carps (sub-adult stage): (**a**), black carp; (**b**), grass carp; (**c**), silver carp; (**d**), bighead carp.

2.2. Test Procedures and Apparatus

The experimental apparatus consisted of a T-shaped flume with a unilateral sudden expansion, the main body of which was 4.0 m in length, 0.8 m in width, and 0.6 m in height (Figure 2). The lower part of the flume featured a reservoir for water storage. In the experiments, water was pumped from the reservoir at the front of the flume, passing through a grate for energy dissipation and rectification, and entering the test section of the flume. Then, it flew through a straight channel into the rear part of the flume and returned to the reservoir via a drainage outlet. A tailgate was installed in front of the drainage outlet to control the water depth within the flume. When water passed through the test section, a rapid flow area (2.0 m in length and 0.2 m in width) was formed in the straight channel, while a slow flow area was formed in the adjacent space with dimensions of 2.0 m in length and 0.6 m in width, resulting in the spatial differentiation of flow velocities. Fish nets were installed at both the front and rear of the test section to prevent the experimental fish from leaving the test area. A high-definition CCD camera with a field of view covering the entire test section was installed at the top of the test section of the flume.

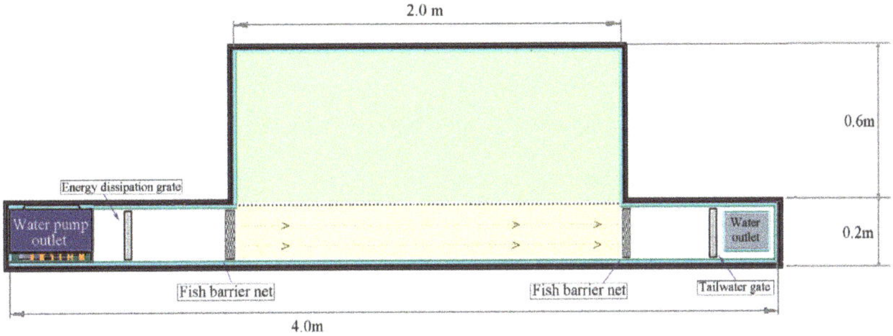

Figure 2. Schematic diagram of the experimental apparatus. The rapid flow area is shown in yellow, and the slow flow area is shown in green.

The experiment was conducted with five inlet flow velocity gradients for each fish species (i.e., the flow velocities when the water entered the test section). These gradients were set at 0, 0.4, 0.8, 1.2, 1.6, and 2.0 times the average body length (BL) of the experimental fish (0.15 m), corresponding to velocities of 0.06 m/s, 0.12 m/s, 0.18 m/s, 0.24 m/s, and 0.30 m/s, respectively. For each velocity tested, the flow velocities of the rapid and slow flow areas were measured by the multi-point velocity measurement method before the experiments started. Specifically, the rapid flow area and slow flow area were all divided into several 10 cm×10 cm grids, and a flow velocity meter (LS1206B, BAIFENG, Qingdao, China) was used to measure the velocity at the center point of each grid. For the center point of each grid, the flow velocity of the upper part (0.05 m below the water surface), the middle part (0.25 m below the water surface), and the bottom part of the water layer were measured. The measurement probe stayed in the water for 30 s during each measurement, and the value of flow velocity was recorded after the numerical display of the instrument stabilized. The flow velocity of each grid was defined as the average of the surface flow velocity, the middle flow velocity, and the bottom flow velocity at the central point of this grid. The flow velocity of the rapid flow area was determined as the average of the flow velocities of all the grids it contains, and the flow velocity of the slow flow area was calculated in the same way. The flow velocities in each area of different experimental groups were shown in Table 1.

Table 1. Average flow velocities at various locations within the experimental area.

Groups	Inlet Flow Velocity (m/s)	Rapid Flow Area (m/s)	Slow Flow Area (m/s)
0 BL	0.000	0.000 ± 0.000	0.000 ± 0.000
0.4 BL	0.062	0.051 ± 0.006	0.012 ± 0.005
0.8 BL	0.123	0.100 ± 0.022	0.023 ± 0.005
1.2 BL	0.178	0.133 ± 0.023	0.042 ± 0.013
1.6 BL	0.239	0.193 ± 0.017	0.051 ± 0.016
2.0 BL	0.301	0.227 ± 0.031	0.076 ± 0.023

The flume was filled with water before the experiment, and the water depth was controlled at 0.5 m using the tailgate. Then, the water pump was stopped to allow the water in the flume to reach a static state. In each experiment, a single fish was carefully moved to the test section and allowed to freely swim within it. After a 10 min adaptation period, the CCD camera at the top of the flume began to continuously record the behavior of the experimental fish for 5 min when the water was still. Subsequently, the water pump was turned on, and its operating power was adjusted according to the velocity gradient settings. The inlet flow velocity was gradually increased in increments, and fish behavior was recorded for 5 min at each velocity until reaching the maximum velocity in the experiment. Then, the water pump was turned off, and the experimental fish were removed from the flume. Five specimens of each species were individually tested (five replicates), and no fish were reused. The fish behavior data were stored in mp4 video file format.

2.3. Detection of the Positions of Fish

2.3.1. Training of the Deep Learning Model

A total of 800 representative images were extracted from the recorded behavior videos. The extracted representative images encompassed the behavior states of the FMCCs at various flow velocities and were used to create the dataset required for model training. The experimental fish in the representative frames were labeled using labelimg 2.0 to generate corresponding label information for each image. The 800 data labels were divided into training and validation sets in a 7:1 ratio, with the training set comprising 700 data labels and the validation set containing 100 data labels. Then, transfer training was carried out on the dataset using the pretrained YOLOv5.pt weight file and the YOLOv5s model. The training results indicate that the mean average precision (mAP, an important metric for evaluating model performance) reached 0.95, and the target prediction results in the validation set were consistent with the manual annotation information in the training set. After training, the weight file used for individual fish detection was obtained.

2.3.2. Testing of the Deep Learning Model

Six images were randomly selected from the behavior videos of each fish species at each velocity group, resulting in a test set comprising 120 images, which was used to assess the performance of the above-trained YOLOv5 model in detecting individual fish in this study. The detection results of part of the test set are shown in Figure 3. It can be observed that the mean average precision (mAP) values for the detection of the four species were all around 0.9. The contours of the experimental fish can be clearly distinguished from the experimental background, and the visual effects are within an acceptable range for human eyes. No false detections or misdetections occurred, highlighting the superior performance of the proposed model for the identification of individual fish.

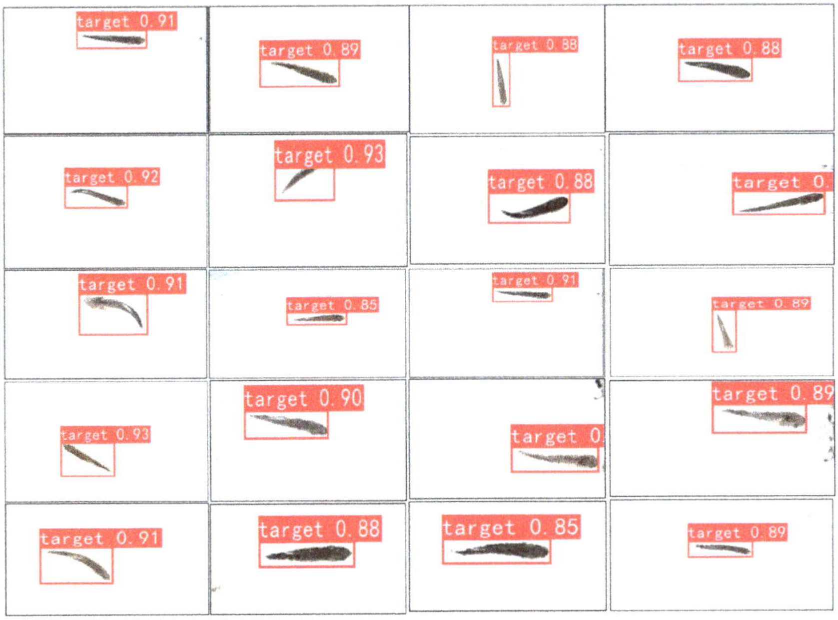

Figure 3. The detection effect of the model on FMCCs.

2.3.3. Detection of the Positions of Fish Using the Deep Learning Model

The weight file generated by the deep learning model was employed to perform object detection on all video data obtained in this study. The position information of the detected experimental fish targets in each frame was acquired and stored. Figure 4 shows the fish-identification process of the YOLOv5 model using the behavior videos. First, the above-mentioned trained weight file was loaded into the model, and then the recorded behavior video was read frame by frame and translated into images, which were used as input. Then, slicing was carried out using the Focus module to initially transform the feature size, converting the original images to 640 $\times$ 480 samples, which were more convenient for the model to operate. The transformed images were inputted into the backbone network composed of multiple convolutional layers, C3 modules, etc. and object features were extracted layer by layer. Then, in the neck network, using convolution and concat, the features at different scales were combined to enhance feature representation. Finally, the head detection network was used for the classification and identification of the experimental fish in the images at multiple scales, outputting the detection results using the position data of each fish species.

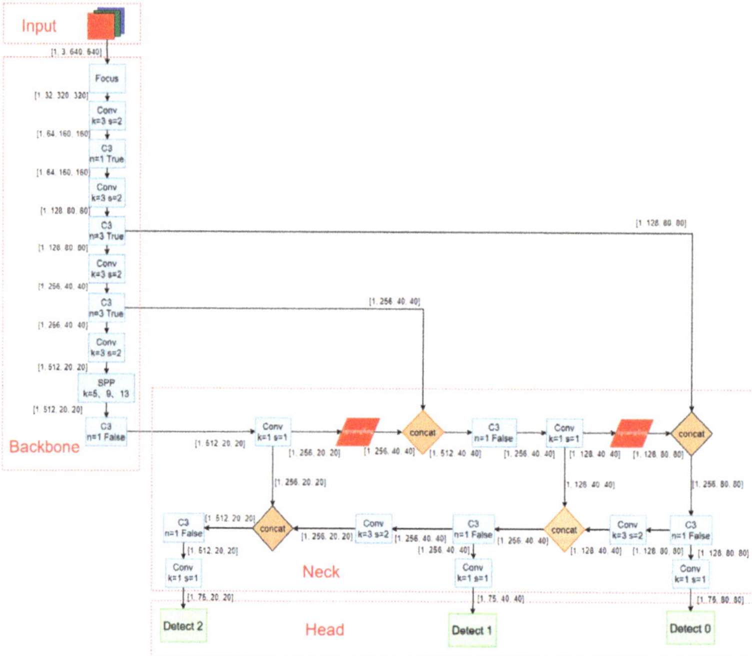

Figure 4. The fish detection process of the YOLOv5 model using behavior videos.

2.4. Data Analysis

According to the position coordinates of the experimental fish, the number of entries per square centimeter into each area during the experiment was counted. All data were analyzed for normal distribution and homogeneity of variance using the Kolmogorov–Smirnov and Levene methods, respectively. An independent-samples *t*-test was employed to analyze the differences in the total frequency of entries into the rapid and slow flow areas for the four species within each experimental group. Additionally, based on the total frequency data of experimental fish entering the rapid flow area across different flow velocity groups, inter-group differences in rheotaxis behavior were analyzed using the LSD (least significant difference) and Tamhane's T2 multiple comparison methods.

3. Results

Figure 5a displays the number of black carp per square centimeter in rapid and slow flow areas across different velocities. No significant differences were found between entries into rapid and slow flow areas under static water (0.0 BL) or at an inlet velocity of 2.0 BL ($p > 0.05$), indicating no directional preference. Occupancy in the rapid flow area was significantly higher than that in the slow flow area when inlet velocities were 0.4 BL, 0.8 BL, 1.2 BL, and 1.6 BL ($p < 0.05$), demonstrating positive rheotaxis. The rheotaxis response progressively decreased as velocity increased beyond 1.2 BL, ultimately resulting in the avoidance of the rapid flow area at the maximum velocity (2.0 BL).

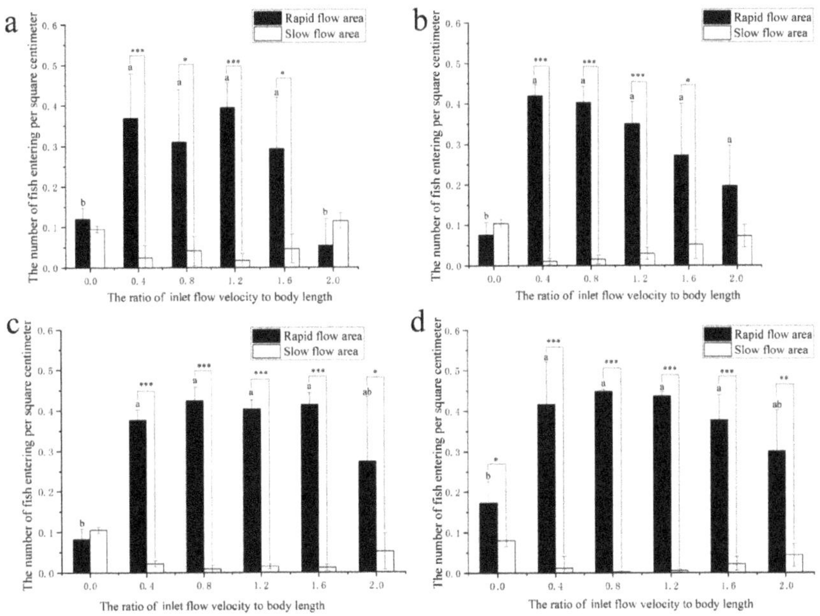

Figure 5. The number of entries per square centimeter into the two areas at the different flow velocities: (**a**), black carp; (**b**), grass carp; (**c**), silver carp; (**d**), bighead carp. For the letters 'a', 'b', and 'ab' above the bar chart, if there is at least one common letter between two groups, it indicates that the difference is not significant ($p > 0.05$). When the marked letters are completely different, the difference is significant ($p < 0.05$). *, **, *** indicates $p < 0.05$, $p < 0.01$, and $p < 0.001$, respectively.

Figure 5b demonstrates grass carp's distribution patterns. No zonal preference was detected under static water ($p = 0.144$). When the inlet flow velocity increased to 0.4 BL, the number of grass carp entering the rapid flow area was significantly higher than that in the slow flow area ($p < 0.001$). However, after the inlet flow velocity increased to 0.4 BL, the number of grass carp entering the rapid flow area gradually decreased with the increase in flow velocity, and they exhibited no obvious preference for either of the two areas at the highest flow velocity of 2.0 BL ($p > 0.05$).

Figure 5c reveals the behavioral responses of silver carp. No significant differences were observed in the number of silver carp entering the two areas ($p = 0.150$) in static water. When the inlet flow velocity increased to 0.4 BL, the number of fish entering the rapid flow area was significantly higher than that in the slow flow area ($p < 0.001$), and the fish consistently preferred the rapid flow area as the inlet flow velocity gradually increased; that is, the number of silver carp entering the rapid flow area was always higher than that in the slow flow area ($p < 0.05$). Nevertheless, when the inlet flow velocity increased to 2.0 BL, the preference of silver carp for the rapid flow area began to exhibit a downward trend.

Figure 5d shows the number of bighead carp per square centimeter in the rapid and slow flow areas at each velocity. All six velocities were associated with significantly higher rapid flow entries ($p < 0.05$). Notably, the number of bighead carp entering the rapid flow area began to decrease with the increase in flow velocity after the inlet flow velocity reached 1.2 BL.

Figure 6 illustrates the fish distribution in the rapid flow area as an indicator of the degree of rheotaxis, which was used for interspecies comparison. At inlet flow velocities of 0 BL, 0.4 BL, and 0.8 BL, no obvious differences were observed in the distribution of the four fish species in the rapid flow area ($p > 0.05$). When the inlet flow velocity was

1.2 BL, the number of grass carp entering the rapid flow area was significantly lower than that of bighead carp ($p = 0.014$). At an inlet flow velocity of 1.6 BL, the total frequency of black carp and grass carp entering the rapid flow area was significantly lower than that of silver carp ($p < 0.05$). At an inlet flow velocity of 2.0 BL, the number of grass carp entering the rapid flow area was significantly lower than that of black carp ($p = 0.048$), silver carp ($p = 0.038$), and bighead carp ($p = 0.027$).

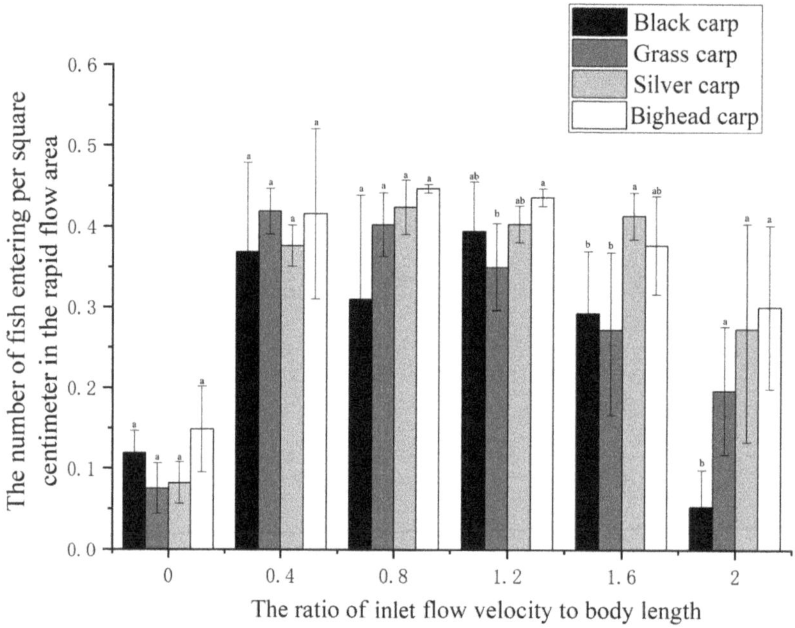

Figure 6. The number of entries per square centimeter in the rapid flow area in each experimental group. For the letters 'a', 'b', and 'ab' above the bar chart, if there is at least one common letter between two groups, it indicates that the difference is not significant ($p > 0.05$). When the marked letters are completely different, the difference is significant ($p < 0.05$).

4. Discussion

According to the experimental results, the four fish species—grass carp, black carp, silver carp, and bighead carp—all exhibited significant rheotaxis. Variations in flow velocity in the environment lead to their preference for faster-flowing waters. Specific differences were observed in the rheotactic behavioral patterns among the four fish species. Overall, their behavioral patterns were relatively similar at low flow velocities below 1.6 body length (BL), but at high flow velocities above 1.6 BL, the rheotaxis of black carp and grass carp was relatively less than that of silver carp and bighead carp. All four fish species preferred rapid flow areas when the overall flow velocity was low but avoided them when the overall flow velocity was high. They exhibited the strongest avoidance of rapid flow areas when the inlet flow velocity was approximately 2.0 BL, with this pattern more evident in black carp and grass carp than in silver carp and bighead carp. The different flow velocity preferences directly affect the efficiency of food acquisition among these four major Chinese carp species. For instance, appropriate flow velocities can facilitate feeding and enhance foraging efficiency, with plankton directly flowing into the mouths of filter-feeding fish when they open their mouths against the current. For herbivorous fish such as grass carp, flowing water may bring aquatic plants to areas where they are more easily accessible. Omnivorous fish may search for various food resources in different flow velocity zones, with relatively flexible flow velocity preferences to adapt to different food source distributions.

Furthermore, compared to black carp and grass carp, silver carp and bighead carp prefer to swim more frequently in still water, resulting in more diverse and dispersed behavioral trajectories. Based on the research team's observations of swimming behavior among various freshwater fish in China, compared to bottom-dwelling fish such as Gobioninae, *Cyprinus* sp. and Cobitidae, surface-dwelling fish such as *Hemiculter leucisculus*, *Culter* sp. and *Hyporhamphus* sp. markedly prefer to swim more frequently. Additionally, the swimming frequency of silver carp and bighead carp may also be related to their feeding habits. Silver carp and bighead carp are filter-feeding fish and thus need to constantly swim in natural waters to better filter out plankton in the water. Previous studies have reported that, among the four major Chinese carp species, silver carp and bighead carp exhibit flow velocity preferences that are closely associated with their feeding habits [20]. However, it has been established that filter-feeding fish have a significantly greater preference for still or slow-moving water than omnivorous and herbivorous fish, which relatively differs from our findings, potentially due to differences in experimental environments and methods. In addition, bighead carp individuals consistently showed a preference for staying upstream near the inlet against the current throughout the experiment. The authors currently cannot provide a reasonable explanation for the specific mechanism of this behavior, and the team will continue to investigate this issue.

In this study, we observed one fish at a time when studying the rheotactic behavior of the four major Chinese carp species. Significant individual behavioral differences were observed among the four fish species, resulting in high standard deviations in experimental data when using individual fish for research. The four major Chinese carp species are schooling fish in both natural and artificial breeding environments, and therefore, due to the presence of the "leader fish effect", behavioral differences among individuals within schools may be masked, resulting in a more consistent behavioral pattern. Studies have shown that when schooling fish are isolated, they may exhibit abnormal and stereotypical behaviors [21]. We estimate that if fish schools were used in this study, the results might differ from those based on individual fish. Therefore, our research team plans to conduct similar rheotactic behavior studies using schools of the four major Chinese carp species to further explore the mechanisms of their rheotactic behavior.

Apart from rheotactic behavior, the results of this study also revealed another behavioral pattern of these species: thigmotaxis. Also known as wall-hugging behavior, thigmotaxis refers to animals' tendency to move around the peripheral areas of their environment and rarely enter the central zone [22,23]. In this study, according to the trajectory distribution of the experimental fish freely swimming in still water, all four fish species exhibited a preference for the edge areas of the tank. As the environmental illuminance and temperature settings used in this experiment are consistent with the conditions during fish acclimation, it is unlikely that they caused a stress response in the fish. Therefore, this behavior is likely inherent to the experimental fish. Although this behavior has been studied in other fish species such as zebrafish (*Danio rerio*) [24] and medaka (*Oryzias latipes*) [12], it has not been reported in the four major Chinese carp species and warrants further investigation.

5. Conclusions

This study revealed the excellent performance of the YOLOv5 convolutional neural network in the detection of experimental fish in behavior experiments. It provides a superior quantitative method for the location identification and data statistics of experimental fish. Subadult black carp preferred rapid flow areas within a flow velocity range of 0.4–1.6 BL (rapid flow velocities of 0.051–0.193 m/s), while grass carp, silver carp, and bighead carp preferred rapid flow areas within a velocity range of 0.4–2.0 BL (rapid flow velocities of 0.051–0.227 m/s). However, this preference decreased when the overall flow velocity increased to a certain threshold, eventually resulting in the avoidance of rapid flow areas. The four species exhibited the highest preference for rapid flow areas at flow velocities of 1.2 BL, 0.4 BL, 0.8 BL, and 0.8 BL, respectively.

Author Contributions: Conceptualization, N.Q.; methodology, N.Q. and Y.Y.; software, W.L. and Y.Y.; validation, N.Q. and W.L.; formal analysis, J.J., G.M. and W.L.; resources, N.Q. and Y.Y.; data curation, W.L. and J.J.; writing—original draft preparation, N.Q. and J.J.; writing—review and editing, N.Q. and S.P.; visualization, N.Q., J.J. and W.L.; supervision, S.P.; project administration, S.P.; funding acquisition, N.Q. All authors have read and agreed to the published version of the manuscript.

Funding: This research was funded by the National Key Research and Development Project (No. 2022YFC3203400), the Guangxi Key Research and Development Program (No. AB24010087), and the Science and Technology Project of the Transportation Department of Qinghai Province (No. JG2024-19).

Institutional Review Board Statement: This research followed the guidelines specified by the research permits from Tianjin Research Institute for Water Transport Engineering, M.O.T (permit no. TKS-EE2024098, 1 December 2024).

Data Availability Statement: The data generated and analyzed during the current study are available from the corresponding author.

Conflicts of Interest: Wenjing Li is employed by China Three Gorges Corporation. The remaining authors declare that this research was conducted in the absence of any commercial or financial relationships that could be construed as a potential conflict of interest.

Abbreviations

The following abbreviations are used in this manuscript:

BL	Body length
CNNs	Convolutional neural networks
YOLO	You only look once
FMCCs	Four major Chinese carps

References

1. Barros, T.F.; Louvise, J.; Caramaschi, R.P. Flow gradient drives morphological divergence in an Amazon pelagic stream fish. *Hydrobiologia* **2019**, *833*, 217–229. [CrossRef]
2. Yuan, X.; Tu, Z.-Y.; Han, J.-C.; Wang, X.-X.; Shi, X.-T.; Liu, G.-Y.; Huang, Y.-P. Effects of flow rate on swimming behavior and energy consumption of *Schizothorax chongi*. *Acta Hydrobiol. Sin.* **2012**, *36*, 270–275. [CrossRef]
3. Castro, V.; Grisdale-Helland, B.; Helland, S.J.; Kristensen, T.; Jørgensen, S.M.; Helgerud, J.; Claireaux, G.; Farrell, A.P.; Krasnov, A.; Takle, H. Aerobic Training Stimulates Growth and Promotes Disease Resistance in Atlantic Salmon (*Salmo salar*). *Comp. Biochem. Physiol. Part A Mol. Integr. Physiol.* **2011**, *160*, 278–290. [CrossRef] [PubMed]
4. Inoue, L.A.K.A.; Hackbarth, A.; Arberláez-Rojas, G.; Moraes, G. Growth Performance and Metabolism of the Neotropical Fish *Piaractus mesopotamicus* under Sustained Swimming. *Aquaculture* **2019**, *511*, 734219. [CrossRef]
5. Gruber, S.J.; Dickson, K.A. Effects of Endurance Training in the Leopard Shark, *Triakis semifasciata*. *Physiol. Zool.* **1997**, *70*, 481–492. [CrossRef]
6. Zhu, Z.; Song, B.; Lin, X.; Xu, Z. Effect of Sustained Training on Glycolysis and Fatty Acids Oxidation in Swimming Muscles and Liver in Juvenile Tinfoil Barb *Barbonymus schwanenfeldii* (Bleeker, 1854). *Fish Physiol. Biochem.* **2016**, *42*, 1807–1817. [CrossRef]
7. Lupandin, A.I. Effect of Flow Turbulence on Swimming Speed of Fish. *Biol. Bull.* **2005**, *32*, 461–466. [CrossRef]
8. Zha, W.; Huang, M.; Zeng, Y. Swimming behavior of crucian carp in an open channel with sudden expansion. *River Res. Appl.* **2019**, *35*, 1499–1510. [CrossRef]
9. Tang, C.; Yan, Q.; Li, W.; Yang, X.; Zhang, S. Impact of dam construction on the spawning grounds of the four major Chinese carps in the Three Gorges Reservoir. *J. Hydrol.* **2022**, *609*, 127694. [CrossRef]

10. Wang, J.N.; Li, C.; Duan, X.B.; Luo, H.H.; Feng, S.X.; Peng, Q.D.; Liao, W.G. The relationship between thermal regime alteration and spawning delay of the four major Chinese carps in the Yangtze River below the Three Gorges Dam. *River Res. Appl.* **2014**, *30*, 987–1001. [CrossRef]
11. Lv, Z.; Wang, G.; Zhang, P.; Ai, X.; Cao, X.; Zheng, W.; Mu, Z.; Yu, B. Optimizing flow regime for the Four Major Chinese Carps by integrating habitat suitability within reservoir operation. *J. Hydrol.* **2023**, *626*, 130226. [CrossRef]
12. Qiu, X.; Nomichi, S.; Chen, K.; Honda, M.; Kang, I.J.; Shimasaki, Y.; Oshima, Y. Short-term and persistent impacts on behaviors related to locomotion, anxiety, and startle responses of Japanese medaka (*Oryzias latipes*) induced by acute, sublethal exposure to chlorpyrifos. *Aquat. Toxicol.* **2017**, *192*, 148–154. [CrossRef] [PubMed]
13. Krizhevsky, A.; Ilya, S.; Geoffrey, E. Imagenet Classification with Deep Convolutional Neural Networks. *Adv. Neural Inf. Process. Syst.* **2012**, *25*, 1097–1105. [CrossRef]
14. Simonyan, K.; Andrew, Z. Very Deep Convolutional Networks for Large-Scale Image Recognition. *arXiv* **2014**, arXiv:1409.1556. [CrossRef]
15. He, K.; Zhang, X.; Ren, S.; Sun, J. Deep Residual Learning for Image Recognition. In Proceedings of the IEEE Conference on Computer Vision and Pattern Recognition, Las Vegas, NV, USA, 27–30 June 2016; pp. 770–778. [CrossRef]
16. Huang, G.; Liu, Z.; Van Der Maaten, L.; Weinberger, K.Q. Densely Connected Convolutional Networks. In Proceedings of the IEEE Conference on Computer Vision and Pattern Recognition, Honolulu, HI, USA, 21–26 July 2017; pp. 4700–4708. [CrossRef]
17. Qiu, C.; Zhang, S.; Wang, C.; Yu, Z.; Zheng, H.; Zheng, B. Improving Transfer Learning and Squeeze- and-Excitation Networks for Small-Scale Fine-Grained Fish Image Classification. *IEEE Access* **2018**, *6*, 78503–78512. [CrossRef]
18. Dos Santos, A.A.; Gonçalves, W.N. Improving Pantanal fish species recognition through taxonomic ranks in convolutional neural networks. *Ecol. Inform.* **2019**, *53*, 100977. [CrossRef]
19. Li, X.; Liu, M.; Zhang, S.; Zheng, R. Fish Trajectory Extraction Based on Object Detection. In Proceedings of the 2020 39th Chinese Control Conference (CCC), Shenyang, China, 27–30 July 2020; IEEE Press: New York, NY, USA, 2020; pp. 6584–6588.
20. Li, M.-N.; Zhu, H.-F.; Jin, Z.-J.; Luo, J.; Xu, J.-W.; Li, Z.-M.; Ma, W.-Z.; Shi, X.-T. The effect of accelerating flow on swimming behavior of juvenile *Aristichthys nobilis* and *Hypophthalmichthus molitrix* during downstream migration. *Acta Hydrobiol. Sin.* **2018**, *42*, 571–577. [CrossRef]
21. Pazmino, S.D.; Kent, M.I.; Ward, A.J. Locomotion and habituation to novel experimental environments in a social fish species. *Behaviour* **2020**, *157*, 1007–1023. [CrossRef]
22. Treit, D.; Fundytus, M. Thigmotaxis as a test for anxiolytic activity in rats. *Pharmacol. Biochem. Behav.* **1988**, *31*, 959–962. [CrossRef]
23. Sharma, S.; Coombs, S.; Patton, P.; De Perera, T.B. The function of wall-following behaviors in the Mexican blind cavefish and a sighted relative, the *Mexican tetra* (Astyanax). *J. Comp. Physiol. A Neuroethol. Sens. Neural Behav. Physiol.* **2009**, *195*, 225–240. [CrossRef]
24. Scherer, E. Behavioural responses as indicators of environmental alterations: Approaches, results, developments. *J. Appl. Ichthyol.* **1992**, *8*, 122–131. [CrossRef]

animals

MDPI

Article

7,8-DHF Modulates Aggressive Behavior in *Sebastes schlegelii*: Phenotype-Dependent Responses in Aggression-Dimorphic Individuals

Shufei Xu [1,2], Xinna Ma [1,2], Yang Xiao [1,2], Tao Zhang [3], Chao Ma [4] and Zhen Ma [1,5,*]

1 Key Laboratory of Environment Controlled Aquaculture, Ministry of Education, Dalian 116023, China; xsf0619@163.com (S.X.); maxinna@dlou.edu.cn (X.M.); xiaoyang@dlou.edu.cn (Y.X.)
2 College of Marine Technology and Environment, Dalian Ocean University, Dalian 116023, China
3 Tianzheng Industrial Co., Ltd., Dalian 116021, China; zht_3000@163.com
4 State Key Laboratory of Coastal and Offshore Engineering, Dalian University of Technology, Dalian 116081, China; machao@dlut.edu.cn
5 College of Fisheries and Life Science, Dalian Ocean University, Dalian 116023, China
* Correspondence: mazhen@dlou.edu.cn

Simple Summary: This study provides the first evidence that the TrkB receptor-specific agonist 7,8-dihydroxyflavone exerts differential regulatory effects on aggression phenotypes in *Sebastes schlegelii* through BDNF-TrkB signaling. Intraperitoneal administration of 2.5 mg/kg 7,8-DHF significantly reduced locomotor acceleration, angular velocity, and activity frequency in high-aggression individuals, while only decreasing angular velocity in low-aggression counterparts. These findings offer novel insights into the neural mechanisms underlying aggressive behavior and informs phenotype-based strategies for aquaculture optimization.

Abstract: Aggressive behavior is regulated by intricate neural circuits and molecular mechanisms, notably through the interaction of brain-derived neurotrophic factor (BDNF) with its receptor, tropomyosin receptor kinase B (TrkB), which influences neuroplasticity and related behavioral phenotypes. We investigate the role of the BDNF signaling pathway in fish aggression using juvenile black rockfish (*Sebastes schlegelii*), which exhibit distinct aggressive phenotypes. The TrkB agonist 7,8-dihydroxyflavone (7,8-DHF) was administered intraperitoneally at doses of 1.25, 2.5, and 5 mg/kg to assess its effects on the behavioral characteristics of high-aggression (H-agg) and low-aggression (L-agg) phenotypes. Our findings indicate the following: (1) The effects of 7,8-DHF are dose-dependent, with 2.5 mg/kg identified as the effective threshold dose for H-agg individuals; (2) in the H-agg group, this dose significantly reduced locomotor acceleration, angular velocity, and activity frequency, while prolonging the first movement latency; (3) in the L-agg group, only angular velocity was significantly decreased with the 2.5 mg/kg treatment, with no significant changes observed in other behavioral parameters. This study provides the first evidence for differential behavioral responses to 7,8-DHF in *S. schlegelii*, demonstrating dose-dependent aggression suppression in H-agg phenotypes and threshold-specific responses in L-agg phenotypes. These insights into the neuro-molecular basis of fish aggression can guide phenotype-specific management in aquaculture, potentially improving stress management, reducing injuries and mortality, and boosting productivity.

Keywords: 7,8-DHF; dose-dependent response; behavioral phenotyping; aggressive behavior; *Sebastes schlegelii*

1. Introduction

Aggressive behavior constitutes an evolutionarily conserved strategy for resource competition, playing vital roles in habitat acquisition, mate selection, and reproductive success within natural ecosystems. However, under high-density intensive aquaculture conditions, the adverse effects of such behaviors are markedly amplified, thereby inducing compounded detrimental consequences [1]. Spatial confinement and restricted behavioral expression in captive environments often lead to frequent physical injuries, secondary infections, and impaired growth performance, collectively undermining both production efficiency and fish welfare standards. Notably, aggressive behavior exhibits considerable inter-individual variability within populations. Empirical research has shown that a minority (15–30%) of individuals with a H-agg phenotype account for over 80% of observed aggressive interactions [2]. This behavioral heterogeneity demonstrates significant environmental stability [3–5], as evidenced by studies on *Oncorhynchus kisutch* (coho salmon), which reveal consistent aggression profiles throughout extended acclimation periods [6]. The persistence of conserved aggression phenotypes strongly suggests that individual differences are primarily regulated by structural neuroplasticity (e.g., stable synaptic remodeling) rather than short-term environmental factors [7].

The maintenance of neuroplastic homeostasis critically depends on brain-derived neurotrophic factor (BDNF) [8], a key regulator of aggressive behavior through tyrosine kinase receptor B (TrkB)-mediated signaling [9–11]. Studies on *Mus musculus* (mice) and *Danio rerio* (zebrafish) have demonstrated that BDNF deficiency enhances neural circuit excitability and increases aggression, thereby revealing the evolutionarily conserved regulation of aggression by the BDNF-TrkB pathway [12,13]. The small-molecule TrkB agonist 7,8-dihydroxyflavone (7,8-DHF) effectively mimics BDNF's neuroregulatory functions while circumventing the pharmacokinetic limitations associated with exogenous BDNF administration, such as its short half-life and poor blood–brain barrier permeability, which restrict its direct behavioral applications [14]. Mechanistically, 7,8-DHF modulates aggression through the activation of downstream MAPK/ERK signaling cascades and exhibits dose-dependent behavioral effects in mammalian models [15,16]. Nie et al. demonstrated the therapeutic potential of 7,8-DHF in ameliorating and even reversing behavioral deficits in mice [17]. Despite these advances, significant knowledge gaps persist regarding the behavioral effects of 7,8-DHF in piscine models. The dose–response relationships and phenotype-specific regulatory mechanisms under aquaculture conditions remain inadequately understood, particularly concerning individuals displaying distinct aggression phenotypes.

Accurate quantification of aggressive behavior is both a vital welfare indicator and a crucial metric for assessing neuromodulatory interventions [18,19]. In Atlantic salmon (*Salmo salar*), Oliveira et al. [20] demonstrated a positive correlation between chasing/nipping behaviors and fin damage. Individuals with high-aggression levels demonstrate significantly greater locomotor acceleration and reduced first movement latency, which have been established as quantitative indicators of aggression [21,22]. Importantly, these behavioral parameters correlate strongly with neural excitability pathways. The administration of fluoxetine, a selective serotonin reuptake inhibitor, significantly prolonged the latency to first movement in mice [23]. This behavioral modulation was closely associated with c-Fos expression, an immediate early gene marker of neuronal activation [24].

The black rockfish, *Sebastes schlegelii*, a species of significant ecological and economic importance in China, Korea, and Japan [25,26], exhibits frequent aggressive behavior that adversely affects production efficiency. Therefore, mitigating this behavior could improve profitability by reducing losses caused by injuries and stress. Recent research on the aggression of *S. schlegelii* has predominantly focused on environmental enrichment and physiological stress responses, while the neural mechanisms underlying its aggressive behavior remain unexplored [27–29]. This study aims to systematically investigate the behavioral responses of *S. schlegelii* with distinct aggression phenotypes to graded doses of 7,8-DHF, addressing three objectives: (1) determining the dose-dependent effects of 7,8-DHF on aggression modulation in juveniles; (2) identifying phenotype-specific differences in response thresholds to 7,8-DHF; (3) elucidating the regulatory patterns of 7,8-DHF across different aggression phenotypes. Our findings will enhance our understanding of the neurobiological basis of fish aggression and may reveal phenotype-dependent neural regulation mechanisms in aggressive behavior.

2. Materials and Methods

2.1. Experimental Fish and Husbandry

Juvenile *S. schlegelii* were obtained from a local hatchery (same genetic background) and maintained in the aquaculture system of the Aquaculture Facilities and Equipment Engineering Center. The recirculating aquaculture system (RAS) maintained all water quality parameters within optimal and stable ranges. Throughout the experimental period, water quality parameters were maintained as follows: water temperature was maintained at 20.0 ± 1.0 °C; salinity at 30 ± 1.0‰; an ammonia nitrogen concentration of 0.22–0.39 mg/L, and dissolved oxygen levels of 6.30–7.98 mg/L. A 12 h/12 h light/dark photoperiod was implemented. The fish were fed three times daily (08:30, 12:30, and 16:30) with commercial feed (2% of body weight per day; crude protein ≥ 44%, crude fat ≥ 5%). Prior to the experiments, the fish were acclimated for at least 7 days to stabilize their physiological conditions and fasted for 24 h. The experimental protocols were approved by the Dalian Ocean University Animal Ethics Committee (Approval No. 2023-47), and all procedures complied with China's National Standards for Laboratory Animal Welfare (GB/T 35892-2018 and GB/T 35823-2018) [30,31].

2.2. Experimental Design

2.2.1. Experimental Fish Tagging

The aggressive behavior experiment employed a random pairing method, ensuring a body weight difference of no more than 5% between paired fish to minimize size-related effects. A total of 48 healthy juveniles (11.12 ± 0.54 cm body length; 18.72 ± 0.92 g body weight) were selected. The fish were anesthetized 72 h prior to the trial using MS-222 (60 mg/L, Sigma-Aldrich, St. Louis, MO, USA), with the duration of anesthesia not exceeding 3 min. Colored plastic discs (d = 5 mm) were attached to the anterior dorsal fin using surgical sutures for identification purposes. The tagged fish were promptly reintroduced to their tanks, with strict adherence to the following criteria for the experimental subjects: the absence of hemorrhage, visible health effects, or abnormal behavior, and normal recovery of mobility within 60 s [32].

2.2.2. Experimental Setup and Distinguishing Aggressive Phenotypes

The experimental tank (40 × 35 × 17.5 cm) contained a 10 cm water depth, with the sidewalls covered by white opaque stickers to minimize external disturbances. A high-definition camera (HIK-DS-2CD3T35D-I5, 1080P@30 fps, Hikvision, Hangzhou, China) was positioned vertically and recorded interactions for 10 min after partition removal (Figure 1) [21]. Aggression phenotyping was conducted using the real opponent paradigm, wherein two size-matched juveniles ($\Delta BW < 5\%$) were introduced into the experimental tank. This better mimicked natural ecological contexts, thereby eliciting ethologically relevant responses and enabling comprehensive quantification of aggressive phenotypes in both focal and opponent fish. Their interactions, including circling, chasing, biting, and withdrawal, were continuously monitored until the first attack–retreat sequence occurred, thereby identifying aggression-dimorphic individuals. Individuals dominating attacks and inducing opponent avoidance were classified as the H-agg phenotype, whereas those consistently exhibiting escape behaviors were designated as the L-agg phenotype. Aggressive behaviors were operationally defined as either sustained chasing (≥3 s/bout) by H-agg individuals toward L-agg individuals or body contact biting, which elicited escape responses in L-agg individuals.

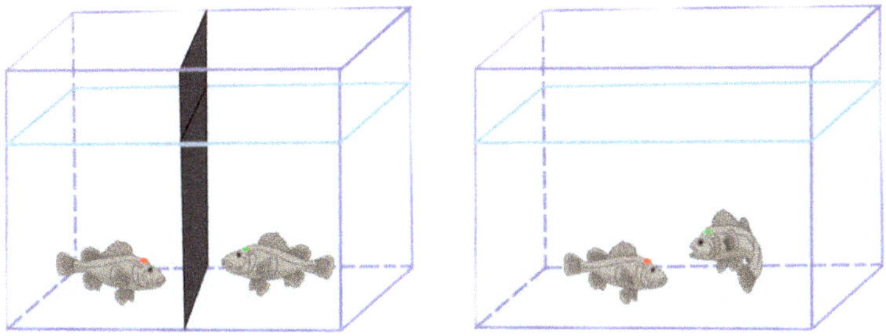

Figure 1. Experimental setup for dyadic confrontations. The high-aggression (H-agg) phenotype is denoted by red markings and the low-aggression (L-agg) phenotype is denoted by green markings.

2.2.3. Experimental Fish Dosing

The experimental design included four groups: three treatment groups receiving 7,8-DHF at concentrations of 1.25, 2.5, and 5 mg/mL in a 30% DMSO-PBS vehicle, and a control group injected with an equivalent 30% DMSO-PBS. Each group consisted of six replicates (N = 6). The dosages were adjusted based on body weight (1.25, 2.5, 5 mg/kg) and administered intraperitoneally with a 100 µL injection using a self-refilling syringe (Socorex 174; Socorex, Ecublens, Switzerland). After a 72 h recovery period post-tagging, the injected fish were acclimated in the test tanks for 1 h before a 10 min behavioral recording session commenced.

2.3. Behavioral Analysis

All trials were video-recorded for 10 min to capture dyadic interactions and individual behavioral signatures. Automated trajectory tracking and parameter extraction were performed using EthoVision XT (v12.0, Noldus Information Technology, Wageningen, The Netherlands). Six behavioral metrics were analyzed: movement distance, locomotor acceleration, angular velocity, first movement latency, activity frequency, and immobility frequency.

Movement distance was quantified as the two-dimensional displacement (cm) of the fish centroid, calculated using the following formula:

$$D_n = \sqrt{(X_n - X_{n-1})^2 + (Y_n - Y_{n-1})^2}$$

where D_n represents the frame-to-frame displacement, and $(X_n - X_{n-1})$ and $(Y_n - Y_{n-1})$ denote the coordinate differences along the X- and Y-axes between consecutive video frames (n and $n-1$).

Locomotor acceleration was mathematically defined as the rate of change in the maximum velocity (cm/s^2):

$$A_n = \frac{v_n - v_{n-1}}{t_n - t_{n-1}}$$

where A_n indicates instantaneous acceleration, $v_n - v_{n-1}$ represents the velocity difference between frames, and $t_n - t_{n-1}$ is the time interval between frames.

Angular velocity was defined as the rate of change in the head direction (°/s):

$$W_n = \frac{T_n}{t_n - t_{n-1}}$$

where w_n denotes the instantaneous turning rate, T_n is the change in head orientation angle between frames, and $t_n - t_{n-1}$ is the sampling interval.

First movement latency was measured as the time (in seconds) from the removal of the partition until the initial displacement exceeded 1.0 cm.

Activity frequency was defined as a 20–60% change in body position between frames (captured at 10 frames per second), while immobility frequency was defined as a change of less than 20% in body position between frames (also at 10 frames per second).

2.4. Statistical Analysis

The data were expressed as arithmetic means ± SEM. All analyses were performed using Microsoft Office Excel 2019 and SPSS 21.0. The normality of the data was assessed using Shapiro–Wilk tests. One-way ANOVA was used to analyze the data that met the assumptions of normality and homoscedasticity, followed by Dunnett's multiple comparison tests. In cases where heterogeneity of variance was detected, non-parametric tests (Kruskal–Wallis) were applied. To compare behavioral parameters between H-agg and L-agg individuals, we conducted two-tailed t-tests on normally distributed data (Shapiro–Wilk test, $p > 0.05$) with homoscedasticity (Levene's test, $p > 0.05$)

3. Results

3.1. Aggressive Behavior Analysis in *S. schlegelii*

In dyadic confrontations, H-agg and L-agg individuals exhibited distinct behavioral phenotypic divergence. Figure 2 illustrates the ethogram of aggressive behaviors in *S. schlegelii*. Circular swimming (Figure 2a) served as a mutual evaluation of combat potential. H-agg individuals initiated 93.60% of offensive acts, primarily manifesting as frequent chasing (Figure 2b) and targeted biting at caudal fins or lateral abdominal regions (Figure 2c). Conversely, L-agg individuals predominantly displayed escape behaviors, rapidly swimming away from attackers and occasionally displaying freezing responses, characterized by fin folding and immobility at the bottom of the tank (Figure 2d). Quantitative analysis of behaviors revealed that aggression constituted 16.84% of the total interaction time during the 10 min observation period. Specifically, chasing accounted for 88.29%, while biting comprised 11.71% of the aggressive interactions (Table 1).

Table 1. Analyses of the duration of particular aggressive interactions (s).

	Ethogram of Aggressive Behavior	Time (s)	Proportion of Phenotypes %	Total Proportion %
Aggressive phenotype	Biting	11.83 ± 1.72	11.71	1.98
	Chasing	89.17 ± 6.23	88.29	14.86
Total		101	100	16.84
Non-aggressive phenotype	Circling	83.50 ± 3.43	16.73	13.91
	Withdrawal	415.50 ± 5.53	83.27	69.25
Total		499	100	83.16

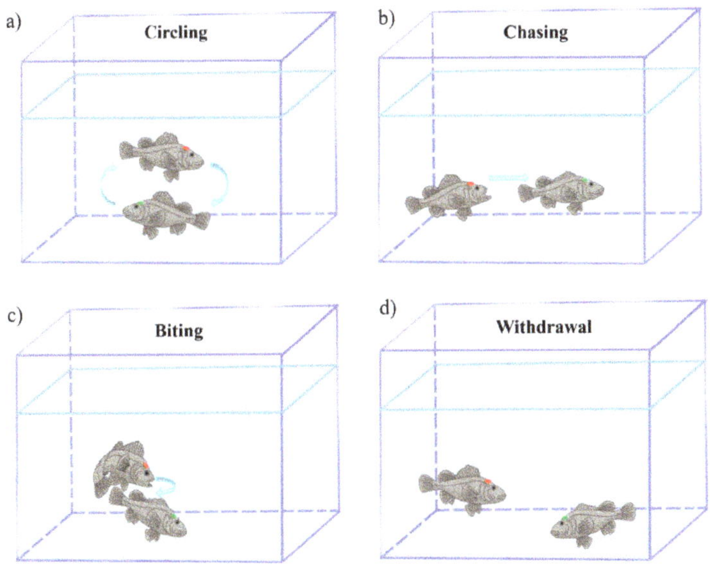

Figure 2. Graphical ethogram of aggressive behavior in *Sebastes schlegelii.* (**a**) Circular swimming; (**b**) aggressive chasing; (**c**) biting; (**d**) submission withdrawal. The high-aggression (H-agg) phenotype is denoted by red markings and the low-aggression (L-agg) phenotype is denoted by green markings.

3.2. Locomotor Parameters of Aggression-Dimorphic Individuals

A quantitative analysis of locomotor parameters across the high- and low-aggression phenotypes revealed significant group differences (Table 2). The H-agg individuals exhibited significantly greater locomotor acceleration compared to their L-agg counterparts ($p < 0.01$). The latency to first movement was substantially shorter in the H-agg individuals than in the L-agg individuals ($p < 0.01$). Conversely, the L-agg individuals exhibited a significantly higher immobility frequency than the H-agg individuals ($p < 0.05$). No significant differences were observed between the groups in terms of movement distance, angular velocity, or activity frequency ($p > 0.05$).

Table 2. Behavioral parameters of H-agg and L-agg in the controls.

	Movement Distance (cm)	Locomotor Acceleration (cm/s^2)	Angular Velocity (°/s)	First Movement Latency (s)	Activity Frequency	Immobility Frequency
H-agg	10.91 ± 1.04	3607.54 ± 156.35 **	48.27 ± 5.62	1.19 ± 0.09 **	191.38 ± 16.50	210.61 ± 33.77 *
L-agg	9.15 ± 0.89	2367.91 ± 169.29	36.15 ± 2.62	1.95 ± 0.16	210.91 ± 23.29	321.39 ± 33.91

Significant differences between H-agg and L-agg are denoted by asterisks (* indicates $p < 0.05$, ** indicates $p < 0.01$).

3.3. Effects of 7,8-DHF on Locomotor States in Aggression-Dimorphic Individuals

The concentration–response relationship of 7,8-DHF on the locomotor activity of fish with distinct phenotypes is shown in Figure 3. Movement distance and locomotor acceleration are considered fundamental aspects of locomotor behavior. The H-agg individuals displayed significant dose-dependent modulation in acceleration (Figure 3a). Specifically, both the 2.5 mg/kg (2633.18 cm/s^2) and 5 mg/kg (2676.71 cm/s^2) treatments resulted in significantly reduced acceleration compared to the controls (3607.54 cm/s^2, $p < 0.05$), with no inter-dose differences ($p > 0.05$). In contrast, the L-agg juveniles exhibited similar locomotor acceleration across all 7,8-DHF concentrations compared to the 0 mg/kg controls

($p > 0.05$, Figure 3b). Movement distance was not significantly affected by the 7,8-DHF treatment in either phenotype ($p > 0.05$, Figure 3c,d).

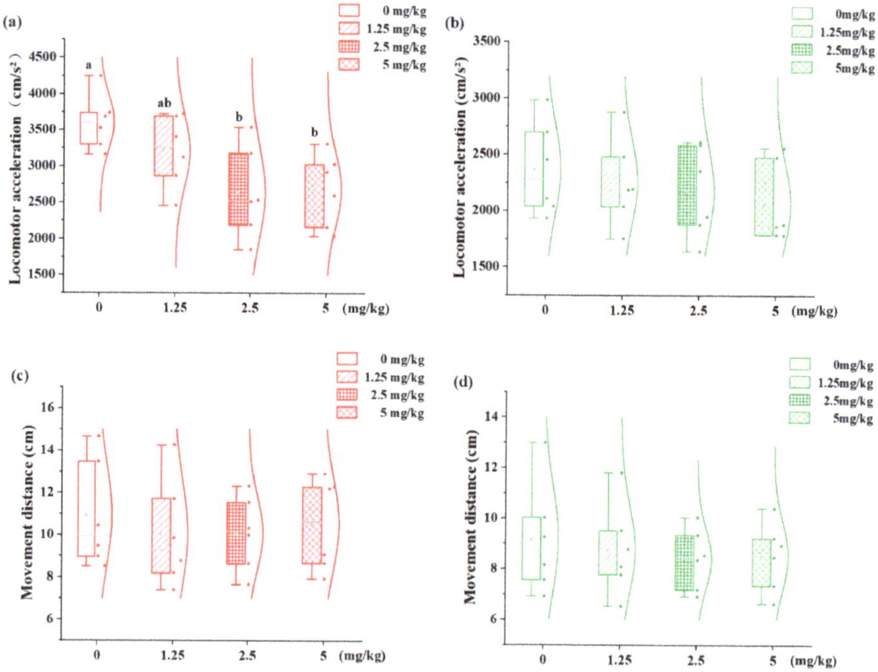

Figure 3. Effects of different concentrations of 7,8-DHF on the locomotor activity of H-agg and L-agg individuals. (**a**) Locomotor acceleration of H-agg individuals; (**b**) locomotor acceleration of L-agg; (**c**) movement distance of H-agg; (**d**) movement distance of L-agg. Values are presented as mean ± SE (n = 6). Different letters indicate significant differences ($p < 0.05$).

3.4. *Effects of 7,8-DHF on Spatial Distribution in Aggression-Dimorphic Individuals*

The concentration–response relationship of 7,8-DHF on spatial distribution patterns is illustrated in Figure 4. Angular velocity and first movement latency indicate spatial preference. The H-agg individuals demonstrated a dose-dependent reduction in angular velocity (Figure 4a). Specifically, the 2.5 mg/kg group (26.56°/s) showed significantly lower values than the control (0 mg/kg: 48.27°/s; $p < 0.05$), with no differences between the 1.25 and 2.5 mg/kg groups. Similarly, the L-agg individuals displayed reduced angular velocity at 2.5 mg/kg (25.85°/s vs. control 36.15°/s, $p < 0.05$), but no inter-group differences at other doses (Figure 4b). In contrast, the H-agg individuals showed dose-dependent increases in first movement latency (Figure 4c). The 2.5 mg/kg group (2.55 s) significantly surpassed all other groups ($p < 0.05$). No such effects were observed in the L-agg individuals across the tested concentrations (Figure 4d).

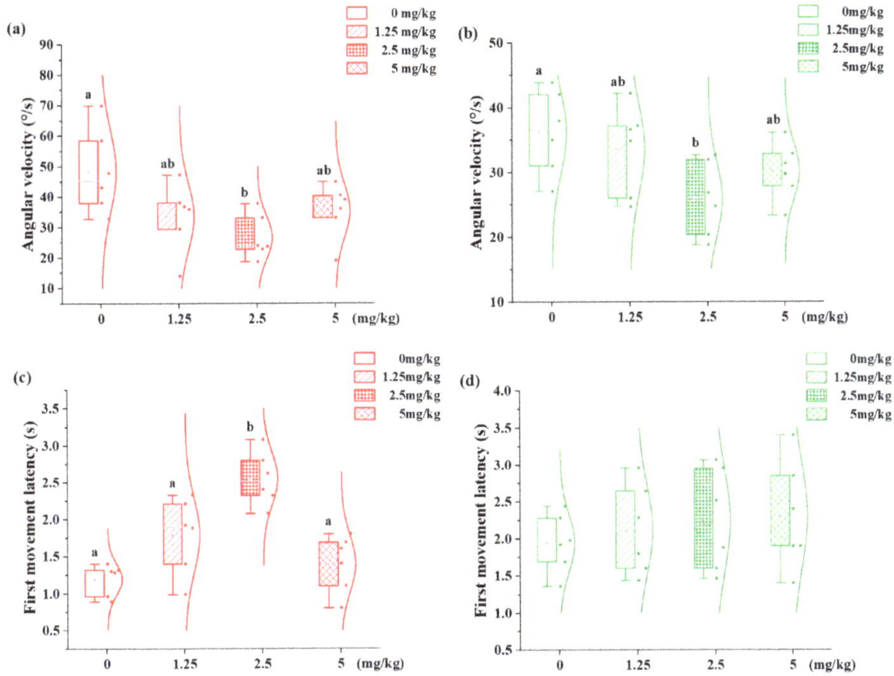

Figure 4. Effects of different concentrations of 7,8-DHF on the positional distribution of H-agg and L-agg individuals. (**a**) Angular velocity of H-agg; (**b**) angular velocity of L-agg; (**c**) first movement latency of H-agg; (**d**) first movement latency of L-agg. Values are presented as mean ± SE (n = 6). Different letters indicate significant differences ($p < 0.05$).

3.5. Effects of 7,8-DHF on Behavioral States in Aggression-Dimorphic Individuals

The dose–response relationship of 7,8-DHF on behavioral states in *S. schlegelii* is demonstrated in Figure 5. Activity frequency and immobility frequency serve as reliable metrics for assessing locomotor activity and behavioral modifications. The H-agg individuals exhibited a significant dose-dependent decrease in activity frequency (Figure 5a): the 2.5 mg/kg group (82.06 bouts) showed a 57.1% reduction compared to vehicle controls (0 mg/kg: 191.38 bouts; $p < 0.05$), with no effects at other concentrations ($p > 0.05$). The L-agg juveniles showed unchanged activity frequencies across all treatment groups ($p > 0.05$, Figure 5b). Immobility duration was not significantly altered in either phenotype at any tested concentrations ($p > 0.05$, Figure 5c,d).

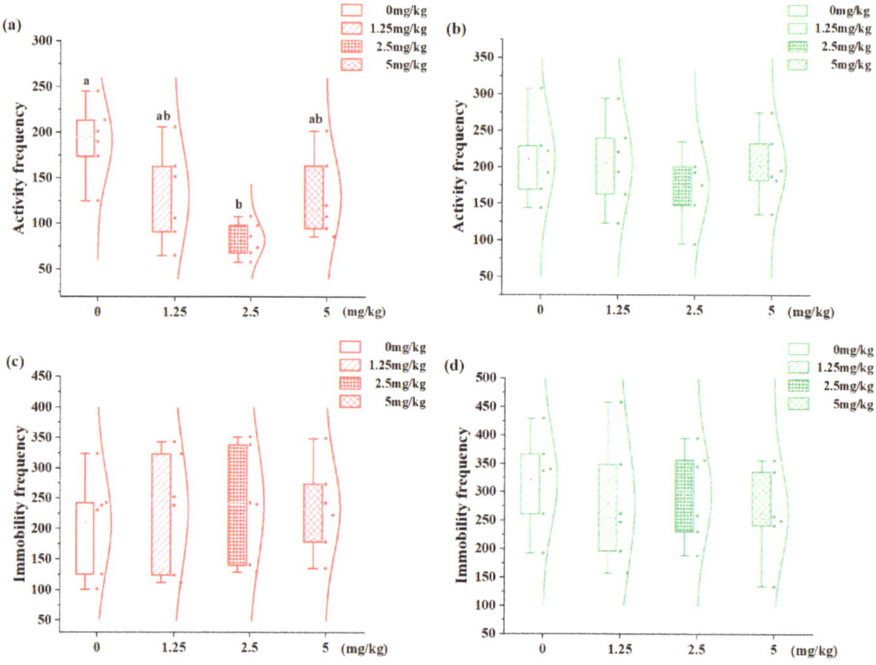

Figure 5. Effects of different concentrations of 7,8-DHF on the physical state of H-agg and L-agg individuals. (**a**) Activity frequency of H-agg; (**b**) activity frequency of L-agg; (**c**) immobility frequency of H-agg; (**d**) immobility frequency of L-agg. Values are presented as mean ± SE (n = 6). Different letters indicate significant differences ($p < 0.05$).

4. Discussion

In dyadic confrontation, juvenile *S. schlegelii* individuals with high aggression and low aggression levels exhibited pronounced marked behavioral dimorphism, indicating a significant divergence in behavioral strategies. H-agg individuals initiated 93.6% of aggressive encounters, characterized by temporally clustered and highly repetitive attacks. These dominant fish predominantly occupied the central zone of the experimental arena, employing proactive approaching [21]. Conversely, L-agg individuals primarily exhibited demonstrated escape and freezing behaviors, showing a strong preference for the peripheral zone, consistent with passive defense mechanisms [33]. This center–periphery dichotomy corresponds to the bold–shy continuum reported in teleost personality studies [34]. Notably, aggressive interactions accounted for only 16.84% of the total observation time (Table 1), suggesting a pattern of "instantaneous high-intensity contest" in intraspecific competition. The chase-to-bite ratio (88.3% chasing vs. 11.7% biting) matched the energetically optimal ratio (4:1) reported in coral reef species [35]. Dominant H-agg individuals maintain social hierarchy through chase-intensive strategies, whereas subordinate L-agg individuals employ motion inhibition and submissive postures to avoid conflict. The greater prevalence of chasing over physical attacks, combined with ritualized circling and fin erection, reflects an evolutionary compromise between asserting dominance and conserving metabolic resources [36].

In the context of behavior classification, the mirror image test is a classical method that offers distinct advantages in terms of standardizing procedures and reducing the risk of physical injury to participants. Nevertheless, in order to elicit a more comprehensive expression of aggressive behaviors and complex behavioral sequences, this study employed the real opponent paradigm [37]. This methodological trade-off enabled the acquisition of more nuanced data through quantitative behavioral analysis. Quantitative analysis revealed that the H-agg individuals exhibited significantly higher locomotor acceleration and shorter first movement latency compared to their L-agg counterparts. These findings align with previous research [21,38], which demonstrates competitive advantages of rapid behavioral initiation in aggressive phenotypes. Notably, the lack of significant differences in movement distance and angular velocity indicates that acceleration and latency parameters more sensitively reflect motivational drive rather than overall locomotor capacity. The data support the neurobehavioral model that dissociates limbic-regulated attack motivation from motor circuit-dependent basic movement capacity [39].

This study provides the first evidence for differential behavioral modulation by 7,8-DHF across aggression phenotypes in *S. schlegelii*. The H-agg individuals exhibited significant dose-dependent reductions in acceleration (45%), angular velocity (28.5%), and activity frequency (57.1%) at optimal doses. TrkB, the high-affinity BDNF receptor, demonstrated increased activity in neural circuits associated with aggression [40], a pattern conserved in rodent and primate models [41,42]. Acting as a TrkB agonist, 7,8-DHF likely attenuates aggressive behaviors through enhanced BDNF/TrkB signaling in H-agg neural circuits. The bell-shaped dose–response curve suggests saturation of the TrkB binding site at 2.5 mg/kg, consistent with receptor occupancy pharmacodynamics [43]. The L-agg resistance to 7,8-DHF suggests fundamental differences in TrkB-expressing neuron ensembles between the phenotypes [44].

The impulsive aggressive responses of the H-agg phenotype appeared to be particularly sensitive to activity-dependent neuroplastic changes mediated by the TrkB signaling pathway. Pharmacological activation of TrkB induced rapid alterations in neuronal activity and neuromuscular transmission, impairing action potential conduction and contractile performance. This impairment is manifested as a reduction in acceleration among juvenile fish [45,46]. 7,8-DHF significantly decreased angular velocity and increased first movement latency in the H-agg phenotypes, without affecting movement distance, indicating a selective suppression of rapid response capability [43] rather than generalized motor function impairment. These effects align with the neuromodulatory action of 7,8-DHF on aggression-associated neural circuits, consistent with findings from mammalian models [47]. Parallel mechanisms occurred in dopamine signaling: D2 receptor antagonists reduced aggression but concurrently impaired motor function [48].

The differential effects of 7,8-DHF on H-agg and L-agg individuals suggest intrinsic differences in their neurotransmitter systems, potentially involving TrkB-dependent regulation of dopaminergic neuron plasticity. In H-agg individuals, excessive dopaminergic activity may drive aggressive behavior, as TrkB activation enhances synaptic plasticity in dopamine neurons [49]. The reduced locomotion in H-agg individuals reflects 7,8-DHF-induced downregulation of dopamine pathways, while L-agg individuals, characterized by lower basal dopamine activity, remained unaffected [50]. Furthermore, aggression phenotype differentiation is consistently associated with altered BDNF gene methylation, which likely underlie reduced basal TrkB signaling in L-agg individuals, thereby elevating their response threshold to exogenous agonists.

Notably, 7,8-DHF exhibited a biphasic (U-shaped) dose–response effect, with maximal suppression of movement initiation, turning speed, and locomotor activity observed in H-agg individuals at 2.5 mg/kg. This implies that supraoptimal TrkB stimulation activates compensatory inhibition mechanisms (e.g., via feedback loops or desensitization), attenuating aggression [51,52]. The suppressed activity and delayed movement initiation at this dose likely reflect metabolic trade-offs, where energy conserved from reduced aggression enhances stress adaptation. These behavioral changes are mediated by conserved neuroregulatory circuits, analogous to cortisol-driven modulation of energy metabolism. The consistency with canonical stress responses indicates that 7,8-DHF functionally emulates these adaptive pathways.

5. Conclusions

BDNF is pivotal in aggression regulation. Our findings show that 7,8-DHF (a TrkB agonist) effectively attenuates aggression in high-aggression *S. schlegelii* juveniles, with its U-shaped dose effect highlighting intricate neuroplasticity–receptor kinetic interplay in behavioral control. The critical dose of 2.5 mg/kg may represent a threshold where the drug's action transitions from specific TrkB receptor activation to broader systemic interference. Subsequent research could explore BDNF-mediated aggression control through CNS expression analysis and mechanistic studies, offering solutions to mitigate fish aggression for enhanced welfare and aquaculture productivity. The present study employed one-way ANOVA to evaluate the independent effects of 7,8-DHF concentration on behavioral outcomes within each phenotype group, without examining potential concentration–phenotype interactions. Future investigations will specifically address these interaction effects to further elucidate 7,8-DHF's behavioral modulation mechanisms.

Author Contributions: Conceptualization, S.X. and Z.M.; methodology, S.X. and Z.M.; formal analysis, S.X., X.M.; investigation, S.X., Y.X., and C.M.; resources, T.Z.; writing—original draft, S.X.; writing—review and editing, Z.M.; supervision, Z.M.; project administration, Z.M.; funding acquisition, Z.M. All authors have read and agreed to the published version of the manuscript.

Funding: This research was funded by the National Key R&D Program of China (No. 2024YFD2400200), the National Natural Science Foundation of China (No. 32473129/32403092), and the Fundamental Scientific Research Project of Liaoning Province (JYTMS20230477/2023-MSLH-019).

Institutional Review Board Statement: The animal study protocol was approved by the Animal Ethics Committee of Dalian Ocean University (Approval No. 2024-36). All experimental procedures strictly complied with China's National Standards for Laboratory Animal Welfare (GB/T 35892-2018 and GB/T 35823-2018).

Informed Consent Statement: Not applicable.

Data Availability Statement: The data supporting the findings of this study are available from the corresponding author upon justified request.

Acknowledgments: The authors gratefully acknowledge the journal editors and peer reviewers for their expert evaluation and constructive comments, which were instrumental in improving this study.

Conflicts of Interest: Author Tao Zang was employed by the company Tianzheng Industrial Co., Ltd. The remaining authors declare that the research was conducted in the absence of any commercial or financial relationships that could be construed as a potential conflict of interest.

References

1. Turnbull, J.F.; Adams, C.E.; Richards, R.H.; Robertson, D.A. Attack site and resultant damage during aggressive encounters in Atlantic salmon (*Salmo salar* L.) parr. *Aquaculture* **1998**, *159*, 345–353. [CrossRef]
2. Carvalho, T.B.; Souza, E.C.M.; Pinheiro-da-Silva, J.; Villacorta-Correa, M.A. Effect of body size heterogeneity on the aggressive behavior of larvae of matrinxã, *Brycon amazonicus* (Characiformes, Bryconidae). *Acta Amaz.* **2018**, *48*, 304–310. [CrossRef]
3. Cavallino, L.; Rincon, L.; Scaia, M.F. Social behaviors as welfare indicators in teleost fish. *Front. Vet. Sci.* **2023**, *10*, 1050510. [CrossRef]
4. Zabegalov, K.N.; Kolesnikova, T.O.; Khatsko, S.L.; Volgin, A.D.; Yakovlev, O.A.; Amstislavskaya, T.G.; Friend, A.J.; Bao, W.; Alekseeva, P.A.; Lakstygal, A.M.; et al. Understanding zebrafish aggressive behavior. *Behav. Process.* **2019**, *158*, 200–210. [CrossRef] [PubMed]
5. Hellinger, J.; Jägers, P.; Spoida, K.; Weiss, L.C.; Mark, M.D.; Herlitze, S. Analysis of the territorial aggressive behavior of the bioluminescent flashlight fish *Photoblepharon steinitzi* in the red sea. *Front. Mar. Sci.* **2020**, *7*, 78. [CrossRef]

6. Rosenau, M.L.; Mcphail, J. Inherited differences in agonistic behavior between two populations of coho salmon. *Trans. Am. Fish. Soc.* **1987**, *116*, 646–654. [CrossRef]
7. Mutschler, C.; Telerman, S.B. Glial plasticity in the zebrafish central nervous system. *Trends Cell Biol.* **2024**, *34*, 531–534. [CrossRef]
8. Kowiański, P.; Lietzau, G.; Czuba, E.; Waśkow, M.; Steliga, A.; Moryś, J. BDNF: A key factor with multipotent impact on brain signaling and synaptic plasticity. *Cell. Mol. Neurobiol.* **2018**, *38*, 579–593. [CrossRef]
9. Meng, L.; Liu, B.; Ji, R.; Jiang, X.; Yan, X.; Xin, Y. Targeting the BDNF/TrkB pathway for the treatment of tumors. *Oncol. Lett.* **2019**, *17*, 2031–2039. [CrossRef]
10. Niu, C.; Yue, X.; An, J.J.; Xu, H.; Xu, B. Genetic dissection of BDNF and TrkB expression in glial cells. *Biomolecules* **2023**, *7*, 549007. [CrossRef]
11. Choo, M.; Miyazaki, T.; Yamazaki, M.; Kawamura, M.; Nakazawa, T.; Zhang, J.; Tanimura, A.; Uesaka, N.; Watanabe, M.; Sakimura, K.; et al. Retrograde BDNF to TrkB signaling promotes synapse elimination in the developing cerebellum. *Nat. Commun.* **2017**, *8*, 195. [CrossRef] [PubMed]
12. Ito, W.; Chehab, M.; Thakur, S.; Li, J.; Morozov, A. BDNF-restricted knockout mice as an animal model for aggression. *Genes Brain Behav.* **2011**, *10*, 365–374. [CrossRef] [PubMed]
13. Liu, Z.H.; Li, Y.W.; Hu, W.; Chen, Q.L.; Shen, Y.J. Mechanisms involved in tributyltin-enhanced aggressive behaviors and fear responses in male zebrafish. *Aquat. Toxicol.* **2020**, *220*, 105408. [CrossRef]
14. Wurzelmann, M.; Romeika, J.; Sun, D. Therapeutic potential of brain-derived neurotrophic factor (BDNF) and a small molecular mimics of BDNF for traumatic brain injury. *Neural Regen. Res.* **2017**, *12*, 7–12. [PubMed]
15. Agrawal, R.; Noble, E.; Tyagi, E.; Zhuang, Y.; Ying, Z.; Gomez-Pinilla, F. Flavonoid derivative 7,8-DHF attenuates TBI pathology via TrkB activation. *Biochim. Et Biophys. Acta* **2015**, *1852*, 862–872. [CrossRef]
16. Dhaliwal, N.; Dhaliwal, J.; Chopra, K. 7, 8-Dihydroxyflavone ameliorates cholinergic dysfunction, inflammation, oxidative stress, and apoptosis in a rat model of Vascular dementia. *Neurochem. Res.* **2024**, *49*, 1137–1149. [CrossRef]
17. Nie, S.; Ma, K.; Sun, M.; Lee, M.; Tan, Y.; Chen, G.; Zhang, Z.; Zhang, Z.; Cao, X. 7,8-Dihydroxyflavone protects nigrostriatal dopaminergic neurons from rotenone-induced neurotoxicity in rodents. *Park. Dis.* **2019**, *3*, 9193534. [CrossRef]
18. Yang, T.; Bayless, D.W.; Wei, Y.; Landayan, D.; Marcelo, I.M.; Wang, Y.; DeNardo, L.A.; Luo, L.; Druckmann, S.; Shah, N.M. Hypothalamic neurons that mirror aggression. *Cell* **2023**, *186*, 1195–1211.e19. [CrossRef]
19. Loveland, J.L.; Fernald, R.D. Differential activation of vasotocin neurons in contexts that elicit aggression and courtship. *Behav. Brain Res.* **2017**, *317*, 188–203. [CrossRef]
20. MacLean, A.; Metcalfe, N.B.; Mitchell, D. Alternative competitive strategies in juvenile Atlantic salmon (*Salmo salar*): Evidence from fin damage. *Aquaculture* **2000**, *184*, 291–302. [CrossRef]
21. Li, H.; Wang, J.; Zhang, X.; Hu, Y.; Liu, Y.; Ma, Z. Comparing behavioral performance and physiological responses of *Sebastes schlegelii* with different aggressiveness. *Fish Physiol. Biochem.* **2022**, *48*, 1333–1347. [CrossRef] [PubMed]
22. Barbosa, H.P.; Lima-Maximino, M.G.; Maximino, C. Acute fluoxetine differently affects aggressive display in zebrafish phenotypes. *Aggress. Behav.* **2019**, *45*, 62–69. [CrossRef] [PubMed]
23. Peeters, D.; Rietdijk, J.; Gerrits, D.; Rijpkema, M.; de Boer, S.F.; Verkes, R.J.; Homberg, J.R. Searching for neural and behavioral parameters that predict anti-aggressive effects of chronic SSRI treatment in rats. *Neuropharmacology* **2018**, *143*, 339–348. [CrossRef]
24. Ariyasiri, K.; Choi, T.I.; Gerlai, R.; Kim, C.H. Acute ethanol induces behavioral changes and alters c-fos expression in specific brain regions, including the mammillary body, in zebrafish. *Prog. Neuro-Psychopharmacol. Biol. Psychiatry* **2021**, *109*, 110264. [CrossRef] [PubMed]
25. Li, R.; Cho, S.H. Fish meal replacement by chicken by-product meal in diet: Impacts on growth and feed availability of juvenile rockfish (*Sebastes schlegelii*), and economical analysis. *Animals* **2025**, *15*, 80. [CrossRef]
26. Li, R.; Cho, S.H. Substitution impact of tuna by-product meal for fish meal in the diets of rockfish (*Sebastes schlegelii*) on growth and feed availability. *Animals* **2023**, *13*, 3586. [CrossRef]
27. Zhang, Z.; Xu, X.; Wang, Y.; Zhang, X. Effects of environmental enrichment on growth performance, aggressive behavior and stress-induced changes in cortisol release and neurogenesis of black rockfish *Sebastes schlegelii*. *Aquaculture* **2020**, *528*, 735483. [CrossRef]
28. Xu, X.; Zhang, Z.; Guo, H.; Qin, J.; Zhang, X. Changes in aggressive behavior, cortisol and brain monoamines during the formation of social hierarchy in black rockfish (*Sebastes schlegelii*). *Animals* **2020**, *10*, 2357. [CrossRef]
29. Jia, Y.; Gao, Y.; Wan, J.; Gao, Y.; Li, J.; Guan, C. Altered physiological response and gill histology in black rockfish, *Sebastes schlegelii*, during progressive hypoxia and reoxygenation. *Fish Physiol. Biochem.* **2021**, *47*, 1133–1147. [CrossRef]
30. *GB/T 35892-2018*; Laboratory Animal—Guideline for Ethical Review of Animal Welfare. The National Standards of the People's Republic of China: Beijing, China, 2018.
31. *GB/T 35823-2018*; Laboratory Animals—General Requirements for Animal Experiment. The National Standards of the People's Republic of China: Beijing, China, 2018.

32. Cooke, S.J.; Hinch, S.G.; Wikelski, M.; Andrews, R.D.; Kuchel, L.J.; Wolcott, T.G.; Butler, P.J. Biotelemetry: A mechanistic approach to ecology. *Trends Ecol. Evol.* **2004**, *19*, 334–343. [CrossRef]
33. Benhaïm, D.; Ferrari, S.; Chatain, B.; Bégout, M.L. The shy prefer familiar congeners. *Behav. Process.* **2016**, *126*, 113–120. [CrossRef]
34. Grunst, A.S.; Grunst, M.L. Animal personality in multiple stressor environments: The evolutionary ecology of among-individual differences in responses to stressor suites. *Proc. R. Soc. B Biol. Sci.* **2024**, *291*, 20241620.
35. Geange, S.W. An evaluation of prior residency and habitat effects on the persistence of settling reef fishes. Ph.D. Thesis, Victoria University of Wellington, Wellington, Australia, 2010.
36. Scarsella, G.E.; Duque, K.S.; Wong, S.C.; Sivaraman, B.; Earley, R.L. Hormonal responses to noncontact aggression in convict cichlid fish. *J. Exp. Zool. Part A Ecol. Genet. Physiol.* **2016**, *325*, 219–230. [CrossRef]
37. Hubená, P.; Horký, P.; Slavík, O. Test-dependent expression of behavioral syndromes: A study of aggressiveness, activity, and stress of chub. *Aggress. Behav.* **2020**, *46*, 412–424. [CrossRef] [PubMed]
38. Axling, J.; Vossen, L.E.; Peterson, E.; Winberg, S. Boldness, activity, and aggression: Insights from a large-scale study in Baltic salmon (*Salmo salar* L.). *PLoS ONE* **2023**, *18*, e0287836. [CrossRef] [PubMed]
39. Tan, S.; Xiao, Y.; Yin, H.H.; Chen, A.I.; Soong, T.W.; Je, H.S. Postnatal TrkB ablation in corticolimbic interneurons induces social dominance in male mice. *Proc. Natl. Acad. Sci. USA.* **2018**, *115*, e9909–e9915. [CrossRef]
40. Adachi, M.; Autry, A.E.; Mahgoub, M.; Suzuki, K.; Monteggia, L.M. TrkB signaling in dorsal raphe nucleus is essential for antidepressant efficacy and normal aggression behavior. *Neuropsychopharmacology* **2017**, *42*, 886–894. [CrossRef]
41. Chen, L.X.; Qi, Z.; Shao, Z.J.; Li, S.S.; Qi, Y.L.; Gao, K.; Liu, S.X.; Li, Z.; Sun, Y.S.; Li, P.Y. Study on antidepressant activity of pseudo-ginsenoside HQ on depression-Like behavior in mice. *Molecules* **2019**, *24*, 870. [CrossRef]
42. Cortese, G.P.; Barrientos, R.M.; Maier, S.F.; Patterson, S.L. Aging and a peripheral immune challenge interact to reduce mature brain-derived neurotrophic factor and activation of TrkB, PLCgamma1, and ERK in hippocampal synaptoneurosomes. *J. Neurosci.* **2011**, *31*, 4274–4279. [CrossRef]
43. Guarino, A.; Bettegazzi, B.; Aziz, N.; Barbieri, M.; Bochicchio, D.; Crippa, L.; Marino, P.; Sguizzato, M.; Soukupova, M.; Zucchini, S.; et al. Low-dose 7,8-Dihydroxyflavone administration after status epilepticus prevents epilepsy development. *Neurotherapeutics* **2022**, *19*, 1951–1965. [CrossRef]
44. Andreska, T.; Lüningschrör, P.; Sendtner, M. Regulation of TrkB cell surface expression-a mechanism for modulation of neuronal responsiveness to brain-derived neurotrophic factor. *Cell Tissue Res.* **2020**, *382*, 5–14. [CrossRef] [PubMed]
45. Hoy, K.C.; Strain, M.M.; Turtle, J.D.; Lee, K.H.; Huie, J.R.; Hartman, J.J.; Tarbet, M.M.; Harlow, M.L.; Magnuson, D.S.K.; Grau, J.W. Evidence that the central nervous system can induce a modification at the neuromuscular junction that contributes to the maintenance of a behavioral response. *J. Neurosci.* **2020**, *40*, 9186–9209. [CrossRef] [PubMed]
46. Schlinger, B.A.; Paul, K.; Monks, D.A. Muscle, a conduit to brain for hormonal control of behavior. *Horm. Behav.* **2018**, *105*, 58–65. [CrossRef] [PubMed]
47. García-Díaz Barriga, G.; Giralt, A.; Anglada-Huguet, M.; Gaja-Capdevila, N.; Orlandi, J.G.; Soriano, J.; Canals, J.M.; Alberch, J. 7,8-dihydroxyflavone ameliorates cognitive and motor deficits in a Huntington's disease mouse model through specific activation of the PLCγ1 pathway. *Hum. Mol. Genet.* **2017**, *26*, 3144–3160. [CrossRef]
48. Dai, B.; Zheng, B.; Dai, X.; Cui, X.; Yin, L.; Cai, J.; Zhuo, Y.; Tritsch, N.X.; Zweifel, L.S.; Li, Y.; et al. Experience-dependent dopamine modulation of male aggression. *Nature* **2025**, *639*, 430–437. [CrossRef]
49. Kim, H.I.; Lee, S.; Lim, J.; Chung, S.; Koo, T.S.; Ji, Y.G.; Suh, Y.G.; Son, W.S.; Kim, S.H.; Choi, H.J. ERRγ ligand HPB2 upregulates BDNF-TrkB and enhances dopaminergic neuronal phenotype. *Pharmacol. Res.* **2021**, *165*, 105423. [CrossRef]
50. Yang, S.; Zhu, G. 7,8-Dihydroxyflavone and neuropsychiatric disorders: A translational perspective from the mechanism to drug development. *Curr. Neuropharmacol.* **2022**, *20*, 1479–1497. [CrossRef]
51. Devi, L.; Ohno, M. 7,8-Dihydroxyflavone, a small-molecule TrkB agonist, reverses memory deficits and BACE1 elevation in a mouse model of Alzheimer's disease. *Neuropsychopharmacology* **2012**, *37*, 434–444. [CrossRef]
52. Minichiello, L. TrkB signalling pathways in LTP and learning. *Nat. Rev. Neurosci.* **2009**, *10*, 850–860. [CrossRef]

animals

MDPI

Article

Sexual Selection in Mosquitofish: Differences in the Use of Mating Cues Between Sexes

Jiefei Wei [1,†], Bowen Feng [2,†], Chenglong Dong [3], Bojian Chen [4] and Kai Liu [1,2,*]

1 Fisheries College, Jimei University, Xiamen 361021, China; jfwei1113@163.com
2 State Key Laboratory of Mariculture Breeding, Fisheries College, Jimei University, Xiamen 361021, China; fengbowen0323@163.com
3 College of Animal Science and Technology, Northwest A&F University, Yangling 712100, China; dongcl1996@nwafu.edu.cn
4 Department of Biology, Queen's University, Kingston, ON K7P 3N6, Canada; bjc18@queensu.ca
* Correspondence: liukai1218@jmu.edu.cn
† These authors have contributed equally to this work.

Simple Summary: Mate choice plays a crucial role in evolution, with individuals assessing potential mates based on various traits. The Western mosquitofish (*Gambusia affinis*), a sexually dimorphic species, provides an ideal model for studying these preferences. This study explored how mating cues influence mate choice in both sexes using morphological data, computer-simulated animations, and preference tests. Results showed that females preferred males with resting-phase gonopodia, likely to avoid forced copulation. Conversely, males preferred younger females, with preference strength increasing with male body size. Additionally, males favored females with larger gravid spots, indicating fecundity, but avoided larger, older females, possibly due to lower reproductive potential. These findings suggest that females prioritize avoiding harassment, while males consider multiple factors in mate selection, highlighting the complexity of mating strategies in *G. affinis*.

Abstract: Sexual selection is a major driver of speciation and evolution, with mate choice being a key component. Individuals assess mate quality by integrating various mating cues. The Western mosquitofish (*Gambusia affinis*), a species exhibiting pronounced sexual dimorphism in body size and secondary sexual traits, serves as an ideal model for studying mate choice. This study examines the impact of mating cues on mate choice in different sexes of *G. affinis* through a combination of morphological parameter database construction, computer-simulated animations, and dichotomous association preference tests. The results showed that male gonopodium status significantly affects female mate choice. Females exhibited a preference for males with resting-phase gonopodia, suggesting their aversion to forced copulation and sexual harassment in coercive mating systems. Furthermore, males preferred younger females, with this preference being positively correlated with male body size. This suggests that males are sensitive to sperm competition intensity and may base their choice on social rank. Geometric morphometric analysis and simulation experiments showed that males preferred females with larger gravid spots, regardless of age, suggesting that gravid spot size reflects female fecundity. Male preference for younger females with streamlined bodies and smaller abdomens was significant, but body size did not affect mate choice in general. Our findings highlight that female and male *G. affinis* employ different mate choice strategies, with females prioritizing male harassment avoidance and males considering multiple mating cues, not solely one dominant characteristic, in their mate choice decisions. These findings demonstrate that mate choice in *G. affinis* involves balancing conflicting preferences for traits associated with reduced harassment risk (e.g., resting-phase gonopodium in males) and those linked to reproductive potential (e.g., large gravid spot in females), highlighting the nuanced decision-making processes in both sexes.

https://doi.org/10.3390/ani15101489

Keywords: mate choice; coercive mating system; mating cues; *Gambusia affinis*

1. Introduction

Mate choice plays a critical role in shaping sexual selection, as individuals evaluate the quality of potential mates and make corresponding mating decisions based on the degree to which they weigh various sources of information. This process is also closely related to an individual's reproductive fitness [1–4]. Several factors influence mate choice decisions, including morphological traits [5,6], behavioral characteristics including personality traits [7,8], and environmental factors [9–11]. Both sexes gather and analyze different types of information to make the most beneficial mating decisions. Therefore, the differences in the selection of mate selection cues between different genders have always been a hot topic in this field.

The purpose of animal mating is to pass on their own genetic material to their offspring by mating with a high-quality mate. In general, certain specific morphological characteristics of an individual can directly reflect certain physical conditions and survival abilities of the animal, such as the suitor's resistance to parasites and diseases, predation ability, defense ability, and brooding ability [12,13]. Female guppies (*Poecilia reticulate*) prefer to mate with brightly colored males, especially males with orange spots on the sides of their bodies, indicating their physical health [14]. The morphological characteristics of females in many species are related to their fertility. For example, female two-spotted gobies (*Gobiusculus flavescens*) have bright orange-yellow abdomens. Male fish clearly prefer females with orange-yellow abdomens. The reason for this preference may be that the orange-yellow abdomen of female fish indicates their fertility [12]. Female *Heterandria formosa* prefers males with large body size [15], while another study did not find such a preference [16], suggesting that male size may not be the only factor affecting female mate selection. In addition, female Eastern mosquitofish (*Gambusia holbrooki*) generally prefer males with a medium mating frequency, which can avoid excessive sexual harassment from males and mating damage caused by forced mating, and obtain sufficient sperm to reproduce offspring [7]. Therefore, mate selection depends on the attractiveness of individuals of the same sex with different phenotypic characteristics to individuals of the opposite sex, including morphology, behavior, and the internal physiology and genes of the animal. This selection can be made directly for the animal itself or through the evaluation of the environmental conditions in which the animal is located. In short, in the process of mate choice, organisms integrate and evaluate various aspects of information about their potential mates to achieve the purpose of selecting the most suitable mate.

Due to reproductive investment differences, gamete disparity, and sexual selection pressures, females are generally choosier than males [3,17]. While many species exhibit female preferences for certain male traits, coercive mating systems deviate from this pattern. In such systems, males often use aggressive or coercive strategies, like sexual harassment or forced copulation, to secure mating opportunities and suppress female choice, shifting the evolutionary dynamics from mutual attraction to male-driven reproductive tactics [18]. These systems are common in species with high male density or intense competition for females, particularly in the Poeciliidae family, where males employ tactics such as trailing and gonopodium thrusting [19]. The gonopodium, a specialized sexually dimorphic anal fin, typically exhibits two positional phases in males. In the resting phase, the gonopodium remains laterally positioned adjacent to the abdominal region. During the active phase, the gonopodium shifts to a protruding forward position relative to the body. This is often followed by gonopodium movements, including lateral swings for courtship displays and thrusts for insemination [20–23]. In coercive systems, the gonopodium plays a crucial role as a specialized organ for sperm transfer, subject to strong sexual selection, with its morphology and length often indicating male health and competitiveness [24]. In the contexts where coercion is prevalent, males with relatively larger gonopodium may gain a reproductive advantage, as these traits enable successful copulation even when females resist [25]. However, females are not passive participants and may evolve adaptive traits, both morphological and behavioral, to resist male coercion, such as changes that make forced copulation more difficult [26]. While sexual coercion may increase male reproductive success, it often comes at significant costs to females, including time and energy expenditure, and sometimes even injury or death [27–29]. Therefore, the interaction between male mating cues (especially the gonopodium status) and the dynamics of coercive mating systems highlight the complex evolutionary pressures in fish mate choice.

Although studies on sexual selection have traditionally focused on male competition and female mate choice, increasing evidence suggests that males also exhibit mate choice behaviors, which are common in nature and play an important role in the evolution of mating systems [30–32]. In nature, females, especially, tend to prefer larger mates [2,33]. Larger individuals may have more advantages in survival and reproduction, such as better access to food resources, enhanced parenting and guarding abilities, and better protection from predators or male harassment [6,34,35]. Numerous studies have also shown that males prefer larger females [36], especially when female size is more strongly correlated with reproductive potential [6,37]. Theoretically, males aim to maximize mating success and prefer females with higher reproductive potential or those exhibiting traits associated with lower sperm competition intensity [38]. Research has shown that males of *H. formosa* prefer to mate with smaller (younger) females, as the risk and intensity of sperm competition are lower with these females, who are more likely to be virgins. The "virgin preference" is important in male mate choice behaviors. In poecilids, females can store sperm from multiple males, making sperm competition an important factor in mating systems [39]. Studies indicate that males tend to avoid sperm competition and prefer females who have recently mated with fewer or no other males [37,40]. Compared to relying on olfactory cues, the visual influence of morphological traits may be more important in determining female mating status [41]. In the life history of poecilids, male growth nearly ceases after sexual maturity, while females continue to grow throughout their lives, with increasing reproductive potential as they grow [6,20,42]. As females age, their morphological traits change dynamically, including increasing body size and changes in specific features such as gravid spots [20,43,44]. These mating cues are important indicators that males use in mate choice.

In poecilids, visual signals are an important means of obtaining external information [45] and play a significant role in mate choice, with morphological traits being one of the most important influencing factors. Poecilids, due to their marked sexual dimorphism, are ideal model organisms for mate choice studies. It has been reported that male poecilids exhibit clear mate preferences [37,46], suggesting that their sexual selection is driven by both sexes. While numerous studies have focused on mate choice related to morphological traits in poecilids, research on the effects of gonopodium status and age-related mating cues remains limited. Western mosquitofish (*Gambusia affinis*), a representative species of Poeciliidae, follows a promiscuous mating system with no parental care or territorial behavior by males [47]. Given that male size ceases to increase after sexual maturity [21,48], males maturing earlier tend to be smaller, while those maturing later are relatively larger. This study focuses on *G. affinis* to explore two key questions through binary mate choice experiments: (1) How does the status of the male gonopodium affect female mate choice? (2) How does female age affect male mate choice? This research contributes to understanding "bidirectional mate choice" and provides a theoretical basis for understanding the complex and variable mating strategies in natural populations.

2. Materials and Methods

2.1. *Origin and Maintenance of Test Fish*

Wild-caught Western mosquitofish *G. affinis* were collected from Ankang City, Shaanxi Province (120°15.58′ N, 30°27.70′ E), all of which were sexually mature individuals. The fish were maintained in several aerated and filtered 200 L aquaria equipped with aquatic plants, driftwood, and rocks for shelter. They were fed twice a day (at 10:00 a.m. and 4:00 p.m.) ad libitum amounts of bloodworms, brine shrimp, and commercial fish flakes. The stocking density in the aquariums was 40 fish per tank, with a sex ratio of approximately 1:1. The water temperature in the tanks was maintained at 25 ± 0.1 °C, with a photoperiod of 12 h per day. Half of the water was replaced every two weeks, with both the rearing and experimental water being sourced from dichlorinated tap water. All animals were allowed to acclimate under laboratory conditions for 2 weeks before experimentation.

2.2. *Test 1- Influence of Male Gonopidium Status on Mate Choice in Females*

2.2.1. Creation of Male *G. affinis* Animations

According to the fish image database created by Chen et al. [49] for *G. affinis*, 20 images were randomly selected from a total of 24 male fish images. These images were animated using Macromedia Flash 8.0 and converted into .flv files (resolution: 1024 × 768; 60 frames/s). The final set of animations consisted of virtual male fish pairs, where each pair showed different gonopodium phase (e.g., active/resting phase) to reflect relative exposure. The only difference between the two was the gonopodium status, and other morphological traits, such as body length (22.30 mm), body color, swimming speed (2.63 cm/s), and background, remained consistent across all pairs. Each animation displayed a virtual male fish swimming horizontally from the left to the right side of the screen against the light gray background, and the male performed a back-and-forth swimming motion. Before changing direction, the virtual male would swim an additional body length distance off-screen, out of the view of the experimental fish, to simulate an invisible turn. To avoid pseudo-replication, the study utilized the 20 high-resolution images of female *G. affinis* as templates to create 20 distinct animation pairs (Supplementary Materials, Figure S1). Additionally, we maintained proportional scaling using Adobe Photoshop CC 2022 (23.0)'s "Content-Aware Scale" tool to avoid distortion.

2.2.2. Female Mate Choice Test

Prior to the experiment, male fish were placed in isolated 1.5 L transparent plastic bottles for 24 h. The isolation tanks were covered with black plastic on three sides and the bottom, and the plastic bottles were suspended inside the isolation tanks, each with small holes for water and gas exchange. The water temperature in the isolation tanks was maintained at 25 ± 0.1 °C, with a 12 h light/dark cycle and sufficient oxygenation. A dichotomous association preference test was conducted following the setup described by Chen et al. [49]. The standard length (SL) of each fish was measured with a precision of 0.1 mm at the end of the experiment. Water in the experimental tank was replaced after every four trials to maintain water quality. Finally, the total time spent in each preference zone was calculated, and the strength of mate preference (SOP) was determined. A total of 33 female fish (Mean ± SE, SL: 30.42 ± 4.66 mm) were tested in the experiment.

2.3. Test 2- Influence of Female Age on Mate Choice in Males

2.3.1. Acquisition of High-Resolution Images of Young and Old Female Fish

This study involved the collection of high-resolution images of female fish captured at different time points to create computer animations. The "older" females were captured in Ankang City and were kept in the laboratory for at least one year before the images were taken. The "younger" females were subadult females, and they reached sexual maturity within one month of being kept in the laboratory. Studies have shown that the natural lifespan of wild *G. affinis* typically does not exceed 12–15 months, although individuals raised under controlled conditions can live up to 18 months [50,51]. Thus, the "older" females were at least one year old, while the "younger" females were no more than 0.5 years old, with a minimum age difference of six months between the two groups. Following the methodology of Chen et al. [49], a total of 40 high-resolution images of wild-caught and laboratory-raised female *G. affinis* were obtained (Mean ± SE, $SL_{Young\ female}$: 28.15 ± 4.57 mm, $n = 20$; $SL_{Old\ female}$: 37.30 ± 2.84 mm, $n = 20$).

2.3.2. Quantify the Relative Gravid Spot Area of Young and Old Female Fish

Using the software Morpho J 1.08.02., the areas of the gravid spot and the lateral body surface of the *G. affinis* were measured from 40 high-resolution photographs [52], and the ratio between the two areas was calculated. Comparison revealed that the relative gravid spot area of young females (gravid spot area/lateral body surface: Mean ± SE = 0.84 ± 0.11%) was significantly smaller than that of older females (1.87 ± 0.29%, independent sample *t*-test, $t_{25.03} = -3.41$, $p = 0.002$, Figure 1c). This indicates that young females have relatively smaller gravid spots, while older females have relatively larger gravid spots, with a significant difference between the two groups. The geometric morphometric analysis followed three steps [53]: (1) the photos were converted to tps format using tpsUtil 1.82 software, and 17 landmark points were digitized with tpsDig 2.05 software. These landmarks provided a complete representation of the lateral body contour of the female fish (Figure 1a,b); (2) a Procrustes fit analysis was performed using Morpho J, and data reduction was applied during shape extraction; and (3) 10 principal components (RW) were extracted from the shape features of the females, accounting for 94.03% of the total shape variation (Table 1), with the first two principal components visualized (total shape variation accumulated to 69.36%; Table 1). Our results showed significant differences in the spine and abdominal morphology between young and older females. Specifically, young females exhibited a streamlined spine and smaller abdominal area, while older females had a "Z"-shaped spine and larger abdominal area (Figure 1d).

Table 1. Results of a principal component analysis on morphological variation.

Relative Warps (RW)	Eigen Value	Variance Explained (%)
RW1	1.91×10^{-3}	60.90
RW2	2.65×10^{-4}	8.46
RW3	2.34×10^{-4}	7.46
RW4	1.13×10^{-4}	3.59
RW5	1.06×10^{-4}	3.39
RW6	8.74×10^{-5}	2.79
RW7	7.95×10^{-5}	2.54
RW8	5.90×10^{-5}	1.88
RW9	5.51×10^{-5}	1.76
RW10	3.95×10^{-5}	1.26

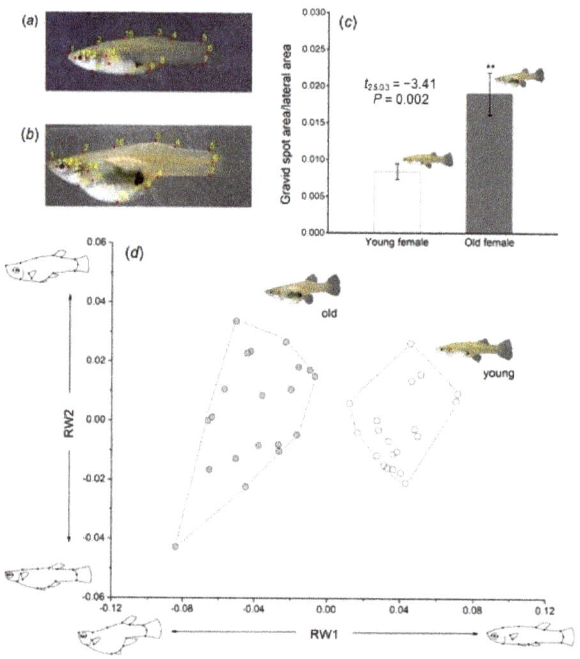

Figure 1. Morphological landmarks, as well as differences in body shape and coloration between young and old female *G. affinis*. Dots and numbers indicate the 17 landmarks used for morphometric analyses in (**a**) young and (**b**) old females. (**c**) Ratio of 'gravid spot area/lateral area' in young and old females with the results of an independent-samples *t*-test. (**d**) Body shape variation along RW1 and RW2. ** refers $p<0.05$.

2.3.3. Creation of Female *G. affinis* Animations

During the creation of computer-simulated animations, the swimming speed of all female fish (2.71 cm/s) and the background (light gray) were kept consistent. Except for in the experiments described in Section 2.3.4, the standard body length of the virtual female fish remained constant across all experimental conditions (Mean $\pm$ SD, SL: 29.1 $\pm$ 4.67 mm). Apart from in the experiments described in Section 2.3.4, the animations displayed two images of the same individual from both sides, with differences in specific morphological traits. Each animation depicted a virtual female fish swimming horizontally from the left side of the light gray background screen to the right, with a return movement. Before changing the swimming direction, the virtual female fish executed an invisible turn equivalent to one body length. To avoid pseudoreplication, we used 40 high-resolution photographs of female *G. affinis* as templates, generating at least 20 pairs of different animations for each experimental group. The images were processed using Macromedia Flash 8.0 and converted into .flv files (resolution: 1024 $\times$ 768; 60 frames/s).

2.3.4. Male Preference for Females of Different Ages

The images of the 40 female fish were randomly paired in pairs to create animations representing both young and aged morphologies. Each animation pair consisted of a virtual young and a virtual old female fish (Supplementary Materials, Figure S2). The standard body length of all virtual female fish was adjusted to 22.30 mm.

2.3.5. Male Preference for Females with Different Relative Areas of Gravid Spots

The ratio of the gravid spot area to the body side area indicates that young females have relatively smaller gravid spots. Therefore, to test whether males can distinguish between young and old females based on the relative area of the gravid spot, this study created two sets of animated pairs ("small" and "large" gravid spot area/body side area ratios) using images of young and old females, respectively. The "small" ratio was defined as the average ratio of gravid spot area to body side area in young females, while the "large" ratio was defined as the average ratio in old females. Additionally, Adobe Photoshop CS5 was used to adjust the gravid spot area in each image to a predefined value, thereby creating the corresponding animations.

2.3.6. Male Preference for Females with Different Morphological Traits (Spine and Abdomen Morphology)

The results of the geometric morphometric analysis indicate that the morphological differences between young and old females are primarily related to the shape of the spine and abdomen. Based on these two key characteristics, a full-factorial design binary mate choice experiment was conducted (six experimental groups: a, b, c, d, e, and f), with 20 pairs of animated images produced for each group. The female fish images used in the animations were derived from young and old female fish. To adjust the morphology of the spine and abdomen of the female fish, Adobe Photoshop CS5 was employed with three parameters as anchor points: θ1, θ2, and the ratio of b-line/a-line (R) (Figure 2). θ1 is the angle between the head and back of the female fish, formed by the connection of landmark points 1, 2, and 2, 3 (mean values: $\theta1_{young} = 197.2°$, $\theta1_{old} = 162.4°$). θ2 represents the angle between the back and tail, formed by the connection of landmark points 2, 3, and 3, 4 (mean values: $\theta2_{young} = 163.4°$, $\theta2_{old} = 151.1°$). The combination of θ1 and θ2 represents the spinal morphology of females of different ages. The a-line is the distance between landmark points 2 and 3, while the b-line is the distance between landmark points 16 and 17. The R-value can represent the abdominal morphology of young (Mean $R_{young} = 0.53$) or old (Mean $R_{old} = 0.86$) females.

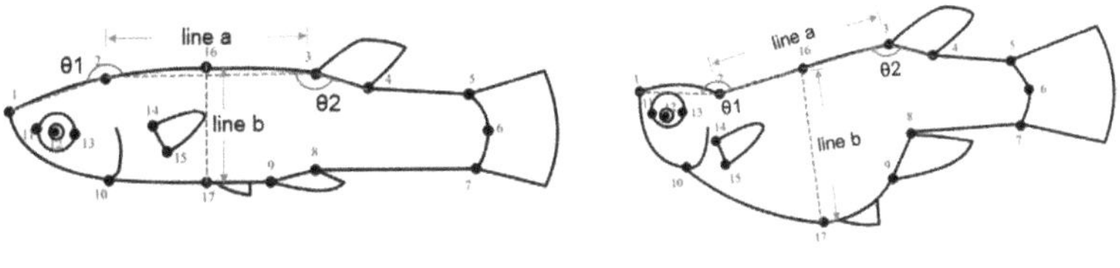

Figure 2. Spinal and abdominal shape manipulation reference for computer-animated female stimuli.

In experiments a–c, the control animations for each pair were derived from images of young female fish, while the experimental animations were generated by modifying the images of young female fish to exhibit the morphological characteristics of older female fish. These modifications included changes to spinal shape (experiment a: using Adobe Photoshop CS5 to adjust the body angles $\theta1_{young}$ and $\theta2_{young}$ to resemble θ1old and θ2old of older female fish), spinal shape combined with abdominal morphology (experiment b: adjusting $\theta1_{young}$, $\theta2_{young}$ and R_{young} to $\theta1_{old}$, $\theta2_{old}$ and R_{old}), and abdominal shape (experiment c: adjusting R_{young} to R_{old}).

In experiments d–f, the animation creation method was similar to that of experiments a–c, with the distinction that in these cases, the control animations were based on images of older female fish, and the experimental animations were processed to exhibit the morphological traits of younger female fish. Specifically, experiment d involved changes to the spinal shape, experiment e combined changes to both spinal and abdominal morphology, and experiment f focused on changes to abdominal shape.

2.4. Statistical Analyses

All statistical analyses in this study were performed using SPSS 19. First, the strength of the female fish mate preference for males with either a continuously active/resting-phase gonopodium was measured, referred to as the mate preference strength (SOP), calculated as follows: SOP = (time spent in the preference area for the virtual male with the continuously active-phase gonopodium) − (time spent in the preference area for the virtual male with the resting-phase gonopodium)/total time spent in both preference areas.

Thus, the SOP value ranges from 1 (indicating a preference for males with continuously active-phase gonopodia) to −1 (indicating a preference for males with resting-phase gonopodia). A one-sample *t*-test was used to compare the SOP values to a hypothesized constant (SOP = 0), testing the overall preference of the females and the experimental conditions. Subsequently, a general linear model (GLM) was conducted with SOP as the dependent variable, including animation number as a random factor and female standard body length as a covariate.

Then, in the experiments investigating male fish preference for female fish of different ages, male preference for female fish with varying relative areas of pregnancy spots, and male preference for female fish with different morphological traits (spinal and abdominal morphology), raw data were used for statistical analysis. Unless otherwise specified, all descriptive statistics are expressed as Mean ± SE.

To determine the strength of mate preference (SOP) for young female fish (characterized by traits such as gravid spot size, spinal, and abdominal morphology), we used the following formula: SOP = (time spent near the preference zone of young trait − time spent near the preference zone of old trait)/total time spent in both preference zones. In the gravid spot and body shape tests, body size was not standardized due to image manipulation constraints. We therefore used SOP value adjustments to account for size bias.

To analyze the overall preference of each group and the experimental results, a one-sample *t*-test was used to analyze the overall SOP values of the experimental fish. A general linear model (GLM) analysis was then performed with SOP as the dependent variable. In the experiments on male preference for female fish of different ages, male preference for female fish with varying relative areas of pregnancy spots, and male preference for female fish with different morphological traits (spinal and abdominal morphology), when processing the three sets of experimental data, animation type (animation type: animations of young or old female fish) was set as a fixed factor, and the standard body length of the experimental male fish (male body size) was included as a covariate in the GLM analysis. In the experiment on male preference for female fish with different morphological traits (spinal and abdominal morphology), when conducting the GLM analysis, animation type, spinal morphology, and abdominal morphology were treated as fixed factors, and the standard body length of the experimental male fish (male body size) was treated as a covariate. A subset analysis was then performed on the data from the young and old female fish animations, excluding the animation type factor. Initially, animation ID was treated as a random factor in the GLM analysis. However, since the *p*-value for animation ID was non-significant ($p > 0.19$) in all analyses, it was excluded from the final model.

3. Results

3.1. Female Preference for Males with Different Gonopodium Status

A general linear model revealed no significant differences in the random variable "Animation ID" and the covariate "Standard Length (SL) of female fish" (Table 2). In assessing the female fish's preference for males with different gonopodial status, it was found that the time spent near males with resting-phase gonopodia was significantly longer than the time spent near males with active-phase gonopodia. By calculating the strength of mate preference and performing a one-sample *t*-test, it was determined that female fish showed a significant aversion to males with persistently active-phase gonopodia (mean SOP = −0.13, $t_{32} = -2.228$, $p = 0.03$, Figure 3).

Table 2. Female preference for males with active-phase gonopodia.

Source	*df*	F	*p*	Wilks' Partial η_p^2
Animation ID	20	1.37	0.28	0.66
SL	1	0.13	0.72	0.009
Error	30			

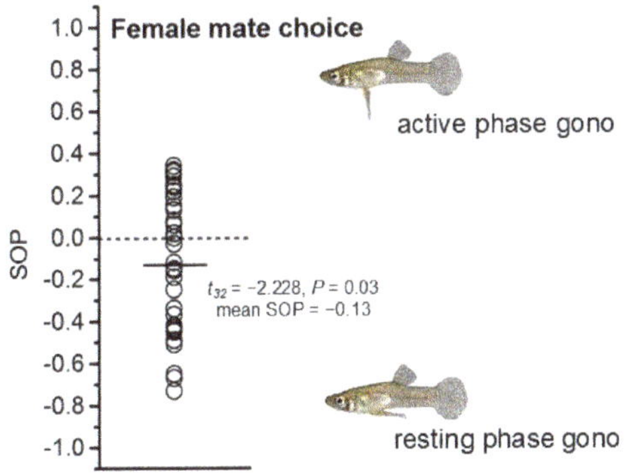

Figure 3. Distribution of individual strength of preference (SOP) values derived from dichotomous association: female choice for male with active-phase gonopodium.

3.2. Male Preference for Females of Different Ages

Compared to the time spent near the animations of older female fish, male fish spent significantly more time near the animations of younger female fish. A one-sample *t*-test on the corresponding SOP values revealed a significant difference ($t_{35} = 2.54$, $p = 0.016$; Figure 4a). A general linear model indicated that male body size significantly influenced male preference for younger female fish ($F_{1,34} = 4.99$, $p = 0.032$), with a positive relation between mate preference and body size ($R^2 = 0.13$; Figure 4b).

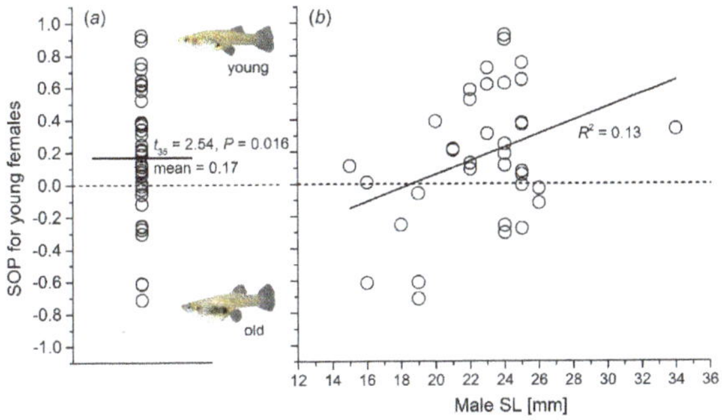

Figure 4. Male mating preference for young female phenotypes in *G. affinis*. (**a**) Distribution of individual strength of preference (SOP) values derived from dichotomous association: male choice for young and old females; (**b**) Relationships between male body size (SL) and SOP for young females.

3.3. Male Preference for Females with Different Relative Areas of Gravid Spots

The results were contrary to the prediction, as males spent more time near the animation of females with larger gravid spot areas (Figure 5). When using the young female fish image as a template to create computer animations, the experimental results showed that males spent significantly more time near the animation of females with larger gravid spot areas compared to those with smaller gravid spot areas. A one-sample *t*-test of the corresponding SOP values revealed a significant difference ($t_{22} = -2.95$, $p = 0.007$; Figure 5a). When using the older female fish image as a template, the results showed that males spent significantly more time near the animation of females with larger gravid spot areas than near those with smaller gravid spot areas. A one-sample *t*-test of the corresponding SOP values showed a significant difference ($t_{23} = -2.49$, $p = 0.021$; Figure 5b). A general linear model showed it was neither the animation model nor the body size of the experimental fish had a significant effect on the corresponding SOP values (Table 3).

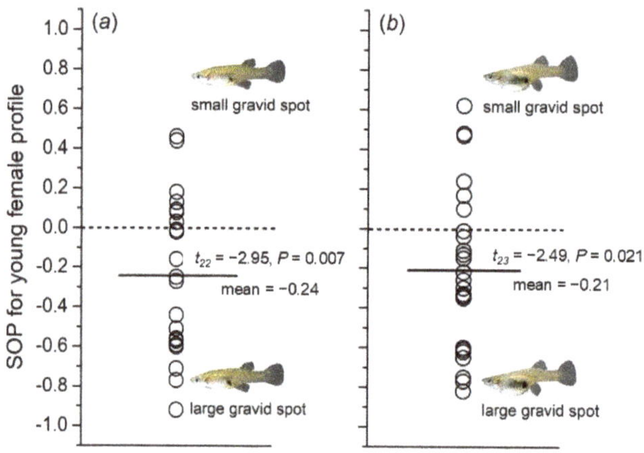

Figure 5. Male mate choice based on gravid spot size. Computer animations showed stimulus females with a large or a small gravid spot and were created using images of (**a**) young and (**b**) old females.

Table 3. Results of general linear model (GLM) using SOP values for gravid spot size as dependent variables; Animation pairs were derived from images of young or old females.

Factor	*df*	*F*	*p*	Wilks' Partial η_p^2
Animation type	0	-	-	-
Animation ID	18	0.92	0.57	0.40
Male body size	1	0.55	0.47	0.022
Animation type × male body size	1	4.21	0.051	0.14
Error	25			

3.4. Male Preference for Females with Different Morphological Traits (Spine and Abdomen Morphology)

In six experimental groups, significant differences in SOP values were observed only when the animation templates were derived from young female fish images, and both spinal and abdominal morphologies were altered to simulate those of older female fish. In this case, the time spent by male fish in the preference area of the young female fish animation was significantly longer than the time spent in the preference area of the virtual older female fish animation. The corresponding SOP values were calculated, and a one-sample *t*-test revealed significant differences ($t_{32} = 2.14$, $p = 0.04$; Figure 6b). In contrast, the results from the other five experimental groups (Figure 6a,c–f) showed no significant preference by male fish for any of the female fish animations. Analysis using a general linear model indicated that none of the factors—animation type (whether the animation was based on young or old female fish), spinal shape, abdominal morphology, or the body size of the experimental fish—nor their interactions, had a significant effect on the results (Table 4).

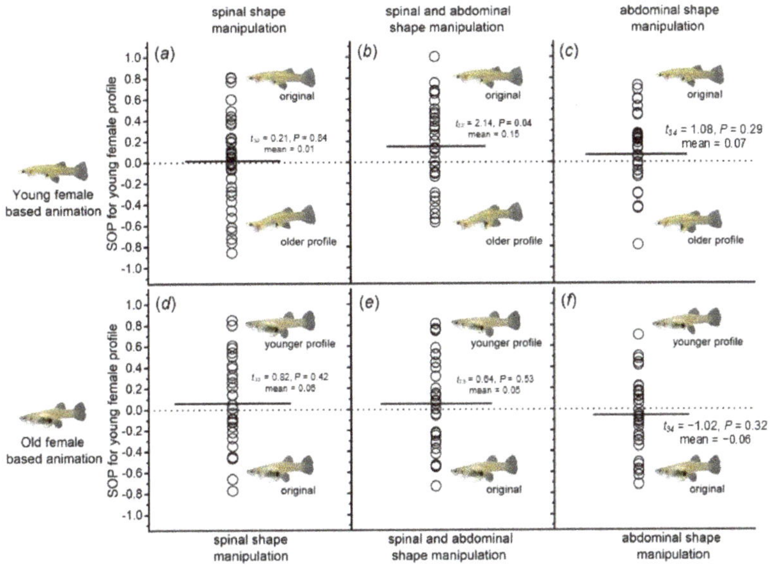

Figure 6. Distribution of individual strength of preference (SOP) values derived from dichotomous association preference tests involving (virtual) spinal and abdominal shape manipulations in stimulus females.

Table 4. Results of general linear model (GLM) using SOP values for spinal and abdominal shape as dependent variables; Animation pairs were derived from images of young or old females.

Source	*df*	*F*	*p*	Wilks' Partial η_p^2
		(a) Both animation types combined		
Animation type	1	0.20	0.66	0.001
Spinal shape	1	0.05	0.83	0.0002
Abdominal shape	1	1.38	0.24	0.01
Male body size	1	1.88	0.17	0.01
Error	206			
		(b) Animations derived from young females		
Spinal shape	1	0.81	0.37	0.01
Abdominal shape	1	2.21	0.14	0.02
Male body size	1	0.11	0.74	0.001
Error	104			
		(c) Animations derived from old females		
Spinal shape	1	1.78	0.19	0.02
Abdominal shape	1	0.03	0.86	0.0003
Male body size	1	3.05	0.084	0.03
Error	99			

4. Discussion

Our study utilized computer-generated animations to simulate male *G. affinis* with different reproductive traits and conducted mate preference experiments to assess female responses. The male fish were kept consistent in various traits, including body size, swimming behavior (such as swimming speed and direction), body coloration, and reproductive fin size. The only variation was in the status of the gonopodia, which were presented in either an active or a resting phase form. The results revealed that female fish significantly preferred males with resting-phase gonopodia, showing a marked aversion to males whose fins remained constantly in an active phase status. This behavior aligns with previous findings in species where males invest relatively little in offspring, opting instead to increase their reproductive success through high-frequency mating. By maximizing copulation attempts, males aim to produce more offspring, often at the expense of female well-being [2]. In many sexually dimorphic species, males engage in sexual harassment as a strategy to force copulation and secure fertilization [54–56]. Male *G. affinis* often engage in sexual harassment of females [57] and force copulations to achieve fertilization. When males sexually harass females, the females' reproductive investment significantly increases, potentially leading to injuries such as genital damage [58]. As a result, females tend to resist most of the males' mating attempts. During the initial mating attempt, males swing their gonopodia to the side and rear of the female to insert their fins into the female's genital opening for ejaculation, thus raising the fins into an active phase status [20]. Female rejection of males with consistently active-phase gonopodia may be a response to their aversion to forced copulations and sexual harassment. Frequent mating can cause significant harm to females, such as genital injury [58] and reduced foraging efficiency [29,59]. Moreover, sexual harassment increases an individual's exposure to predators, thereby elevating the risk of predation [60], which reduces both survival and reproductive fitness. Notably, the lack of response to a cue does not imply a lack of perception but may indicate differing decision thresholds.

In the animal kingdom, mate choice is often seen as a process where females exert dominant influence, especially when it comes to the selection of high-quality mates. However, our study showed males can also exhibit a highly selective preference in their mate choices. Actually, male mate choice does not contradict general patterns but illustrates sex-specific decision strategies. Specifically, we found that male *G. affinis* showed a strong preference for younger females, a preference that is closely linked to the male's body size. The strength of this mate preference (SOP) was positively correlated with the male's body size, suggesting that larger males might be more capable of distinguishing subtle differences in female age and body traits, which could signal higher reproductive potential. This finding is noteworthy because it implies that males do not merely choose mates based on physical attributes like size or coloration, but also on age-related traits that may be indicative of fertility or overall health [61]. The preference for younger females may be related to the fact that younger individuals are typically at their peak reproductive age, which could lead to more successful and healthier offspring [62,63]. Additionally, it is important to consider the role of sperm competition, a well-documented phenomenon in many species. Virgin females, due to the lower intensity of sperm competition within their reproductive tract, may offer a more favorable environment for sperm fertilization. This could explain why males might preferentially choose younger females, as they are likely to be virgins and therefore experience less sperm competition, ensuring a higher likelihood of successful fertilization. As virgin females tend to experience lower levels of sperm competition, many species exhibit a preference for virgin mates [64]. For example, males from the species *H. formosa* prefer smaller females, possibly because they offer a more favorable environment for sperm success due to reduced sperm competition [16]. Hence, male *G. affinis* may not only be selecting mates based on direct physical or age-related traits but also strategically considering the potential benefits of reduced sperm competition. This selective behavior could be an adaptive strategy to maximize reproductive success and ensure that the male's genetic material has a better chance of fertilizing the female's eggs.

Morphological analysis of young and old female fish revealed significant differences in the size of the gravid spot (color pattern), as well as spine and abdominal morphology (morphological traits). When the relative area of the gravid spot was used as a variable, males consistently preferred females with a larger relative area of the gravid spot, with female age becoming a potential influencing factor. In most fish species, the size of the gravid spot is used as a marker to identify the developmental stage of females [44]. Studies have shown that the gravid spot undergoes dynamic changes during different developmental stages [20,43,44]. Research on the life history of certain mosquito fish species indicates that the size and pigment intensity of the gravid spot may be linked to the female's reproductive cycle—spots grow larger and become more intensely colored with age [20,65]. While body size in female fish represents their reproductive capacity, the relative area of the gravid spot may provide a more intuitive indication of their fertility. Therefore, males consistently prefer females with a larger gravid spot, regardless of the female's age. When spine and abdominal morphology were used as variables in a full-factorial design and binary mate choice experiments were conducted, it was found that males showed a significant preference for females with "young abdominal and spine morphological traits". The apparent contradiction between this preference and the males' strong preference for females with a larger gravid spot (an indicator of older age) suggests that male *G. affinis* do not rely on a single morphological characteristic to select potential mates. In studies on the impact of body size on male mate choice, males showed no significant preference for females with larger body sizes, whether the females were young or old. This result contrasts with the earlier finding of males' preference for younger females, suggesting that body size is a potential preference trait but is overshadowed by age. Research has shown that with age, female body length increases, along with reproductive capacity [20]. However, since males did not prefer larger females, it suggests that body size does not intuitively reflect a female's reproductive potential. Additionally, as female body size increases with age, older females may have lower overall fitness.

In natural environments, mate choice in many species does not rely on a single trait but rather on the integration of multiple cues. This strategy of multi-cue integration is thought to enhance the flexibility and adaptability of decision making, particularly in contexts where signals are variable or environmental conditions are unpredictable [66]. For *G. affinis*, factors such as body size, gravid spot size, age, and harassment risk may all serve as parallel sources of information with differing weights. Males and females are likely to integrate and balance these cues when making mate choices, leading to individual differences in preference patterns [4]. Therefore, we interpret the subtle deviations in SOP values observed in our study as reflecting a complex cognitive evaluation process in *G. affinis*.

5. Conclusions

In summary, we found that during the mate choice process of *G. affinis*, females exhibited a strong aversion to males with continuously active phase genital fins, indicating a clear rejection of forced mating and male sexual harassment. In the male mate choice process, males preferred to mate with younger females, possibly due to the lower intensity of sperm competition associated with younger females. Additionally, larger males showed a stronger rejection of older females, likely because larger males have higher social status and more mating opportunities, making them more "choosy" in mate choice compared to smaller males. Furthermore, male *G. affinis* showed a significant preference for females with larger gravid spots (an indicator of older morphological traits) and those with streamlined bodies and smaller abdomens (indicating younger morphological traits). This suggests that no single mating cue dominates male mate choice. Future research could explore the plasticity of mate choice in this species during its invasive expansion, particularly under extreme environmental changes, and investigate the evolutionary patterns of genital fin morphology and behavioral traits in sexual selection, revealing the adaptive mechanisms behind the reproductive behaviors of this species.

Supplementary Materials: The following supporting information can be downloaded at: https://www.mdpi.com/article/10.3390/ani15101489/s1, Figure S1: Images of males (M1–M20) used to generate computer animations; Figure S2: Images of females (Pair1–Pair20) used to generate computer animations.

Author Contributions: Investigation, Data curation, Formal analysis, Writing—original draft: J.W.; Writing—review and editing: B.F.; Investigation, Data curation: C.D.; Conceptualization, Methodology, Investigation, Data curation, Writing—review and editing: B.C.; Writing—original draft, Writing—review and editing, Funding acquisition: K.L. All authors have read and agreed to the published version of the manuscript.

Funding: This research was funded by the Natural Science Foundation of Fujian Province, China, Grant No. 2024J01705.

Institutional Review Board Statement: This study met all relevant ethical standards for experimental research, approved by the Animal Care and Use Committee of the Fisheries College of Jimei University (Animal Ethics No. 1067).

Informed Consent Statement: Not applicable.

Data Availability Statement: The original contributions presented in the study are included in the article; further inquiries can be directed to the corresponding author.

Acknowledgments: We sincerely thank the reviewers for their insightful comments to improve the manuscript.

Conflicts of Interest: The authors declare no conflicts of interest.

References

1. Darwin, C. *The Descent of Man, and Selection in Relation to Sex*; D. Appleton: New York, NY, USA, 1872; Volume 2.
2. Andersson, M. *Sexual Selection*; Princeton University Press: Princeton, NJ, USA, 1994.
3. Andersson, M.; Simmons, L.W. Sexual selection and mate choice. *Trends Ecol. Evol.* **2006**, *21*, 296–302. [CrossRef] [PubMed]
4. Rosenthal, G.G. *Mate Choice: The Evolution of Sexual Decision Making from Microbes to Humans*; Princeton University Press: Princeton, NJ, USA, 2017.
5. Cornuau, J.H.; Rat, M.; Schmeller, D.S.; Loyau, A. Multiple signals in the palmate newt: Ornaments help when courting. *Behav. Ecol. Sociobiol.* **2012**, *66*, 1045–1055. [CrossRef]
6. Herdman, E.J.; Kelly, C.D.; Godin, J.G.J. Male mate choice in the guppy (*Poecilia reticulata*): Do males prefer larger females as mates? *Ethology* **2004**, *110*, 97–111. [CrossRef]
7. Sommer-Trembo, C.; Plath, M.; Gismann, J.; Helfrich, C.; Bierbach, D. Context-dependent female mate choice maintains variation in male sexual activity. *R. Soc. Open Sci.* **2017**, *4*, 170303. [CrossRef]
8. Ryan, M.J.; Rand, A.S. The sensory basis of sexual selection for complex calls in the túngara frog, *Physalaemus pustulosus* (sexual selection for sensory exploitation). *Evolution* **1990**, *44*, 305–314.
9. Heuschele, J.; Mannerla, M.; Gienapp, P.; Candolin, U. Environment-dependent use of mate choice cues in sticklebacks. *Behav. Ecol.* **2009**, *20*, 1223–1227. [CrossRef]
10. Candolin, U.; Salesto, T.; Evers, M. Changed environmental conditions weaken sexual selection in sticklebacks. *J. Evol. Biol.* **2007**, *20*, 233–239. [CrossRef]
11. Järvenpää, M.; Lindström, K. Water turbidity by algal blooms causes mating system breakdown in a shallow-water fish, the sand goby Pomatoschistus minutus. *Proc. R. Soc. Lond. Ser. B Biol. Sci.* **2004**, *271*, 2361–2365. [CrossRef]

12. Amundsen, T.; Forsgren, E. Male mate choice selects for female coloration in a fish. *Proc. Natl. Acad. Sci. USA* **2001**, *98*, 13155–13160. [CrossRef]
13. López, S. Parasitized female guppies do not prefer showy males. *Anim. Behav.* **1999**, *57*, 1129–1134. [CrossRef]
14. Harmison, J.; Ogborn, M.; Pett, J.; Tolman, D. Female Mate Choice in Guppies Based on Color Contrast and Sensory Exploitation. *Biology Poster* 2013. Available online: https://digitalcommons.usu.edu/biology_posters/98 (accessed on 18 May 2025).
15. Aspbury, A.S.; Basolo, A.L. Repeatable female preferences, mating order and mating success in the poeciliid fish, Heterandria formosa. *Behav. Ecol. Sociobiol.* **2002**, *51*, 238–244. [CrossRef]
16. Ala-Honkola, O.; Säilä, L.; Lindström, K. Males prefer small females in a dichotomous choice test in the poeciliid fish Heterandria formosa. *Ethology* **2010**, *116*, 736–743. [CrossRef]
17. Clutton-Brock, T. Reproductive competition and sexual selection. *Philos. Trans. R. Soc. B Biol. Sci.* **2017**, *372*, 20160310. [CrossRef]
18. Parker, G.A. Sexual conflict over mating and fertilization: An overview. *Philos. Trans. R. Soc. B Biol. Sci.* **2006**, *361*, 235–259. [CrossRef] [PubMed]
19. Bisazza, A.; Vaccari, G.; Pilastro, A. Female mate choice in a mating system dominated by male sexual coercion. *Behav. Ecol.* **2001**, *12*, 59–64. [CrossRef]
20. Pyke, G.H. A review of the biology of Gambusia affinis and G. holbrooki. *Rev. Fish Biol. Fish.* **2005**, *15*, 339–365. [CrossRef]
21. Hughes, A.L. Male size, mating success, and mating strategy in the mosquitofish *Gambusia affinis* (Poeciliidae). *Behav. Ecol. Sociobiol.* **1985**, *17*, 271–278. [CrossRef]
22. Wang, S.; Cummings, M.; Kirkpatrick, M. Coevolution of male courtship and sexual conflict characters in mosquitofish. *Behav. Ecol.* **2015**, *26*, 1013–1020. [CrossRef]
23. Martinez-Rivera, N.; Serrano-Velez, J.L.; Torres-Vazquez, I.I.; Langerhans, R.B.; Rosa-Molinar, E. Are superficial neuromasts proprioceptors underlying fast copulatory behavior? *Front. Neural Circuits* **2022**, *16*, 921568. [CrossRef]
24. Simmons, L.W. Sexual selection and genital evolution. *Austral Entomol.* **2014**, *53*, 1–17. [CrossRef]
25. Gonzalez-Voyer, A.; Fitzpatrick, J.L.; Kolm, N. Sexual selection determines parental care patterns in cichlid fishes. *Evolution* **2008**, *62*, 2015–2026. [CrossRef] [PubMed]
26. Jennions, M.D.; Petrie, M. Why do females mate multiply? A review of the genetic benefits. *Biol. Rev.* **2000**, *75*, 21–64. [CrossRef] [PubMed]
27. Grayson, K.L.; De Lisle, S.P.; Jackson, J.E.; Black, S.J.; Crespi, E.J. Behavioral and physiological female responses to male sex ratio bias in a pond-breeding amphibian. *Front. Zool.* **2012**, *9*, 24. [CrossRef]
28. Mühlhäuser, C.; Blanckenhorn, W.U. The costs of avoiding matings in the dung fly Sepsis cynipsea. *Behav. Ecol.* **2002**, *13*, 359–365. [CrossRef]
29. Magurran, A.E.; Seghers, B.H. A cost of sexual harassment in the guppy, *Poecilia reticulata*. *Proc. R. Soc. Lond. Ser. B Biol. Sci.* **1994**, *258*, 89–92.
30. Swierk, L.; Myers, A.; Langkilde, T. Male mate preference is influenced by both female behaviour and morphology. *Anim. Behav.* **2013**, *85*, 1451–1457. [CrossRef]
31. Edward, D.A.; Chapman, T. The evolution and significance of male mate choice. *Trends Ecol. Evol.* **2011**, *26*, 647–654. [CrossRef]
32. Rosenqvist, G. Male mate choice and female-female competition for mates in the pipefish Nerophis ophidion. *Anim. Behav.* **1990**, *39*, 1110–1115. [CrossRef]
33. Morina, D.L.; Demarais, S.; Strickland, B.K.; Larson, J.E. While males fight, females choose: Male phenotypic quality informs female mate choice in mammals. *Anim. Behav.* **2018**, *138*, 69–74. [CrossRef]
34. Blanckenhorn, W.U.; Mühlhäuser, C.; Morf, C.; Reusch, T.; Reuter, M. Female choice, female reluctance to mate and sexual selection on body size in the dung fly Sepsis cynipsea. *Ethology* **2000**, *106*, 577–593. [CrossRef]
35. Schlupp, I.; McKnab, R.; Ryan, M. Sexual harassment as a cost for molly females: Bigger males cost less. *Behaviour* **2001**, *138*, 277–286. [CrossRef]
36. Wong, R.Y.; So, P.; Cummings, M.E. How female size and male displays influence mate preference in a swordtail. *Anim. Behav.* **2011**, *82*, 691–697. [CrossRef]
37. Dosen, L.D.; Montgomerie, R. Female size influences mate preferences of male guppies. *Ethology* **2004**, *110*, 245–255. [CrossRef]
38. Bonduriansky, R. The evolution of male mate choice in insects: A synthesis of ideas and evidence. *Biol. Rev.* **2001**, *76*, 305–339. [CrossRef]
39. Pilastro, A.; Mandelli, M.; Gasparini, C.; Dadda, M.; Bisazza, A. Copulation duration, insemination efficiency and male attractiveness in guppies. *Anim. Behav.* **2007**, *74*, 321–328. [CrossRef]
40. Wong, B.B.; McCarthy, M. Prudent male mate choice under perceived sperm competition risk in the eastern mosquito fish. *Behav. Ecol.* **2009**, *20*, 278–282. [CrossRef]
41. Deaton, R. Factors influencing male mating behaviour in *Gambusia affinis* (Baird & Girard) with a coercive mating system. *J. Fish Biol.* **2008**, *72*, 1607–1622.

42. Reznick, D.; Hrbek, T.; Caura, S.; De Greef, J.; Roff, D. Life history of Xenodexia ctenolepis: Implications for life history evolution in the family Poeciliidae. *Biol. J. Linn. Soc.* **2007**, *92*, 77–85. [CrossRef]
43. Fleuren, M.; Quicazan-Rubio, E.M.; van Leeuwen, J.L.; Pollux, B.J. Why do placentas evolve? Evidence for a morphological advantage during pregnancy in live-bearing fish. *PLoS ONE* **2018**, *13*, e0195976. [CrossRef]
44. Norazmi-Lokman, N.H.; Purser, G.; Patil, J.G. Gravid spot predicts developmental progress and reproductive output in a livebearing fish, *Gambusia holbrooki*. *PLoS ONE* **2016**, *11*, e0147711. [CrossRef]
45. Kodama, I.; Yamanaka, A.; Endo, K.; Koya, Y. Role of the yellow spot around the urogenital opening of female mosquitofish (*Gambusia affinis*) as a cue for copulation. *Zool. Sci.* **2008**, *25*, 1199–1204. [CrossRef] [PubMed]
46. Bisazza, A.; Marconato, A.; Marin, G. Male mate preferences in the mosquitofish *Gambusia holbrooki*. *Ethology* **1989**, *83*, 335–343. [CrossRef]
47. Winkler, P. Thermal preference of *Gambusia affinis* affinis as determined under field and laboratory conditions. *Copeia* **1979**, *1979*, 60–64. [CrossRef]
48. Vargas, M.; De Sostoa, A. Life history of *Gambusia holbrooki* (Pisces, Poeciliidae) in the Ebro delta (NE Iberian peninsula). *Hydrobiologia* **1996**, *341*, 215–224. [CrossRef]
49. Chen, B.-j.; Liu, K.; Zhou, L.-j.; Gomes-Silva, G.; Sommer-Trembo, C.; Plath, M. Personality differentially affects individual mate choice decisions in female and male Western mosquitofish (*Gambusia affinis*). *PLoS ONE* **2018**, *13*, e0197197. [CrossRef]
50. Swenton, D.M.; Kodric-Brown, A. Habitat and life history differences between two species of *Gambusia*. *Environ. Biol. Fishes* **2012**, *94*, 669–680. [CrossRef]
51. Pyke, G.H. Plague minnow or mosquito fish? A review of the biology and impacts of introduced *Gambusia* species. *Annu. Rev. Ecol. Evol. Syst.* **2008**, *39*, 171–191. [CrossRef]
52. Klingenberg, C.P. MorphoJ: An integrated software package for geometric morphometrics. *Mol. Ecol. Resour.* **2011**, *11*, 353–357. [CrossRef]
53. Ouyang, X.; Gao, J.; Xie, M.; Liu, B.; Zhou, L.; Chen, B.; Jourdan, J.; Riesch, R.; Plath, M. Natural and sexual selection drive multivariate phenotypic divergence along climatic gradients in an invasive fish. *Sci. Rep.* **2018**, *8*, 11164. [CrossRef]
54. McKinney, F.; Evarts, S. Sexual coercion in waterfowl and other birds. *Ornithol. Monogr.* **1998**, 163–195. [CrossRef]
55. Réale, D.; Boussès, P.; Chapuis, J.-L. Female-biased mortality induced by male sexual harassment in a feral sheep population. *Can. J. Zool.* **1996**, *74*, 1812–1818. [CrossRef]
56. Bisazza, A. Male competition, female mate choice and sexual size dimorphism in poeciliid fishes. *Mar. Freshw. Behav. Physiol.* **1993**, *23*, 257–286. [CrossRef]
57. Plath, M.; Makowicz, A.M.; Schlupp, I.; Tobler, M. Sexual harassment in live-bearing fishes (Poeciliidae): Comparing courting and noncourting species. *Behav. Ecol.* **2007**, *18*, 680–688. [CrossRef]
58. Peters, G.; Mäder, B. Morphologische Veränderungen der Gonadenausführgänge sich fortpflanzender Schwertträgerweibchen (*Xiphophorus helleri* Heckel). *Zool. Anz.* **1964**, *173*, 243–257.
59. Pilastro, A.; Benetton, S.; Bisazza, A. Female aggregation and male competition reduce costs of sexual harassment in the mosquitofish *Gambusia holbrooki*. *Anim. Behav.* **2003**, *65*, 1161–1167. [CrossRef]
60. Pocklington, R.; Dill, L.M. Predation on females or males: Who pays for bright male traits? *Anim. Behav.* **1995**, *49*, 1122–1124. [CrossRef]
61. Watkins, C.D. Mate assessment based on physical characteristics: A review and reflection. *Biol. Rev.* **2025**, *100*, 113–130. [CrossRef]
62. Kenrick, D.T.; Keefe, R.C. Age preferences in mates reflect sex differences in human reproductive strategies. *Behav. Brain Sci.* **1992**, *15*, 75–91. [CrossRef]
63. Sutter, D.A.; Suski, C.D.; Philipp, D.P.; Klefoth, T.; Wahl, D.H.; Kersten, P.; Cooke, S.J.; Arlinghaus, R. Recreational fishing selectively captures individuals with the highest fitness potential. *Proc. Natl. Acad. Sci. USA* **2012**, *109*, 20960–20965. [CrossRef]
64. Guevara-Fiore, P.; Skinner, A.; Watt, P. Do male guppies distinguish virgin females from recently mated ones? *Anim. Behav.* **2009**, *77*, 425–431. [CrossRef]
65. Edwards, T.M.; Guillette, L.J., Jr. Reproductive characteristics of male mosquitofish (*Gambusia holbrooki*) from nitrate-contaminated springs in Florida. *Aquat. Toxicol.* **2007**, *85*, 40–47. [CrossRef] [PubMed]
66. Hebets, E.A.; Papaj, D.R. Complex signal function: Developing a framework of testable hypotheses. *Behav. Ecol. Sociobiol.* **2005**, *57*, 197–214. [CrossRef]

biology

MDPI

Review

Aquatic Organisms in Response to Salinity Stress: Ecological Impacts, Adaptive Mechanisms, and Resilience Strategies

Tariq Dildar [1,2,3], Wenxiao Cui [1,2,3], Mhd Ikhwanuddin [2,3,4] and Hongyu Ma [1,2,3,*]

1 Guangdong Provincial Key Laboratory of Marine Biotechnology, Shantou University, Shantou 515063, China; 21tariq@stu.edu.cn (T.D.); wxcui@stu.edu.cn (W.C.)
2 International Joint Research Center for the Development and Utilization of Important Mariculture Varieties Surrounding the South China Sea Region, Shantou University, Shantou 515063, China; ikhwanuddin@umt.edu.my
3 STU-UMT Joint Shellfish Research Laboratory, Shantou University, Shantou 515063, China
4 Higher Institute Centre of Excellence (HICoE), Institute of Tropical Aquaculture and Fisheries, Universiti Malaysia Terengganu, Kuala Nerus 21030, Terengganu, Malaysia
* Correspondence: mahy@stu.edu.cn; Tel.: +86-0754-86503471

Simple Summary: In aquatic ecosystems, abiotic factors, such as temperature, pH, and salinity, play crucial roles in shaping biological communities. Compared to temperature and pH, salinity is particularly significant because of its profound impact on the osmoregulatory capacities of organisms across a variety of aquatic habitats, ranging from freshwater to seawater. Understanding how organisms adapt to salinity variability is essential for biodiversity conservation and the sustainable management of aquatic resources. This review first examines the physiological, molecular, and behavioral adaptations of aquatic organisms in response to survive and thrive under salinity stress. Then, the mechanisms of osmotic regulation, ion transport, and oxidative stress responses, especially key signaling pathways involved in adaptive processes, are explored. Moreover, genetic and epigenetic modifications and the interactions between organisms and microbiomes are emphasized for their contribution to resilience. Finally, the review suggests potential strategies for mitigating salinity stress, such as nutritional interventions and the development of salinity-resistant varieties in aquaculture.

Abstract: Salinity stress presents a major ecological challenge for aquatic organisms, particularly in environments where salinity levels fluctuate. These fluctuations are becoming more pronounced due to climate change, further destabilizing aquatic ecosystems. Understanding how organisms adapt to such variability is essential for biodiversity conservation and the sustainable management of aquatic resources. This review examines the physiological, molecular, and behavioral adaptations that enable aquatic organisms to survive and thrive under salinity stress. Specifically, it explores mechanisms of osmotic regulation, ion transport, and oxidative stress responses, highlighting key signaling pathways—such as AMP-activated protein kinase (AMPK), Phosphatidylinositol 3-kinase–protein kinase (PI3K-AKT), Mitogen-activated protein kinase (MAPK), and the Hippo pathway—that facilitate these adaptive processes. The review also emphasizes the genetic and epigenetic modifications that contribute to resilience, underscoring the importance of genetic diversity for species survival in fluctuating salinity conditions. Furthermore, the interactions between host organisms and their microbiomes are discussed as critical factors influencing resilience. The review addresses the impact of salinity fluctuations on species distribution and biodiversity, with a focus on the implications of climate change for aquatic ecosystems. Finally, strategies for mitigating salinity stress, such as nutritional interventions and the development of salinity-resistant varieties, are explored, particularly in aquaculture. Overall, this review consolidates current knowledge on organismal adaptations, molecular mechanisms, and environmental challenges, offering valuable insights for ecological research and aquaculture practices in the face of climate change.

Keywords: salinity stress; adaptive mechanisms; aquatic organisms; osmoregulation; microbiome interaction; genetic diversity

1. Introduction

In aquatic ecosystems, both the mean and variance of several abiotic factors, such as temperature, pH, and salinity, play crucial roles in shaping biological communities [1]. While temperature and pH are recognized as key abiotic factors influencing aquatic environments, salinity is particularly significant due to its profound impact on species' osmoregulatory capacities across a variety of aquatic habitats, ranging from freshwater to hypersaline ecosystems [2]. Climate change is exacerbating these salinity concerns by elevating temperatures and altering precipitation patterns, resulting in more intense and unpredictable salinity variations in both freshwater and marine systems [3]. Organisms' ability to tolerate specific salinity levels is intrinsically linked to their osmoregulatory mechanisms, which are shaped by their life history traits. To maintain osmotic balance, organisms expend considerable energy regulating ion uptake and secretion in response to hypoosmotic and hyperosmotic conditions, thereby ensuring stable internal fluid concentrations [4]. In hypoosmotic environments, organisms lose ions to the surrounding water. To maintain osmotic balance, ion transporters such as Na^+/K^+-ATPase (NKA) and $Na^+/K^+/2Cl^-$ co-transporters (NKCC1) actively expel Na^+ and retain K^+ In contrast, in hyperosmotic environments, organisms must avoid excessive ion buildup. In this case, NKCC1-mediated Na^+ inflow is increased to support ion absorption, whereas NKA promotes the extrusion of surplus Na^+ to maintain homeostasis [5]. These energy-dependent systems are necessary for organisms' survival in environments with changing salinity, ensuring that cells can adjust to the osmotic stress imposed by the surrounding medium. However, climate-driven salinity fluctuations are pushing these adaptive systems to their limits, as indicated by increased metabolic and oxidative stress in aquatic species exposed to rapidly changing salinity [6].

Despite the growing understanding of the impact of salinity on aquatic organisms, a significant research gap remains regarding the species-specific mechanisms of osmoregulation across diverse environments, particularly in the context of rapidly changing climate conditions. This study addresses this gap by exploring the adaptive strategies employed by various species to cope with fluctuating salinity levels and the broader implications of these adaptations for ecosystem stability and biodiversity. Salinity shifts can change species distributions by modifying organisms' physiological and behavioral attributes, which influence dynamics of competition, predator–prey relationships, and community structure [7]. These changes can have a cascading effect on ecological processes like nutrient cycling, food web interactions, and habitat availability, eventually influencing aquatic ecosystem stability and function [8].

The evolution of osmoregulatory mechanisms has enabled aquatic organisms to colonize diverse ecological niches [9], with adaptations occurring at the molecular, physiological, ecological, and behavioral levels to cope with varying salinity conditions [10]. At the behavioral level, organisms may modify their feeding or movement patterns to avoid stressful conditions or seek more favorable environments. At the molecular level, they respond to salinity stress through changes in gene expression, activation of signaling pathways, and epigenetic modifications [11]. Physiologically, osmoregulatory processes involving ion transport are essential for maintaining internal homeostasis across environmental salinity gradients [3].

For species to effectively adapt to salinity stress, they must exhibit appropriate genetic diversity and population structure [12]. For instance, organisms with greater genetic diversity typically exhibit a wider range of physiological responses to fluctuations in salinity, enhancing their survival in environments with variable salinity levels [13]. Furthermore, genetic diversity in gill function, osmoregulatory processes, and the production of antioxidant enzymes plays a critical role in determining species' ability to cope with salt-induced oxidative stress [14]. Beyond intrinsic traits, interactions between host organisms and their microbiomes can also influence their tolerance to salinity stress [15]. Specifically, by modulating host physiology [16] and enhancing stress responses [17], the microbiome can significantly affect how aquatic organisms respond to fluctuating salinity levels.

Despite the recognized importance of salinity adaptation for the survival of aquatic species, significant gaps remain in our understanding of the ecological impacts, adaptive mechanisms, and molecular responses of aquatic organisms, particularly among crustaceans and fish. This knowledge is increasingly critical in the context of global environmental changes, as anthropogenic activities not only alter salinity levels and introduce other stressors [8] but also affect the population and genetic structures of species, which in turn influence their capacity to respond to these changes [18]. This review aims to address these gaps by understanding species-specific mechanisms of osmoregulation under fluctuating salinity stress, with a particular focus on aquatic organisms such as crustaceans and fish. While existing research has explored the impact of salinity on marine life, the roles of microbiomes and genetic diversity in enhancing salinity tolerance have been insufficiently studied. Further, our objective is to investigate how adaptive strategies at the molecular, physiological, and behavioral levels contribute to survival in environments characterized by fluctuating salinity. Additionally, we aim to explore the potential role of epigenetic modifications and the complex interactions between host organisms and their microbiomes in coping with salinity stress. In addition, this study examines innovative strategies for mitigating salinity stress in aquatic organisms, such as nutritional interventions and selective breeding programs designed to enhance salinity tolerance.

2. Ecological Consequences of Salinity Stress on Aquatic Organisms

Aquatic organisms have developed a range of adaptive strategies to cope with the ecological pressures imposed by salinity stress, which allows them to thrive in environments with fluctuating salinity levels.

2.1. Salinity Ranges in Aquatic Environment

Aquatic ecosystems exhibit a wide range of salinity levels, from less than 1 ppt in freshwater environments to over 400 ppt in hypersaline habitats [3]. In estuarine regions, where freshwater meets seawater, salinity fluctuates from 0.5 to 35 ppt on a daily or seasonal basis, influenced by factors such as tides, river inflows, and evaporation [19]. These dynamic salinity gradients create distinct ecological zones, with transitional areas that support both freshwater and marine species. These zones often exhibit greater biodiversity compared to strictly freshwater or marine systems [20]. Hypersaline environments, such as shallow lagoons and arid coastal regions, may reach salinities of up to 350 ppt, as seen in areas like the Gulf of Mexico, Mediterranean, and Red Sea [21]. These lakes are renowned for their hypersaline conditions, which significantly influence the physiological responses of the organisms inhabiting them. Characterized by high concentrations of salts like sodium chloride (NaCl), sodium sulfate (Na_2SO_4), and other soluble minerals, these hypersaline lakes present unique challenges to the resident biota. For instance, in the Aral Sea, increasing salinity, particularly from sodium sulfate as the lake shrinks, has altered the ionic composition of the water, impacting the osmoregulatory processes of aquatic organisms [22]. Similarly, the Dead Sea, with its extreme salinity, primarily composed of chloride salts, supports only specialized extremophiles such as *Artemia salina* and other halophilic species. These organisms must adapt to high concentrations of chloride and sulfate ions, requiring significant physiological modifications in metabolic and ion-transport mechanisms to maintain homeostasis. Furthermore, the chemical composition of these hypersaline lakes, especially the dominance of chloride and sulfate ions, affects the bioavailability of essential nutrients, influencing ion exchange and osmotic regulation. Studies on the Dead Sea and similar hypersaline environments highlight how these extreme conditions drive adaptations at the biochemical and cellular levels, such as alterations in ion-transport and osmoregulatory strategies, which are critical for the survival of organisms in these challenging environments [23].

The significance of these ecological regions depends on the complex interactions between salinity and other environmental factors such as temperature, oxygen, and nutrient availability. For example, salinity-induced stratification in coastal ecosystems can lead to the emergence of separate layers in the water column, influencing the distribution of species and their access to nutrients [24]. These layers, controlled by freshwater influxes and melting ice, provide niches where certain species can thrive, while others may struggle to survive [25]. Furthermore, salinity changes shape ecotones, which are transition zones between various biological communities, such as those found at the freshwater–saltwater ecosystem interface [26]. These ecotones are centers of biodiversity, with a wide range of species, including euryhaline and stenohaline organisms. Euryhaline species, such as *Mozambique tilapia*, are particularly adapted to a wide range of salinities, allowing them to thrive in environments with frequent salinity changes [27]. These species perform a vital role in maintaining ecological stability by enabling species migration and fostering biodiversity in transitional conditions [28]. In contrast, stenohaline species, such as *Cyprinus carpio*, are confined to stable environments with narrow salinity ranges and are more vulnerable to salinity changes in transitional zones [29].

2.2. *Effect of Salinity Variability on Species Distribution and Biodiversity*

The effect of salinity variability on species distribution and biodiversity in estuarine ecosystems is significant, as species exhibit varying levels of tolerance to fluctuating salinity levels. Euryhaline species, capable of thriving across a broad range of salinity conditions, contrast with stenohaline species, which are restricted to either freshwater or marine environments with stable salinity. Fluctuations in salinity can induce physiological stress, particularly in invertebrates, which typically show less tolerance compared to fish [30]. Such stress can drive significant shifts in community composition, especially within invertebrate populations, as many species are replaced by lentic taxa better adapted to rising salinity [31].

Extreme cases of salinity fluctuation are found in hypersaline environments, where salinity levels exceed that of seawater. These environments tend to exhibit reduced biodiversity due to their harsh conditions. However, some species have evolved remarkable adaptations to survive in hypersaline conditions. For example, *Artemia* can tolerate salinities exceeding 100 ppt, utilizing advanced osmotic regulatory mechanisms [32]. On the other hand, *Oreochromis mossambicus* and *Oreochromis niloticus* reveal significant tolerance to varying salinities, with the ability to adapt to salinity levels of 0–100 ppt and 0–35 ppt, respectively. In these extreme environments, species richness tends to decline, and community composition shifts toward a small number of highly specialized species [33,34].

As salinity surpasses 35 g/L, only species with exceptional salinity tolerance are able to persist, resulting in the dominance of crustaceans, particularly *Artemia*, which is the most prevalent taxon in these ecosystems [35].

The dominance of specialized species in hypersaline environments underscores the remarkable adaptability of organisms to fluctuating salinity levels. These adaptations are critical for maintaining ecosystem functions such as nutrient cycling and food web dynamics [36]. For example, in hypersaline ecosystems, halophilic invertebrates like *Artemia* thrive at extreme salinities (>120–160 g/L), driving efficient energy transfer from primary production to higher trophic levels, despite the absence of higher consumers [37]. These simplified food webs consist primarily of *Artemia*, and the absence of complex predation pressures results in cascading effects, where *Artemia* reduces phytoplankton populations, significantly altering nutrient cycling and primary production [38]. Vertebrates such as salt-tolerant fish (up to 120 g/L) and specialist waterbirds also influence ecosystem functioning; flamingos and other birds reduce *Artemia* numbers while spreading nutrients across the landscape [39]. This results in a dynamic interplay between bottom-up (invertebrate-mediated) and top-down (avian predation) mechanisms that preserves the particular productivity of hypersaline ecosystems despite their lower biodiversity [39,40].

2.3. *Climate Change Implications on Salinity Fluctuations*

Climate change is altering salinity dynamics in marine ecosystems, primarily through increased evaporation, shifting precipitation patterns, and polar ice melt, all of which contribute to both hyper- and hyposaline conditions [41]. These salinity fluctuations interact with other climate stressors, such as ocean warming and acidification, creating complex challenges for marine organisms. For instance, marine bivalves like *Crassostrea gigas* and *Mytilus edulis* exhibit increased carbonic anhydrase activity under conditions of both low salinity and ocean acidification, suggesting potential negative impacts on biomineralization and shell formation. The compounding effects are particularly severe when temperature and salinity stressors coincide, as seen in mussels that exhibit elevated expression of stress-related genes under both low salinity and high temperature conditions [42].

The physiological impacts of these combined stressors are leading to concern ecological consequences. Some species facing simultaneous temperature increases and salinity fluctuations surpass their physiological tolerance thresholds, potentially leading to decreased fitness, reproductive failure, and increased mortality rates [43]. Freshwater species are especially vulnerable to salinization effects, with studies documenting significant physiological stress responses in fish populations. Geographic spawners face particular challenges, as decreasing stream flows and higher temperatures disrupt their reproductive cycles, sometimes leading to complete spawning failures [44]. These climate-driven changes are further complicated by ocean acidification, which affects behaviors and survival rates in fish larvae [45].

Climate-induced shifts are also driving significant changes in species distributions, leading to disease outbreaks in several species and having wide-ranging impacts on ecosystem dynamics [46]. For example, warming water above 15 °C impairs the immune system of *Homarus americanus* (temperate North Atlantic), making them more susceptible to diseases like gaffkemia and epizootic shell disease. In *Chionoecetes opilio* (Arctic and sub-Arctic regions), thermal stress has triggered outbreaks of bitter crab disease caused by Hematodinium, with prevalence increasing five-fold per 1°C rise in temperature [47]. These shifts disrupt host–pathogen dynamics, often favoring pathogen proliferation over host defenses, particularly in ectothermic species.

To address these challenges, adaptive strategies for management are essential for aquaculture systems. Selective breeding programs focus on developing climate-resilient strains, including salinity-tolerant species [48,49]. Nutrigenomic approaches offer potential for mitigating stress impacts through customized dietary formulations [49], while advanced reproductive technologies like surrogate broodstock help preserve genetic diversity under changing conditions [50].

3. Organismal Adaptations to Salinity Stress

To persist and thrive in environments with variable salinity, aquatic organisms employ a range of adaptive strategies that include both physiological and behavioral mechanisms. This section explores these adaptations in detail, examining the physiological, behavioral, molecular, and cellular mechanisms that enable these organisms to withstand and adapt to such fluctuations.

3.1. Adaptive Strategies for Salinity Tolerance in Aquatic Organisms

Maintaining osmotic homeostasis in environments where salinity fluctuates is crucial for survival. Freshwater species, such as teleost fish, maintain osmotic balance by actively absorbing salt from the environment and excreting diluted urine to counteract passive salt loss and excessive water influx. In contrast, marine species face the challenge of preventing dehydration and excessive salt accumulation. These species conserve water and excrete excess salts primarily through specialized cells in the gills and kidneys [51].

Euryhaline organisms use complex molecular processes to maintain osmotic balance under changing salinity conditions, with aquaporins (AQPs) and ion transporters playing crucial functions in cellular water and ion homeostasis. Euryhaline organisms, such as *O. niloticus*, use similar osmoregulatory systems, such as NKA, to control ion transport and maintain ionic balance. *O. niloticus* can modify their osmoregulatory functions to changing salinities by depending on ion transporters and aquaporins to manage water and ion homeostasis, which is essential for survival in both freshwater and brackish environments [52]. Studies on crustaceans show that AQPs (AQP3, AQP4, and AQP11) are differentially expressed in osmoregulatory organs such as gills, where they control water permeability to prevent excessive efflux during high-salinity stress. These proteins collaborate with ion-transport genes including NKAα, carbonic anhydrase, and V-type H^+-ATPase (VHA) subunits to control ion uptake and excretion. For example, in *Fenneropenaeus chinensis*, coordinated overexpression of AQP3/4 and ion transporters in the gills improves salinity tolerance by regulating cellular hydration and ionic balance [53]. Similarly, *Litopenaeus vannamei* has tissue-specific AQP expression patterns, with acute salinity changes inducing dynamic alterations in AQP3, AQP4, and AQP11 levels to improve water and ion transport [54]. These molecular adaptations enhance physiological mechanisms such as *Macrobrachium* spp. spawning in brackish water and energy-efficient ion regulation mediated by NKA and VHA, allowing survival over an extensive range of saline conditions [55].

Salinity tolerance in *Oreochromis niloticus* (Nile tilapia) is a crucial factor for expanding its aquaculture potential in brackish and saline environments. Recent studies have revealed that Nile tilapia exhibits significant physiological adaptations to higher salinity, particularly in its gill structure and osmoregulatory mechanisms. The proliferation of chloride cells and increased Na^+/K^+ ATPase activity in the gills enable these fish to better regulate ion balance in saline conditions [52]. Additionally, studies on the natural salinity tolerance of different *O. niloticus* haplotypes indicate that selecting strains with inherent salinity tolerance could further optimize aquaculture systems in coastal and brackish water regions [56]. While high salinity can negatively affect reproduction, limiting population growth, this characteristic could be utilized as a population control strategy in aquaculture [43]. Additionally, studies show that proper acclimatization and dietary management can further optimize growth in saline conditions, with brackish water (5–10 ppt) being particularly suitable for farming Nile tilapia in intensive systems [57].

Diadromous fish, which migrate between freshwater and marine environments, exhibit complex life cycles that accommodate transitions between distinct salinities. Their ability to maintain euryhalinity has evolved independently across multiple teleost lineages, demonstrating the remarkable physiological and evolutionary flexibility of these species [58].

Marine elasmobranchs employ a distinct osmoregulatory strategy, using organic osmolytes such as trimethylamine-N-oxide (TMAO) to maintain osmotic balance. They excrete excess NaCl through the rectal gland, a strategy that contrasts with most teleost fish, which rely on different osmoregulatory mechanisms to actively regulate their extracellular fluid osmolality [59]. In contrast, hagfish are osmoconformers, maintaining body fluid NaCl concentrations that closely match seawater, thereby minimizing the need for active osmoregulation [60]. These varied strategies highlight the diverse adaptations among fish species to maintain osmotic balance across different environments.

Crustaceans have evolved isosmotic intracellular regulation, a mechanism that allows cells to adjust their volume in response to osmotic stress. In low salinity environments, cells reduce their regulatory volume to prevent swelling, while in high-salinity environments, they increase their regulatory volume to maintain cellular integrity. This intracellular regulation is favored over isosmotic extracellular regulation due to its greater efficiency in managing cellular volume under varying salinity conditions [55]. Additionally, osmoregulatory capacities can be gender-specific, as seen in species like *Callinectes sapidus*, where such mechanisms may contribute to the species' invasive success [61].

3.2. Behavioral Adaptations to Salinity Fluctuations in Aquatic Organisms

Aquatic organisms living in ecosystems with frequent salinity fluctuations have developed a variety of behavioral adaptations to cope with these changes. A key strategy is selective habitat use. Many species, particularly those in hypersaline or brackish environments, avoid high-salinity areas by seeking out habitats with lower salinity, thus minimizing osmoregulatory stress. This behavior enables them to allocate energy toward fitness enhancing functions [62]. Studies show that different species use different behavioral strategies. For instance, the euryhaline *Kryptolebias marmoratus* increases activity in hypersaline conditions to find lower salinity refuges, whereas *Danio rerio* exhibit significantly reduced locomotor activity at high salinity (5.78 g/L), conserving energy by reducing movement duration and distance traveled during both light and dark phases [63]. Interestingly, *Poecilia latipinna* exhibit more varied responses that are not precisely related to salinity concentration, implying species-specific adaptive processes. Additionally, *P. latipinna* prefers lower salinity habitats, especially when subjected to additional stresses such as predator cues [64]. This variability in tolerance enhances the overall population's ability to survive and thrive under fluctuating environmental conditions.

3.3. Osmotic Stress and Ion-Transport Mechanism Response in Fluctuating Salinity Environments

The physiological response of aquatic organisms to hypo- and hyper-salinity stress emphasizes the complex mechanisms involved in maintaining osmotic balance. When exposed to fluctuating salinity, organisms activate ion-transport mechanisms, such as Na^+/K^+-ATPase, to regulate the movement of ions across their membranes and maintain osmotic equilibrium. This activation comes at a significant cost, however, as it increases the energy demand required to support enhanced transport. To meet these heightened energy requirements, organisms must increase their overall energy expenditure. Additionally, as part of this adaptive response, gill permeability is adjusted to facilitate more efficient ion exchange. Collectively, these physiological adjustments enable organisms to cope with the challenges posed by fluctuating salinity environments, ensuring their survival and maintaining homeostasis under osmotic stress (Figure 1) [3].

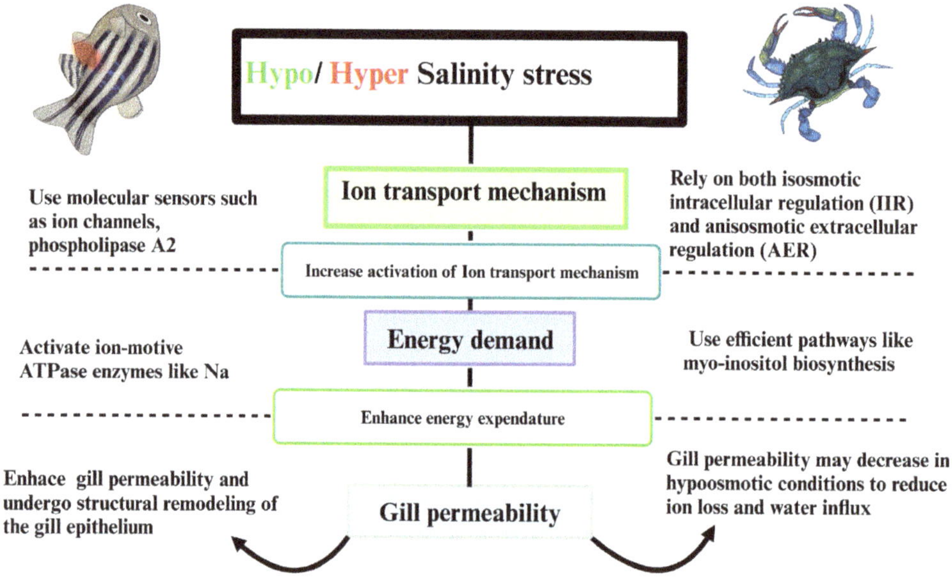

Figure 1. The physiological responses of aquatic organisms to hypo- and hyper-salinity stress, highlighting the activation of ion-transport mechanisms to maintain osmotic balance. This increased ion transport leads to higher energy expenditure, which in turn affects gill permeability, influencing the overall physiological response of aquatic organisms to fluctuating salinities.

Exposure to hyperosmotic conditions triggers specific molecular responses to support osmotic regulation. For example, hyperosmotic stress activates phosphorylation of tight junction proteins by myosin light chain kinase, which modulates the activity of Na^+/Cl^-/taurine co-transporters, helping to maintain ion balance [65]. In hypoosmotic conditions, dephosphorylation of focal adhesion kinase enhances the activity of $Na^+/K^+/2Cl^-$ co-transporters and Cystic Fibrosis Transmembrane Conductance Regulator, aiding ion regulation and restoring osmotic equilibrium [66]. Furthermore, the expression of osmotic stress transcription factor 1 Osteoclast Stimulating Factor 1 is upregulated in response to salinity fluctuations, further modulating these transport mechanisms to help organisms to adopt varying osmotic environments [67].

Different species exhibit unique adaptations to cope with salinity fluctuations. For example, *Sinonovacula constricta* can maintains osmotic balance by modulating free amino acids like taurine and increasing Na^+/K^+-ATPase activity in the gills under fluctuating salinities [68]. Similarly, juvenile *Eriocheir sinensis* display optimal osmoregulatory performance at 8‰ salinity, but their ion transport is disrupted when salinity deviates from range [69].

Some marine fish also exhibit remarkable adaptations to fluctuating salinity. Species such as *O. mossambicus* undergo extensive gill remodeling and adjust their ion-transport systems to meet the high energy demands of fluctuating salinities [70]. Similarly, the *Fundulus heteroclitus* and Tilapia sp. adjust ion transport and use neuroendocrine regulation to cope with salinity changes [71]. The *Acanthopagrus schlegelii* also adapts by changing gene expression in its gills, enhancing ion-transport efficiency and adjusting Na^+/K^+-ATPase activity [72].

Chronic salinity stress requires long-term metabolic adjustments to maintain homeostasis. For example, in response to prolonged salinity exposure, *Salmo salar* undergo metabolic reprogramming that regulates enzymes and pathways for salt metabolism to cope with prolonged salinity exposure [51]. In fish, osmoregulation accounts for 1–50% of total energy expenditure, which increases during extreme salinity stress [73]. Brine shrimp (*Artemia*) are metabolically stable around 35–150 g/L but consume more oxygen below 35 g/L, indicating increased osmoregulatory needs [74]. Organisms maintain homeostasis by implementing metabolic changes, such as switching energy substrates [75] or increasing ion transporters like Na^+/K^+-ATPase to enhance efficiency [76].

3.4. Oxidative Stress Response in Aquatic Organisms Due to Salinity Stress

Salinity stress induces oxidative stress by disrupting redox balance through interactions between osmotic and oxidative pathways (Table 1) [77]. This imbalance increases the production of reactive oxygen species (ROS), leading to cellular damage. The production of ROS occurs as a result of altered cellular respiration and metabolic processes in response to salinity changes [3]. For example, when aquatic organisms experience osmotic stress, this disrupts the balance between the generation and scavenging of ROS [78]. In cells, changes in ion concentrations (such as sodium and chloride ions) can affect mitochondrial function, leading to an increase in electron leakage during cellular respiration. This electron leakage can result in the production of superoxide anions ($O^{2\bullet-}$), a primary ROS [79].

To counter this oxidative stress, organisms activate antioxidant defense mechanisms such as SOD, catalase (CAT), and glutathione peroxidase (GPx). These antioxidants work in concert to neutralize ROS. For instance, SOD catalyzes the dismutation of superoxide into oxygen and hydrogen peroxide, while CAT converts hydrogen peroxide into water and oxygen, and GPx reduces hydrogen peroxide or lipid peroxides to water or corresponding alcohols, respectively [80]. In some species, increased ROS production due to osmotic stress can also lead to alterations in the expression of these antioxidant enzymes.

For example, fluctuating salinities elevate ROS levels in *Scylla serrata*, which triggers the activities of these antioxidant enzymes, such as SOD and CAT, to reduce oxidative damage [81]. On the other hand, rapid salinity reductions in *E. sinensis* compromise oxidative balance by lowering antioxidant enzyme activities, but enhance defenses through an increase in total antioxidant capacity (T-AOC), GSH-PX, and CAT activity [82].

Table 1. Salinity stress-related changes in antioxidant enzyme activity across different aquatic species.

Species	Salinity Type	Duration	Enzyme Activity	References
Brachionus koreanus	High	3–24 h	ROS and GST activity elevated	[83]
Crassostrea gigas	High/Low	4 months	Elevated level of CAT, whereas no effect observed at SOD	[84]
Litopenaeus vannamei	High/Low	24 h	Activities of SOD, GPx, and CAT decreased at both lower and higher salinities	[85]
Eriocheir sinensis	High/Low	24–72 h	At higher salinities, GSH-Px activity is enhanced; works synergistically with SOD and CAT to mitigate lipid peroxidation damage	[86]
Procambarus clarkiis	Low	48 h	Antioxidants like CAT and GPX-PX decrease, while MDA levels increase	[87]

Table 1. *Cont.*

Species	Salinity Type	Duration	Enzyme Activity	References
Scylla olivacea	Medium	Varied	Increase in gill SOD activities helps cope with elevated ROS levels from high metabolic demands	[88]
Lates calcarifer	High/Low	8 weeks	GPX and CAT activities increase in high salinity and decrease in low salinity	[89]
Oreochromis niloticus	Medium/Low	4 weeks	The activity of SOD, GPX, and CAT elevates under medium salinity alongside hypoxia	[90]
Dicentrarchus labrax, *Chanos chanos*	High/Low	4 weeks	SOD level of activity elevated in both fresh and saline water	[91]
Cyprinus carpio	Medium/Low	8 weeks	Exposure to medium salinity reduced lysozyme activity, SOD, CAT, GPx, and increased MDA levels before and after heat stress	[92]

Species can adapt to salinity-induced oxidative stress by altering respiration and metabolic strategies. For instance, *Tigriopus brevicornis* reduces oxygen consumption under stressful conditions, suggesting a different metabolic strategy to conserve energy and mitigate oxidative damage [93]. In bivalves such as *C. gigas*, the activation of antioxidant enzymes is triggered, which helps mitigate oxidative stress [84]. The sensitivity of *Mytilus galloprovincialis* to oxidative stress is particularly evident when salinity fluctuates. At low salinity, markers like protein carbonyls and Thiobarbituric acid reactive substances significantly increase in gill tissues, indicating oxidative stress and cellular damage in *M. galloprovincialis* under fluctuating salinities [94]. These studies support the dynamic nature of oxidative stress responses in bivalves under varying salinity conditions.

Fish also employ diverse strategies to manage oxidative stress caused by salinity fluctuations. For example, *Chanos chanos* upregulate enzymatic antioxidants to maintain redox balance and prevent apoptosis in seawater, but in freshwater, despite the compensatory increase in antioxidants, *C. chanos* still undergoes apoptosis due to an imbalance between ROS production and the capacity of antioxidant defenses to neutralize the oxidative stress [91]. Similarly, antioxidant-rich diets further enhance these enzymatic defenses, as observed in *Salmo trutta*, where dietary supplements improve CAT and GPx activity, thus protecting against oxidative damage [95]. According to these studies, fish employ a range of strategies, including enzymatic regulation and dietary adaptations to mitigate oxidative stress and maintain redox homeostasis under fluctuating salinity conditions.

3.5. Molecular Responses to Salinity Stress

3.5.1. Key Signaling Pathways

Signaling pathways are vital in mediating responses to environmental stressors like salinity fluctuations. These pathways regulate several important processes, including ion transport, osmoregulation, cell survival, metabolism, and immunity. This section examines important signaling pathways—AMPK, PI3K-AKT, MAPK, and the Hippo pathway (Table 2)—that play a fundamental role in maintaining homeostasis. The intricate molecular signaling pathways activated in response to salinity stress highlight the complex mechanisms that enable organisms to cope with fluctuating salinity levels. Upon exposure to salinity stress, two major pathways, AMPK and MAPK, are activated to regulate critical processes such as glycolysis, fatty acid oxidation, and energy metabolism, thereby maintaining cellular homeostasis. Additionally, calcium (Ca^{2+}) signaling plays a pivotal role in triggering intracellular responses. The figure also emphasizes the involvement of the

RAS, P13K-AKT, and YAP/TAZ pathways, which are essential for regulating cell survival, growth, apoptosis, and differentiation under stress. These signaling cascades interconnect in a feedback loop, ensuring a coordinated response. Furthermore, the figure underscores the importance of aquaporins (AQP1 and AQP11) in managing osmotic balance by controlling water and ion transport. Immune responses, inflammation, and apoptosis are also regulated by these pathways to mitigate cellular damage. Collectively, this network of molecular responses allows organisms to adapt to salinity fluctuations, ensuring their survival and maintaining cellular integrity under osmotic stress (Figure 2).

Table 2. Key pathways involved in salinity stress response in aquatic organisms.

Specie	Pathway	Salinity Level	Duration	Effects	References
Oreochromis niloticus	AMPK	8 and 16 psu	8 weeks	Mediates energy metabolism under stress, maintaining cellular energy homeostasis.	[96]
Rachycentron canadum	MAPK	Acute stress (30 ppt to 10 ppt for one hour), Chronic (30 ppt–10 ppt for 4 days)	1 h to 28 days	MAPK pathway upregulated under acute salinity stress, inducing apoptosis and inflammation; chronic stress shows varied expression for long-term adaptation.	[97]
Ictalurus punctatus	MAPK	3 and 7 psu	3 weeks	p38-MAPK mediates gene expression, metabolism, and cellular homeostasis under salinity stress.	[98]
Takifugu fasciatus	MAPK	0, 10, and 20 ppt	3 days	20 ppt salinity led to increased oxidative stress, and apoptosis-related gene expression.	[99]
Scylla paramamosain	AMPK	23 and 4 ppt	6 weeks	Regulates stress responses, gene expression, apoptosis, and oxidative stress management; protects against cell damage via p38-MAPK.	[100]
Scophthalmus maximus	PI3K-AKT	5, 10, and 30 ppt	24 h	Low saline stress downregulated gene expression, protein content, and AKT1 phosphorylation in gill tissues.	[101]
Litopenaeus vannamei	AMPK	3, 20, and 30 psu	0 h to 96 h	AMPK elevated during acute stress, indicating higher energy needs; sustained during chronic stress for osmoregulation and energy balance.	[102]
Litopenaeus vannamei	RAS	25 and 3 ppt	Short-term stress (0 h–96 h), Long-term stress (8 weeks)	RAS pathway downregulated under stress, affecting homeostasis, osmoregulation, and blood glucose levels.	[103]

Table 2. *Cont.*

Specie	Pathway	Salinity Level	Duration	Effects	References
Scylla paramamosain	RAS	3, 20, and 23 ppt	0 h to 168 h	NKA activity increased under stress to regulate osmotic balance, aided by RAS pathway for improved metabolic activity.	[104]
Penaeus monodon	PI3K-AKT	30, 18, and 3 psu	45 days	Prolonged low saline stress downregulated RAS pathway, diverting energy from growth to ionic homeostasis and osmoregulation.	[105]

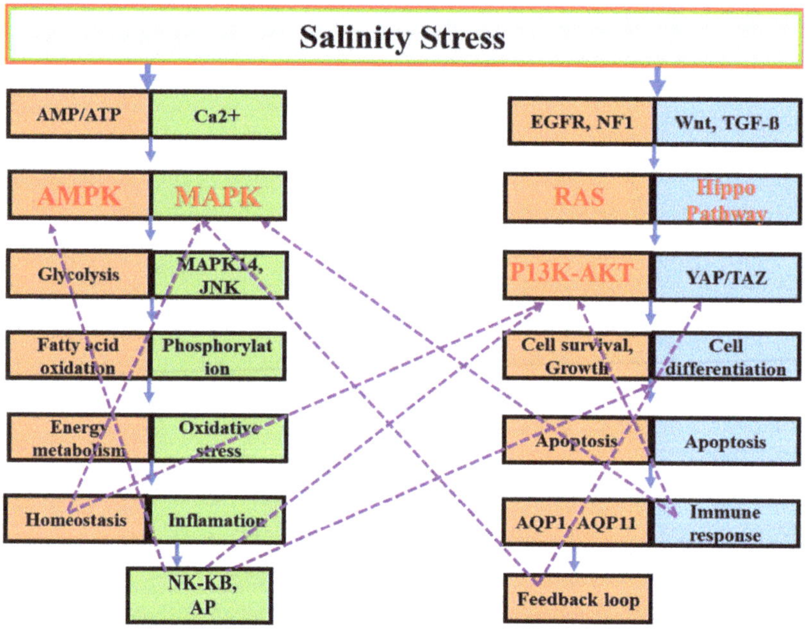

Figure 2. Salinity stress activates the MAPK, RAS, PI3K-AKT, AMPK, and Hippo pathways. In this figure, the red color represents the pathways, highlighting their roles in the stress response, while other colors correspond to the responses related to these pathways. The MAPK pathway induces stress and inflammatory responses through proteins such as MAPK11, JNK, and MKK4. The RAS and PI3K-AKT pathways elevate cell survival and osmoregulation. The AMPK pathway is triggered by AMP/ATP ratio alterations, which improves energy metabolism. The Hippo pathway, in conjunction with YAP/TAZ, modulates cell proliferation and immunological responses.

MAPK Signaling Pathway Response in Aquatic Species Under Salinity Stress

The mitogen-activated protein kinase (MAPK) signaling pathway regulates response to salinity stress by transmitting environmental signals to the nucleus for adaptive changes [106,107]. Salinity-induced increases in cytosolic calcium (Ca^{2+}) activate the MAPK pathway, one of several signaling pathways responding to osmotic stress. This activation leads to the phosphorylation of key proteins involved in stress responses, including transcription factors and enzymes that help the organism adapt to changes in salinity [108]. For example, in the *Scophthalmus maximus*, p38 MAPK upregulation induces downstream phosphorylation events that mediate osmoregulation [106]. Similarly, in *Lateolabrax maculatus* hyperosmotic shock triggers MAPK11 (p38β) expression, further supporting the MAPK pathway role in salinity stress responses [109]. Additionally, MAPK14a has been associated to oxidative damage in *Ictalurus punctatus* under extreme salinity stress, highlighting the role of MAPK pathways in managing both osmotic and oxidative stress [98].

Among MAPK subfamilies, c-Jun N-terminal kinase (JNK) plays a prominent role in managing osmotic and oxidative stress, as well as regulating metabolism, inflammation, and apoptosis. For instance, in *L. maculatus*, JNK activation facilitates osmotic adaptation by regulating mitogen-activated protein kinase 4 (MKK4) [109]. A similar mechanism is observed in the fish *Rachycentron canadum*, where extreme salinity fluctuations lead to changes in MAPK gene expression, particularly in the gills and kidneys, underscoring their involvement in ion transport and osmotic regulation during low salinity stress [97].

In addition to osmotic stress, salinity combined with elevated carbonate alkalinity activates the MAPK pathway, leading to increased ROS production and apoptosis in *E. sinensis* [110]. Similarly, in *Penaeus monodon*, JNK activation regulates MKK4 during low-salinity adaptation [111]. In *Chlamys farreri*, JNK expression increases sharply within hours of environmental stress, highlighting the rapid role of JNK in molluscan stress responses [98].

PI3K-AKT Signaling Pathways in Response to Salinity Stress

The PI3K-AKT signaling pathway, often activated by RAS, plays a vital role in physiological regulation and cellular resilience. In crustaceans, this pathway is integral not only to cellular resilience under stress but also to the innate immune response [112]. The central components of this pathway, phosphatidylinositide 3-kinases (PI3Ks) and AKT (protein kinase B), regulate essential cellular processes. For instance, they inhibit pro-apoptotic proteins, such as the BCL2-antagonist of cell death, and prevent NF- κB nuclear translocation from cells under stress [113].

In response to low salinity, the PI3K-AKT pathway modulates gene expression and protein synthesis to maintain tissue integrity and remove damaged cells through apoptosis in *S. maximus* [96]. Notably, low-salinity stress downregulates AKT1 phosphorylation and gene expression, particularly in the gill, highlighting tissue-specific responses [101]. Additionally, this pathway mediates the expression of aquaporin genes such as AQP1 and AQP11, and the PI3K-AKT pathway facilitates water and ion exchange essential for maintaining osmotic balance in fluctuating environments [114].

Brain transcriptome profiling analysis of *Oreochromis niloticus* under prolonged hypersaline stress revealed that the PI3K-AKT signaling pathway plays a key role in the upstream regulation of osmoregulatory processes in response to salinity stress [113]. Key genes within this pathway, including G protein-coupled receptors and solute carrier families, exhibit differential expression under salinity fluctuations [115]. Furthermore, inhibition of this pathway confirms its pivotal role in modulating ion channel activity, which is essential for maintaining osmotic balance under stress [101].

AMPK Signaling Pathway and Energy Regulation in Response to Salinity Stress

AMP-activated protein kinase helps to maintain energy homeostasis by monitoring the AMP/ATP ratio. When this ratios increase, AMPK is activated through phosphorylation at threonine 172 in the AMPKα subunit, triggering metabolic adjustments to restore energy balance [116]. Under salinity stress, AMPK modulates energy metabolism to meet the heightened demands of osmoregulation [105].

Building on this general mechanism, crustaceans exhibit unique metabolic adaptations under salinity stress. In *L. vannamei*, salinity stress alters lipid metabolism, influencing fatty acid biosynthesis and glycosphingolipid metabolism to meet energy demands [117]. In *S. paramamosain*, low salinity stress increases the AMP/ATP ratio in the hepatopancreas, thereby enhancing ATP production and upregulating genes involved in glycogen and lipid metabolism. These metabolic changes sustain energy needs during stress and facilitate osmotic pressure regulation [100].

In addition to crustaceans, fish also exhibit AMPK-mediated metabolic responses to salinity stress. In *Paralihthys olivaceus*, salinity fluctuations activate AMPK through phosphorylation and allosteric AMP binding, leading to increased glycolysis and fatty acid oxidation that provides the necessary energy to manage osmotic challenges [118]. Similarly, in *Ctenopharyngodon Idella*, activation of AMPK by 5-aminoimidazole-4-carboxamide ribonucleotide reduces serum glucose levels and regulates lipogenesis-related proteins, effectively reallocating energy resources under stress [119].

The Role of the Hippo Signaling Pathway in Salinity Stress

The Hippo signaling pathway plays a role in maintaining tissue and cellular integrity by integrating environmental cues and regulating cellular processes [120]. This pathway plays a role in stress responses and innate immunity in aquatic species like *L. vannamei*. Under optimal conditions, it regulates cell proliferation and anti-apoptotic mechanisms; however, during stress, such as salinity stress, the pathway shifts to promote apoptosis and cell cycle arrest to preserve cellular homeostasis [54]. In bivalves, the pathway normally promotes cell proliferation and anti-apoptotic mechanisms under optimal conditions; however, under stress, such as salinity stress, it shifts towards triggering apoptosis and cell cycle arrest to maintain cellular homeostasis [121].

Beyond its direct effects, interactions with other signaling pathways further enhance its role in managing environmental stress. Its crosstalk with the Wnt pathway, via β-catenin-mediated gene activation, supports tissue growth and repair. Further interactions with bone morphogenetic proteins and the transforming growth factor-β superfamily are vital for muscle growth and innate immunity in both crustaceans and fish [122]. The regulation of metabolism and regeneration by YAP/TAZ further emphasizes Hippo's pivotal role in balancing apoptosis and cell survival during stress and repair. This function is vital for adaptation and resilience in stressful environments [123]. Together, these mechanisms highlight the pathway's pivotal role in enabling adaptive responses to changing salinity conditions in aquatic organisms.

The complex interplay of these pathways regulates cellular responses to salt stress via synergistic interaction and antagonistic regulation. Under moderate salinity stress, these pathways interact to promote cell survival and adaptation. The PI3K-AKT pathway stimulates Hippo-YAP/TAZ signaling by inhibiting LATS1/2 kinases, enabling tissue repair and cell proliferation [121]. Simultaneously, MAPK (especially JNK/p38) and Hippo regulate an integrated response wherein JNK-mediated phosphorylation of YAP/TAZ increases antioxidant defenses while retaining the ability to induce apoptosis if stress becomes extreme [124]. However, under extreme salinity conditions, these pathways typically possess antagonistic interactions. AMPK activation combats PI3K-AKT signaling by suppressing mTORC1 via TSC2 phosphorylation, effectively favoring energy conservation for osmoregulation over growth activities [101]. The pathways also show tissue-specific interactions, with gill tissues preferring AMPK-PI3K synergy for ion regulation and hepatopancreatic tissues using AMPK-Hippo antagonism to manage metabolic demands [125]. The immune osmotic balance is regulated by PI3K-AKT and MAPK interaction in NF-κB activation [126]; however, this is counterbalanced by PI3K-AKT's repression of JNK-induced apoptosis in crustaceans like *P. monodon* [127]. These dynamic interactions impose a complex regulatory network, enabling organisms to precisely regulate their stress responses based on the magnitude of salinity fluctuations.

3.5.2. Stress-Responsive Genes

Stress-responsive genes play pivotal roles in regulating processes such as osmoregulation, metabolism, immune function, and apoptosis under salinity stress [128]. Heat shock proteins (HSPs), particularly HSP70 and HSP90, are molecular chaperones that stabilize cellular proteins under osmotic stress. In fish species such as *Luciobarbus capito*, high salinity triggers the expression of stress-responsive genes, including, HSP70 and HSP90 in the spleen, which helps stabilize cellular proteins under osmotic stress [129]. Transcription factors such as Forkhead box O proteins further mediate stress responses by regulating genes involved in stress resistance, metabolism, and apoptosis, enhancing adaptability during osmotic challenges [130]. In *Acanthopagrus schlegelii* the upregulation of Osteoclast Stimulating Factor 1 regulates downstream genes for ion transport and stress response [128]. Likewise, caspase genes, which are key regulators of apoptosis, are modulated by salinity stress. In *Takifugu fasciatus*, caspase 3, 7, and 9 are upregulated under high salinity, promoting apoptosis, while their expression decreases at moderate salinity, favoring cell survival [99]. Similar patterns are observed in the *Syngnathus typhle*, where salinity stress compromises immune defense [131]. Cell cycle-related genes, such as p53, are activated to promote the proliferation of ionocytes, enhancing the ion exchange capacity of gills in response to changing salinity [132]. This adjustment facilitates the organism's immediate survival under stressful conditions.

In addition to apoptosis regulation, other genes, such as encoding aquaporins (AQPs), play a significant role in osmoregulation during salinity stress. In *L. vannamei*, the expression of LvAQP3, LvAQP4, and LvAQP11 decreases across tissues under salinity stress, except in the hepatopancreas, where they facilitate osmoregulation through amino acid metabolism [133]. Similar to the effects observed in *L. vannamei*, *Fenneropenaeus chinensis* also shows downregulation of AQP3 and AQP4 in the gills to minimize water loss and regulate ion balance [53]. These results indicate the adaptive role of AQPs in maintaining osmotic and ionic balance.

Beyond water transport, metabolic adjustments are essential for energy allocation during salinity stress. In *S. paramamosain*, enzymes like citrate synthase are upregulated to enhance energy production through the TCA cycle, while glycogen synthase and hexokinase are downregulated, indicating a shift from energy storage to immediate use [134]. Lipid metabolism also adopts accordingly, with reduced fatty acid synthesis via fatty acid synthase and increased lipid mobilization to meet energy demands [135]. Under both hypoosmotic and hyperosmotic conditions, organisms adjust the synthesis and degradation of organic osmolytes to maintain osmotic balance. Enzymes like ornithine decarboxylase are upregulated to boost the production of polyamines, while glutamine synthetase expression is downregulated, fine-tuning osmolyte metabolism and supporting cellular homeostasis [136].

3.5.3. Epigenetic Modifications Under Salinity Fluctuations

Epigenetic modifications such as DNA methylation, histone modifications, and non-coding RNA activities enable organisms to adapt to salinity stress. These heritable molecular changes dynamically regulate gene expression, facilitating both immediate responses and long-term resilience to salinity fluctuations without altering the DNA sequence [137]. By modulating gene activity, epigenetic mechanisms help organisms cope with environmental changes in ways that are heritable and adaptive over time. Figure 3 illustrates the critical role of epigenetic modifications and non-coding RNAs (ncRNAs) in regulating gene expression in response to salinity stress. Non-coding RNAs, including microRNAs (miRNAs), regulate genes involved in osmoregulation, metabolism, and stress response, particularly those crucial for ion transport and water balance. These ncRNAs enable organisms to fine-tune gene expression in response to fluctuating salinity levels, ensuring optimal survival. Additionally, DNA methylation at gene promoters typically leads to transcriptional repression, allowing organisms to downregulate non-essential genes and conserve energy, especially under stress conditions. Histone modifications, such as the methylation of lysine residues on histone H3, further regulate gene expression by altering chromatin structure, thus controlling the accessibility of transcription factors to DNA. Together, these epigenetic mechanisms enable organisms to efficiently adapt to salinity stress, maintaining essential physiological functions while minimizing energy expenditure on non-critical processes.

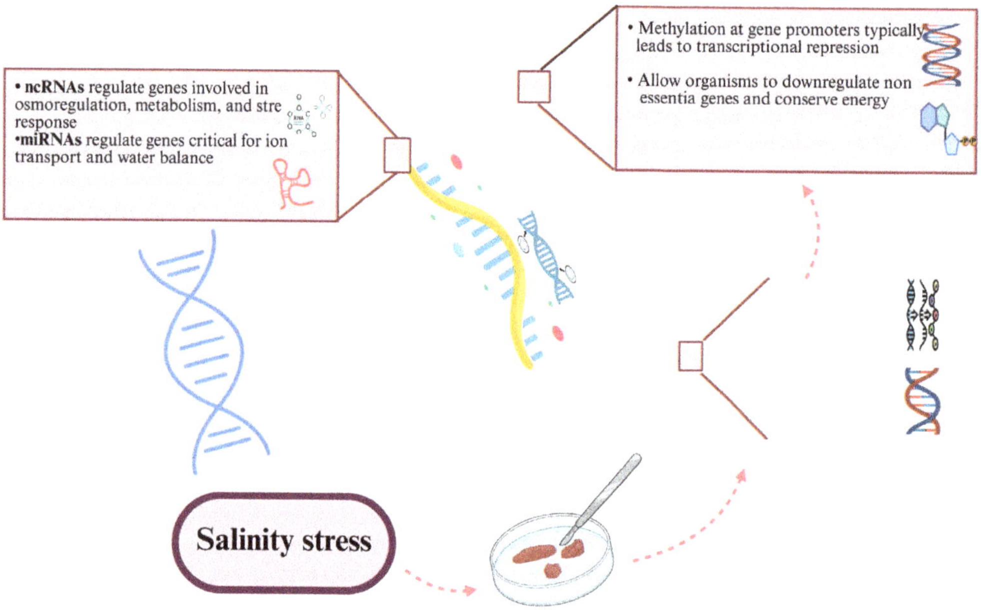

Figure 3. The role of non-coding RNAs (ncRNAs) and microRNAs (miRNAs) in regulating gene expression under salinity stress. It emphasizes how methylation at gene promoters leads to transcriptional repression, allowing organisms to conserve energy by downregulating non-essential genes. It also highlights the regulation of gene expression through histone modifications, affecting chromatin structure and transcription factor access to DNA.

In the context of salinity fluctuations, studies have shown that low salinity induces tissue-specific methylation changes in organisms. For instance, in *Portunus trituberculatus*, low salinity increases methylation in gill tissues, while decreasing it in muscle and gills [138]. Likewise, in *C. gigas*, salinity fluctuations leads to DNA methylation changes in genes associated with nucleic acid metabolism, tropomyosin, and cellular transport, highlighting the role of these modifications in osmoregulatory adaptation [139].

In addition to DNA methylation, histone modifications and non-coding RNAs also contribute to osmoregulatory responses. In *Daphnia magna*, miRNAs are involved in transgenerational inheritance of salinity tolerance, where sustained hypomethylation of stress-relevant genes persists across generations [140]. Such epigenetic memory prepares offspring for similar environmental challenges, highlighting the evolutionary significance of these mechanisms.

Several studies have reported evidence for the heritability or transgenerational transmission of epigenetic adaptation under environmental stresses [141]. However, there is little research on transgenerational inheritance under saline stress. For example, in *D. magna*, high-salinity exposure in the parental F0 generation resulted in distinct low methylation patterns in six protein-coding genes involved in DNA repair, cytoskeleton organization, and protein synthesis in the unexposed F1, F2, and F3 generations [142]. Some studies have examined transgenerational plasticity in response to salinity changes. For instance, research on the *Gasterosteus aculeatus* has investigated its adaptability to changing salinity conditions. The results show that the outcomes of transgenerational plasticity may vary depending on life stage, with early life stages being especially vulnerable to increased salinity [143]. Similarly, research on *Crassostrea virginica* evaluated transgenerational plasticity in response to low salinity. The findings revealed that transgenerational plasticity possessed no substantial impact on low-salinity tolerance, with parental genotype playing a significant role in determining larval size variation. This suggests that evolutionary adaptation is the primary strategy by which *C. virginica* can tolerate future salinity reductions. In some marine fish species, low salinity triggers distinct epigenetic response. Likewise, prolonged exposure to low salinity in *Cynoglossus semilaevis* alters H3K4me3 patterns influencing growth and liver structure. It also upregulates demethylase genes and DNA methyltransferases, enhancing methylation in renal tissues and regulating immune-related genes like *MASP1* to balance energy use and immune responses under stress [144]. These studies indicate that epigenetic modifications enable organisms to adaptively respond to environmental stressors like salinity fluctuations, ensuring both immediate and long-term survival by synchronizing physiological and metabolic processes across generations.

4. Microbiome Interactions and Response Under Salinity Stress

Host–microbiome interactions play a significant role in salinity tolerance as the gut microbiota play a role in metabolism, immune regulation, and disease resistance [145]. Changes in salinity can lead to shifts in the microbial community composition, often favoring opportunistic microbial taxa like *Proteobacteria* and *Vibrio* spp. under higher salinity conditions, while lower salinity levels tend to favor beneficial bacteria such as *Actinobacteria* and *Lactobacillus* [146]. These patterns are consistent across some species, as demonstrated in *Penaeus monodon*, where altered salinity destabilized gut microbial networks, elevated pathogenic potential (particularly *Vibrio*), and compromised immune functions, with 35 microbial biomarkers accurately diagnosing salinity stress [15]. Similarly, in fish species like *Gambusia* and *Paracheirodon sphenops*, salinity variations disrupted their gut microbiome, adversely impacting their health [147]. Meanwhile, gut microbiome of *Oxygymnocypris stewartia* did not show any microbiome change in response to stress [148], indicating a species-specific response of the microbiome under salinity stress.

The observed microbiome shifts under salinity stress may contribute to changes in salinity tolerance rather than being stress-induced. For example, in the case of *Nibea albiflora*, high salinity favors harmful *Vibrio* species, whereas intermediate salinity promotes probiotic taxa, indicating improved immunity and growth, showing that the microbiome is functionally adapted to salinity variations rather than merely a stress response [149]. Similarly, in *Coilia nasus*, salinity mitigation reduces harmful microbes, such as *Escherichia coli*, while enhancing good taxa such as *Actinobacteria* and *Corynebacterium*, which could be associated with improved health and stress tolerance under variable salinity conditions [150]. These studies indicate that microbiome shifts are more than just consequences of salinity stress; they may also play an active role in modifying the organism's ability to withstand such stress.

Salinity stress induces species-specific shifts in gut microbiota that mechanistically compromise gut health, immune function, and nutrient absorption in aquatic organisms. In *Micropterus salmoides*, elevated salinity increases *Bacillus* abundance, which may enhance osmolyte production and stress tolerance [151]. In contrast, *Salmo salar* exhibits distinct microbial profiles in freshwater (*Firmicutes/Actinobacteria*) versus seawater (*Proteobacteria*), with the latter associated with pro-inflammatory responses through TLR/NF-κB signaling [152]. These microbial changes directly impact gut barrier integrity, as demonstrated in *Macrobrachium rosenbergii*, where hypotonic stress (≤8‰) promotes pathogenic *Vibrio* and *Flavobacterium* proliferation while reducing antioxidant capacity (↓SOD, ↑MDA/NO). This oxidative stress downregulates tight junction proteins and elevates iNOS activity, impairing intestinal barrier function and increasing susceptibility to inflammation [17]. Similarly, hypo-salinity in *Sebastes schlegelii* reduces beneficial *Photobacterium* (SCFA producers) while increasing *Mycoplasma*, leading to bile acid dysmetabolism and downregulation of nutrient transporters (SGLT-1/PepT1). These changes, coupled with suppressed NKA activity, disrupt ion-dependent nutrient absorption and exacerbate metabolic inefficiency [153].

In *L. vannamei*, hypoosmotic conditions led to Vibrio overgrowth, reduced microbial richness, and altered lipopolysaccharide biosynthesis [16]. Similarly, in mud crabs, salinity fluctuations increased intestinal permeability, promoted pathogen translocation, and activated inflammatory pathways, exacerbating microbial imbalance [46]. Figure 4 demonstrates how salinity stress induces changes in the gut microbiota of aquatic organisms, leading to gut microbiota dysbiosis. Under fluctuating salinity conditions, the microbial composition in the gut becomes imbalanced, with beneficial microbes potentially decreasing and less beneficial or harmful microbes increasing. This dysbiosis disrupts the functional capabilities of the microbiome, impairing essential processes such as nutrient absorption, immune system regulation, and overall gut health. The figure highlights that these disruptions can compromise the organism's ability to adapt to salinity stress, potentially affecting its survival and physiological functions. Overall, this underscores the critical role of a balanced microbiome in maintaining the health and stress resilience of aquatic organisms exposed to changing salinity environments.

Functional studies reveal that the gut microbiota play an important role in fish osmoregulation via distinct pathways. In Medaka, microbial ecotone manufacturing in seawater-adapted fish was associated with increased Na^+/K^+-ATPase and cftr expression in gills following ecotone treatment, establishing a direct relationship between microbiota activity and osmoregulatory gene regulation. Under acute salinity stress, Nile tilapia exhibited microbial enrichment of osmoregulatory genes (myo-inositol-1-monophosphatase, glutathione S-transferase) and ABC transporters, while carbohydrate-active enzymes were depleted, indicating microbiota-mediated metabolic reprogramming for osmotic balance [154]. Transcriptomic investigations in teleosts validated microbiota's influence on ion-transport genes and osmolyte production pathways during salinity adaptation [155]. These studies provide evidence that gut microbiota plays a functional role in osmoregulation by regulating gene expression and producing metabolites under salinity stress.

At the molecular level, salinity-induced microbiome shifts modulate the expression of stress response genes, antimicrobial peptides, and the prophenoloxidase system, thereby influencing how invertebrates maintain intestinal homeostasis [156]. Under low salinity, structural proteins such as mucins and peritrophins are upregulated, fortifying the intestinal barrier and favoring beneficial taxa [157]. However, hypoosmotic conditions also elevate oxidative stress markers, which can damage intestinal tissues and further disrupt host–microbiome interactions [158]. Thus, salinity stress significantly alters host–microbiome interactions by shifting the microbial community composition, disrupting metabolic functions, and compromising immune and intestinal homeostasis, ultimately impacting host health and stress tolerance.

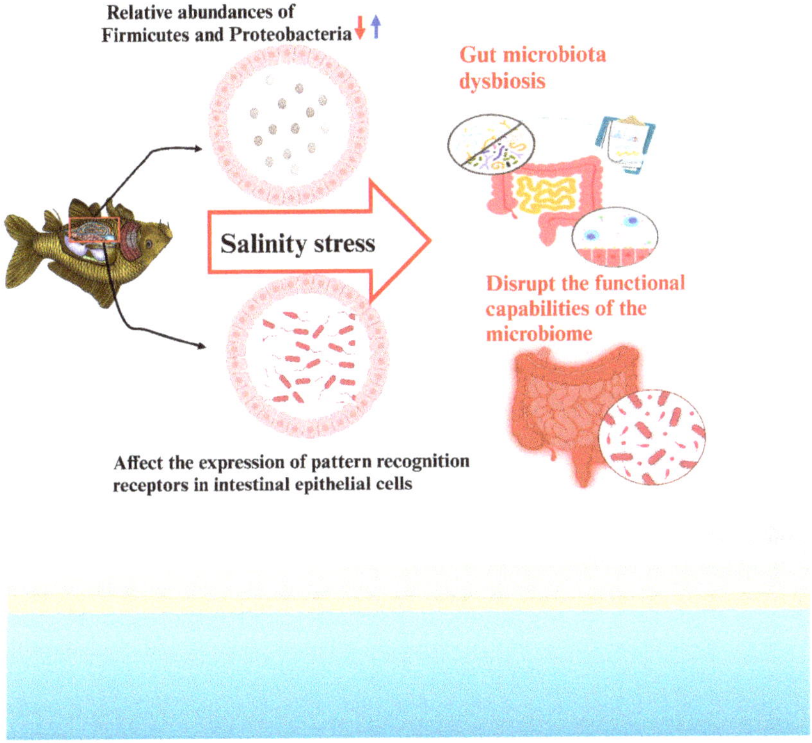

Figure 4. The impact of salinity stress on gut microbiota in fish, leading to dysbiosis, or imbalance in microbial communities. This dysbiosis can disrupt the functional capabilities of the microbiome, affecting the overall health and metabolic functions of the fish. The visual emphasizes the connection between environmental stressors and alterations in gut health.

Microbiome modifications owing to salinity stress in aquaculture can be efficiently addressed using probiotics or microbial consortia. The disturbance of gut microbiota under varied salinity conditions, resulting in lower species richness and increased disease susceptibility, can be addressed by introducing beneficial microbial communities. For example, microbial consortia containing ammonia-oxidizing bacteria, nitrite-oxidizing bacteria, and denitrifying bacteria have been shown to reduce total ammonia nitrogen levels in shrimp farms, improving water quality and reducing salinity stress [159]. Probiotics, on the other hand, have been shown to balance gut microbiota, boost immunological responses, and improve pathogen resistance in response to environmental stress [158]. These microbiome-based techniques provide a sustainable and cost-effective strategy to reduce salt stress, enhance shrimp health, and preserve productivity in aquaculture systems.

5. Genetic Diversity and Resilience of Aquatic Species

Genetic diversity is fundamental to aquatic organisms' resistance to environmental stresses, which include both chemical contaminants and abiotic factors such as salinity changes. In the crab *Portunus armatus*, genetic traits impose tolerance to elevated temperatures and salinity changes [160], whereas in *Chironomus riparius*, populations exposed to long-term mercury contamination developed elevated mercury tolerance without genetic erosion, implying directional selection for resistant traits. In addition, in both situations, communities with greater baseline genetic diversity performed better under stress conditions [161]. These findings highlight the dual significance of genetic diversity as a reservoir for adaptive traits under directional selection and as a barrier against environmental obstacles [61].

However, extreme variations in salinity can negatively impact genetic diversity. For example, shifts in salinity have been linked to noticeable changes in the genetic structure of *Artemia* populations, leading to a decline in their diversity [162]. This highlights the negative impact that drastic environmental changes can have on genetic stability, which in turn affects the resilience of populations. Despite this, species with higher genetic diversity tend to fare better under fluctuating conditions, underlining the importance of maintaining genetic variation for adaptive capacity [163].

Salinity also influences population structure and gene flow within aquatic species. For example, the salinity barrier created by the Amazon-Orinoco plume causes genetic differences between Atlantic fiddler crabs populations from the Caribbean and Brazil, showing how environmental factors can separate populations and limit gene flow [164]. On the other hand, the more adaptable *Callinectes danae* exhibits less genetic differentiation and more gene flow, indicating its ability to better cope with salinity fluctuations [165]. This comparison highlights the differential adaptability of species to salinity fluctuations, emphasizing the subsequent effects on genetic diversity.

Hybridization studies on tilapia show that hybrid fish have faster growth rates and greater tolerance to salinity fluctuations compared to their purebred counterparts, demonstrating how hybridization can increase genetic diversity and improve the ability to cope with environmental stress [166]. Similarly, *F. heteroclitus* exhibit significant genetic variations in genes associated with environmental stress responses, such as the aryl hydrocarbon receptor, which enhance their ability to tolerate normally lethal levels of stress [137]. In addition, *Esox lucius* fish populations have evolved flexible reproductive strategies, with certain fish species able to produce more embryos in saline environments than their freshwater counterparts, further enhancing their ability to survive in variable habitats [167]. These reproductive adaptations are another way in which genetic diversity supports resilience in dynamic environments. Future research should focus on evaluating the long-term evolutionary impacts of hybridization on population genetics, particularly the persistence of hybrid vigor and the potential for genetic assimilation over multiple generations.

6. Strategies for Mitigating Salinity Stress in Aquatic Organisms

Effective strategies to mitigate salinity stress in aqua cultural systems include nutritional interventions and the development of salinity-tolerant strains.

6.1. Nutritional Intervention

Dietary adjustments play a key role in enhancing osmoregulatory efficiency and stress resilience. For example, in *L. vannamei* protein requirements increase under higher salinities to optimize growth and osmoregulatory functions [168]. Supplementing with essential amino acids, such as methionine and tryptophan, further supports physiological stability under salinity stress [169]. In addition to proteins, lipid metabolism also supports energy demands for osmoregulation. Supplementing the diet with lipids enhances energy availability and osmoregulatory functions in shrimps, especially when bile salt production is insufficient [170]. Furthermore, n-3 polyunsaturated fatty acids (PUFAs, e.g., EPA and DHA) not only enhance growth and immune responses in aquatic organisms, but also contribute to stress resilience by maintaining membrane fluidity, reducing oxidative damage, and modulating inflammatory pathways crucial for counteracting osmotic stress [171]. Other key nutrients such as myo-inositol, choline, and vitamin D3 improve antioxidative function, thereby boost salinity stress resilience in *L. vannamei* [170]. For instance, dietary supplementation of vitamins E and C enhanced growth performance and salinity tolerance in *L.*

vannamei [172]. Long-term supplementation with nutrients including vitamins (vitamins C and E) and minerals (selenium) has been revealed to improve stress tolerance in aquatic animals, including higher osmoregulatory ability and immunological responses under chronic stress [173]. Using fish oil and soybean oil in the diet of mud crabs improves nutritional value, taste, and odor. Meanwhile, low salinity negatively affected taste and odor, indicating that the presence of fish oil might reduce these consequences [174]. Carotenoids, including astaxanthin, further enhance osmotic stress resistance in crustaceans, strengthening their ability to cope with salinity fluctuations [175]. Increased calcium levels shorten molt intervals and accelerate exoskeletal calcification, optimizing crab production [176]. Additionally, nucleotides supplementation has shown considerable effects on osmoregulatory competence and enhanced survival rates during salinity transitions in fish species such Atlantic salmon and red sea bream [177]. These results highlight the importance of a balanced diet, including proteins, lipids, vitamins, nucleotides, and minerals, in supporting osmoregulatory function and maintaining the health of aquatic species under fluctuating salinity conditions. In species like *O. niloticu*, salinity-induced morphological changes in the intestine can disrupt nutrient absorption, highlighting the need for a tailored amino acid profile to sustain protein digestibility [178]. Fish populations also require flexible reproductive strategies to thrive in saline environments. Supplementation with essential fatty acids, such as DHA, has been shown to improve salinity tolerance in fish larvae [179]. Furthermore, carbohydrates can act as protein-sparing agents under metabolic stress in fish, improving overall nutrient utilization. For example, diets containing 15–20% carbohydrates benefit fish in osmoregulatory processes [180].

In addition to traditional nutrients, commercial products such as Optimun (Chemoforma, Augst, Switzerland) have been found to reduce plasma cortisol and glucose levels in rainbow trout and red drum, thereby enhancing acute stress tolerance [181]. Vannagen™ (Chemoforma, Augst, Switzerland) has been shown to improve stress resilience in fish, with studies on sole confirming its efficacy in lowering stress-induced cortisol levels [182]. Another product, Maxi-Gen™ Plus Canadian Bio-Systems Inc. (Calgary, AB, Canada), has been successfully incorporated into the diets of Atlantic salmon to improve hypo-osmoregulatory ability and reduce cortisol levels during smoltification, thus increasing growth performance and stress resilience [183].

6.2. Development of Salinity-Resistant Varieties

Selective breeding has emerged as a promising strategy to enhance salinity tolerance in aquatic species. For instance, the Sukamandi strain of tilapia, a hybrid between *O. niloticus* and *O. aureus*, demonstrates increased salinity tolerance through gene introgression from *O. aureus*. Selective breeding for these traits has reinforced the salinity tolerance of this strain, making it better suited for environments with fluctuating salinity levels [184]. Similarly, selective breeding programs targeting *P. hypophthalmus* have successfully enhanced growth and survival in saline environments [185]. In shrimp aquaculture, the identification of genetic markers linked to salinity tolerance has facilitated the distinction between salinity-tolerant families and those more susceptible to environmental stress. This enables more efficient selection of brood stock with inherent salinity-resistant traits, improving the overall effectiveness of breeding programs aimed at enhancing resilience to salinity fluctuations [186].

However, intensive selective breeding for salinity tolerance entails risks that must be carefully analyzed. One important concern is the population's low genetic diversity, which may hinder the species' capacity to respond to other environmental stressors such as disease outbreaks or temperature variations [163]. Another major issue is inbreeding depression, which can cause decreases in vital traits including growth, reproduction, and overall fitness, rendering the population more vulnerable to various challenges [187]. Furthermore, intensive selective breeding may result in the loss of significant genetic features that are important for future breeding demands or in response to unexpected environmental changes [188]. These risks highlight the significance of balancing selective breeding efforts with techniques that preserve genetic diversity and ensure the population's long-term resilience.

To implement strategies for mitigating salinity stress while supporting biodiversity preservation in natural habitats, it is essential to balance interventions with ecological sustainability. Nutritional interventions and selective breeding should be carefully managed to avoid disrupting natural food webs and genetic diversity. For instance, while nutrient supplementation can enhance osmoregulatory functions in aquaculture [174], similar approaches in natural ecosystems should use native, ecologically appropriate resources to support local species without introducing non-native elements. Selective breeding efforts aimed at enhancing salinity tolerance should be paired with genetic studies to preserve wild genetic diversity, preventing the risks of inbreeding and loss of adaptability.

7. Shortcomings

Despite significant progress in understanding the effects of salinity stress, several research gaps remain that hinder a more comprehensive understanding. Notably, there is a lack of detailed species-specific responses to both acute and chronic salinity fluctuations, which are essential for understanding how different species cope with salinity stress over both short-term and long-term periods. Insufficient long-term ecological studies limit our ability to predict cumulative effects of salinity stress over time, and there is limited knowledge regarding the interactive effects of salinity with other environmental stressors such as temperature and pollution. Additionally, the molecular mechanisms underlying behavioral adaptations to salinity changes remain underexplored, and there is a lack of comparative data on ion-transport mechanisms across diverse species.

The time-dependent nature of salinity stress is especially important, as short-term exposure may lead to acute physiological responses, while long-term stress can trigger chronic adaptations, potentially altering species fitness and survival. Further investigation into the long-term effects and heritability of epigenetic changes in response to salinity stress is needed to assess their potential for transgenerational adaptation. Finally, the decline in genetic diversity, especially in the context of rapid climate change, calls for increased research into how genetic variation can influence species' ability to adapt to environmental changes. Addressing these shortcomings is vital for developing a more comprehensive understanding of salinity stress and for implementing effective strategies to protect aquatic ecosystems amid increasing environmental variability.

8. Conclusions

Salinity stress is a significant environmental factor that shapes aquatic ecosystems, influencing species distribution and altering biodiversity. The ability of species to adapt to fluctuations in salinity levels through diverse physiological, behavioral, and molecular mechanisms is crucial for maintaining ecosystem stability and resilience. Genetic diversity plays a key role in allowing populations to cope with changing salinity, ensuring long-term survival and adaptability. Furthermore, the intricate relationships between aquatic organisms and their microbiomes under salinity stress underscore the importance of microbial balance for overall health and resilience. Strategies such as nutritional interventions and the selective breeding of salinity-resistant strains show promise for enhancing the tolerance of aquaculture species to salinity stress, which also has implications for preserving biodiversity in natural habitats.

Author Contributions: Conceptualization: T.D. and H.M.; methodology: T.D. and H.M.; validation: T.D., W.C., M.I. and H.M.; formal analysis: T.D. and H.M.; investigation: T.D., W.C. and H.M.; data curation: T.D., W.C., M.I. and H.M.; writing—original draft preparation: T.D.; writing—review and editing: W.C., M.I. and H.M.; visualization: H.M.; supervision: H.M.; project administration: H.M.; funding acquisition: H.M. All authors have read and agreed to the published version of the manuscript.

Funding: This research was funded by the National Natural Science Foundation of China (32473150), the Program of Agricultural and Rural Department of Guangdong Province (2024-SPY-00-013), and the Science and Technology Project of Guangdong Province (STKJ202209029).

Institutional Review Board Statement: Not applicable.

Informed Consent Statement: Not applicable.

Data Availability Statement: No new data were created or analyzed in this study.

Conflicts of Interest: The authors declare no conflicts of interest.

References

1. Gattoni, K.; Gendron, E.M.S.; Borgmeier, A.; McQueen, J.P.; Mullin, P.G.; Powers, K.; Powers, T.O.; Porazinska, D.L. Context-Dependent Role of Abiotic and Biotic Factors Structuring Nematode Communities Along Two Environmental Gradients. *Mol. Ecol.* **2022**, *31*, 3903–3916. [CrossRef] [PubMed]
2. Brauner, C.J.; Gonzalez, R.J.; Wilson, J.M. Extreme Environments: Hypersaline, Alkaline, and Ion-Poor Waters. *Fish Physiol.* **2012**, *32*, 435–476.
3. Bal, A.; Panda, F.; Pati, S.G.; Das, K.; Agrawal, P.K.; Paital, B. Modulation of Physiological Oxidative Stress and Antioxidant Status by Abiotic Factors Especially Salinity in Aquatic Organisms: Redox Regulation under Salinity Stress. *Comp. Biochem. Physiol. Part—C Toxicol. Pharmacol.* **2021**, *241*, 108–941.
4. Seale, A.P.; Cao, K.; Chang, R.J.A.; Goodearly, T.R.; Malintha, G.H.T.; Merlo, R.S.; Peterson, T.L.; Reighard, J.R. Salinity Tolerance of Fishes: Experimental Approaches and Implications for Aquaculture Production. *Rev. Aquac.* **2024**, *16*, 1351–1373. [CrossRef]
5. Su, M.; Liu, N.; Zhang, Z.; Zhang, J. Osmoregulatory Strategies of Estuarine Fish *Scatophagus argus* in Response to Environmental Salinity Changes. *BMC Genom.* **2022**, *23*, 545. [CrossRef]
6. Williams, L. The Physiological and Behavioural Response of Mytilus Mussels to Salinity in a Changing Ocean. Ph.D. Thesis, University of Exeter, Exeter, UK, 2024.
7. Kroeker, K.J.; Sanford, E. Ecological Leverage Points: Species Interactions Amplify the Physiological Effects of Global Environmental Change in the Ocean. *Ann. Rev. Mar. Sci.* **2022**, *14*, 75–103. [CrossRef]
8. Röthig, T.; Trevathan-Tackett, S.M.; Voolstra, C.R.; Ross, C.; Chaffron, S.; Durack, P.J.; Warmuth, L.M.; Sweet, M. Human-induced Salinity Changes Impact Marine Organisms and Ecosystems. *Glob. Change Biol.* **2023**, *29*, 4731–4749. [CrossRef]
9. McNamara, J.C.; Faria, S.C. Evolution of Osmoregulatory Patterns and Gill Ion Transport Mechanisms in the Decapod Crustacea: A Review. *J. Comp. Physiol. B Biochem. Syst. Environ. Physiol.* **2012**, *182*, 997–1014. [CrossRef]
10. Anger, K. Neotropical *Macrobrachium* (Caridea: Palaemonidae):On the Biology, Origin, and Radiation of Freshwater-Invading Shrimp. *J. Crustac. Biol.* **2013**, *33*, 151–183. [CrossRef]
11. Xu, W.B.; Zhang, Y.M.; Li, B.Z.; Lin, C.Y.; Chen, D.Y.; Cheng, Y.X.; Guo, X.L.; Dong, W.R.; Shu, M.A. Effects of Low Salinity Stress on Osmoregulation and Gill Transcriptome in Different Populations of Mud Crab *Scylla paramamosain*. *Sci. Total Environ.* **2023**, *867*, 161522. [CrossRef]
12. Pini, J.; Planes, S.; Rochel, E.; Lecchini, D.; Fauvelot, C. Genetic Diversity Loss Associated to High Mortality and Environmental Stress during the Recruitment Stage of a Coral Reef Fish. *Coral Reefs* **2011**, *30*, 399–404. [CrossRef]
13. Leeuwis, R.H.J.; Gamperl, A.K. Adaptations and Plastic Phenotypic Responses of Marine Animals to the Environmental Challenges of the High Intertidal Zone. In *Oceanography and Marine Biology*; CRC Press: Boca Raton, FL, USA, 2022; pp. 625–679.
14. Rivera-Ingraham, G.A.; Barri, K.; Boël, M.; Farcy, E.; Charles, A.-L.; Geny, B.; Lignot, J.-H. Osmoregulation and Salinity-Induced Oxidative Stress: Is Oxidative Adaptation Determined by Gill Function? *J. Exp. Biol.* **2016**, *219*, 80–89. [CrossRef] [PubMed]
15. Chaiyapechara, S.; Uengwetwanit, T.; Arayamethakorn, S.; Bunphimpapha, P.; Phromson, M.; Jangsutthivorawat, W.; Tala, S.; Karoonuthaisiri, N.; Rungrassamee, W. Understanding the Host-Microbe-Environment Interactions: Intestinal Microbiota and Transcriptomes of Black Tiger Shrimp *Penaeus monodon* at Different Salinity Levels. *Aquaculture* **2022**, *546*, 737371. [CrossRef]
16. Chang, Y.-T.; Huang, W.-T.; Wu, P.-L.; Kumar, R.; Wang, H.-C.; Lu, H.-P. Low Salinity Stress Increases the Risk of *Vibrio parahaemolyticus* Infection and Gut Microbiota Dysbiosis in Pacific White Shrimp. *BMC Microbiol.* **2024**, *24*, 275. [CrossRef]

17. Liu, B.; Gao, Q.; Sun, C.; Song, C.; Liu, M.; Zhou, Q.; Zheng, X.; Liu, X. Response of Microbiota and Immune Function to Different Hypotonic Stress Levels in Giant Freshwater Prawn *Macrobrachium rosenbergii* Post-Larvae. *Sci. Total Environ.* **2022**, *844*, 157258. [CrossRef]
18. Wang, L.Y.; Gu, Y.Y.; Zhang, Z.M.; Sun, A.L.; Shi, X.Z.; Chen, J.; Lu, Y. Contaminant Occurrence, Mobility and Ecological Risk Assessment of Phthalate Esters in the Sediment-Water System of the Hangzhou Bay. *Sci. Total Environ.* **2021**, *770*, 144705. [CrossRef]
19. Tweedley, J.R.; Dittmann, S.R.; Whitfield, A.K.; Withers, K.; Hoeksema, S.D.; Potter, I.C. Hypersalinity: Global Distribution, Causes, and Present and Future Effects on the Biota of Estuaries and Lagoons. In *Coasts and Estuaries: The Future*; Elsevier: Amsterdam, The Netherlands, 2019; pp. 523–546.
20. Telesh, I.V.; Khlebovich, V.V. Principal Processes within the Estuarine Salinity Gradient: A Review. *Mar. Pollut. Bull.* **2010**, *61*, 149–155. [CrossRef]
21. Serrano, M.A.; Cobos, M.; Magaña, P.J.; Díez-Minguito, M. Sensitivity of Iberian Estuaries to Changes in Sea Water Temperature, Salinity, River Flow, Mean Sea Level, and Tidal Amplitudes. *Estuar. Coast. Shelf Sci.* **2020**, *236*, 106624. [CrossRef]
22. Slade, J.H.; VanReken, T.M.; Mwaniki, G.R.; Bertman, S.; Stirm, B.; Shepson, P.B. Aerosol Production from the Surface of the Great Lakes. *Geophys. Res. Lett.* **2010**, *37*, L18807. [CrossRef]
23. Abbas, M.S.; Yang, Y.; Zhang, Q.; Guo, D.; Godoi, A.F.L.; Godoi, R.H.M.; Geng, H. Salt Lake Aerosol Overview: Emissions, Chemical Composition and Health Impacts under the Changing Climate. *Atmosphere* **2024**, *15*, 212. [CrossRef]
24. Li, G.; Cheng, L.; Zhu, J.; Trenberth, K.E.; Mann, M.E.; Abraham, J.P. Increasing Ocean Stratification over the Past Half-Century. *Nat. Clim. Change* **2020**, *10*, 1116–1123. [CrossRef]
25. Hughes, D.J.; Alderdice, R.; Cooney, C.; Kühl, M.; Pernice, M.; Voolstra, C.R.; Suggett, D.J. Coral Reef Survival under Accelerating Ocean Deoxygenation. *Nat. Clim. Change* **2020**, *10*, 296–307. [CrossRef]
26. Matela, M.; Obolewski, K. Structural Diagnosis of Benthic Invertebrate Communities in Relation to Salinity Gradient in Baltic Coastal Lake Ecosystems Using Biological Trait Analysis. *Sci. Rep.* **2022**, *12*, 12750. [CrossRef]
27. Stickney, R.R. Tilapia Tolerance of Saline Waters: A Review. *Fish-Culturist* **1986**, *48*, 161–167. [CrossRef]
28. Vargas-Chacoff, L.; Regish, A.M.; Weinstock, A.; McCormick, S.D. Effects of Elevated Temperature on Osmoregulation and Stress Responses in Atlantic Salmon *Salmo salar* Smolts in Fresh Water and Seawater. *J. Fish Biol.* **2018**, *93*, 550–559. [CrossRef]
29. Ingersoll, C.G.; Dwyer, F.J.; Burch, S.A.; Nelson, M.K.; Buckler, D.R.; Hunn, J.B. The Use of Freshwater and Saltwater Animals to Distinguish between the Toxic Effects of Salinity and Contaminants in Irrigation Drain Water. *Environ. Toxicol. Chem.* **1992**, *11*, 503–511. [CrossRef]
30. Day, J.W.; Crump, B.C.; Kemp, W.M.; Yáñez-Arancibia, A. Introduction to Estuarine Ecology. *Estuar. Ecol.* **2012**, 1–18. [CrossRef]
31. Iglesias, M.C.-A. A Review of Recent Advances and Future Challenges in Freshwater Salinization. *Limnetica* **2020**, *39*, 185–211.
32. Zhang, C.; Zhou, L.; Ye, J.; Wang, L.; Huang, K.; Zhai, Q. Effects of Acute Salinity Stress on the Serum Osmolality, Serum Ion Concentrations, and ATPase Activity in Gill Filaments of Japanese Seabass (*Lateolabrax japonicus*) Fed with Diets Containing Different Magnesium Levels. *J. Fish. China* **2012**, *36*, 1425. [CrossRef]
33. Perschbacher, P.W. Historical Use of Tilapia in Intensive Co-culture. In *Tilapia in Intensive Co-Culture*; Wiley-Blackwell: Hoboken, NJ, USA, 2017; pp. 36–49.
34. Yue, G.H.; Ma, K.Y.; Xia, J.H. Status of Conventional and Molecular Breeding of Salinity-tolerant Tilapia. *Rev. Aquac.* **2024**, *16*, 271–286. [CrossRef]
35. Getz, E.; Eckert, C. Effects of Salinity on Species Richness and Community Composition in a Hypersaline Estuary. *Estuaries Coasts* **2023**, *46*, 2175–2189. [CrossRef]
36. Balushkina, E.V.; Golubkov, S.M.; Golubkov, M.S.; Litvinchuk, L.F.; Shadrin, N.V. Effect of Abiotic and Biotic Factors on the Structural and Functional Organization of the Saline Lake Ecosystems. *Zh. Obshch. Biol.* **2009**, *70*, 504–514. [PubMed]
37. Golubkov, S.M.; Shadrin, N.V.; Golubkov, M.S.; Balushkina, E.V.; Litvinchuk, L.F. Food Chains and Their Dynamics in Ecosystems of Shallow Lakes with Different Water Salinities. *Russ. J. Ecol.* **2018**, *49*, 442–448. [CrossRef]
38. Mohebbi, F. The Brine Shrimp *Artemia* and Hypersaline Environments Microalgal Composition: A Mutual Interaction. *Int. J. Aquat. Sci.* **2010**, *1*, 19–27.
39. Saccò, M.; White, N.E.; Harrod, C.; Salazar, G.; Aguilar, P.; Cubillos, C.F.; Meredith, K.; Baxter, B.K.; Oren, A.; Anufriieva, E. Salt to Conserve: A Review on the Ecology and Preservation of Hypersaline Ecosystems. *Biol. Rev.* **2021**, *96*, 2828–2850. [CrossRef]
40. Redón, S.; Vasileva, G.P.; Georgiev, B.B.; Gajardo, G. Exploring Parasites in Extreme Environments of High Conservational Importance: *Artemia franciscana* (Crustacea: Branchiopoda) as Intermediate Host of Avian Cestodes in Andean Hypersaline Lagoons from Salar de Atacama, Chile. *Parasitol. Res.* **2020**, *119*, 3377–3390. [CrossRef]
41. Strang, C.I.; Bosker, T. The Impact of Climate Change on Marine Mega-Decapod Ranges: A Systematic Literature Review. *Fish. Res.* **2024**, *280*, 107165. [CrossRef]

42. Wu, F.; Xie, Z.; Lan, Y.; Dupont, S.; Sun, M.; Cui, S.; Huang, X.; Huang, W.; Liu, L.; Hu, M.; et al. Short-Term Exposure of *Mytilus Coruscus* to Decreased PH and Salinity Change Impacts Immune Parameters of Their Haemocytes. *Front. Physiol.* **2018**, *9*, 166. [CrossRef]
43. Kelly, M.W.; DeBiasse, M.B.; Villela, V.A.; Roberts, H.L.; Cecola, C.F. Adaptation to Climate Change: Trade-offs among Responses to Multiple Stressors in an Intertidal Crustacean. *Evol. Appl.* **2016**, *9*, 1147–1155. [CrossRef]
44. Morales-Marín, L.A.; Rokaya, P.; Sanyal, P.R.; Sereda, J.; Lindenschmidt, K.E. Changes in Streamflow and Water Temperature Affect Fish Habitat in the Athabasca River Basin in the Context of Climate Change. *Ecol. Modell.* **2019**, *407*, 108718. [CrossRef]
45. Mitra, A.; Abdel-Gawad, F.K.; Bassem, S.; Barua, P.; Assisi, L.; Parisi, C.; Temraz, T.A.; Vangone, R.; Kajbaf, K.; Kumar, V. Climate Change and Reproductive Biocomplexity in Fishes: Innovative Management Approaches towards Sustainability of Fisheries and Aquaculture. *Water* **2023**, *15*, 725. [CrossRef]
46. Sohaib Ahmed Saqib, H.; Yuan, Y.; Shabi Ul Hassan Kazmi, S.; Li, S.; Zheng, H.; Zhang, Y.; Ikhwanuddin, M.; Ma, H. Salinity Gradients Drove the Gut and Stomach Microbial Assemblages of Mud Crabs (*Scylla paramamosain*) in Marine Environments. *Ecol. Indic.* **2023**, *151*, 110315. [CrossRef]
47. Shields, J.D. Climate Change Enhances Disease Processes in Crustaceans: Case Studies in Lobsters, Crabs, and Shrimps. *J. Crustac. Biol.* **2019**, *39*, 673–683. [CrossRef]
48. Gibson, D.; Riecke, T.V.; Catlin, D.H.; Hunt, K.L.; Weithman, C.E.; Koons, D.N.; Karpanty, S.M.; Fraser, J.D. Climate Change and Commercial Fishing Practices Codetermine Survival of a Long-lived Seabird. *Glob. Change Biol.* **2023**, *29*, 324–340. [CrossRef]
49. Hakim, M.M.; Ganai, N.A.; Ahmad, S.M.; Asmi, O.; Akram, T.; Hussain, S.; Gora, A.H. Nutrigenomics: Omics Approach in Aquaculture Research to Mitigate the Deficits in Conventional Nutritional Practices. *J. Entomol. Zool. Stud.* **2018**, *6*, 582–587.
50. Yoshizaki, G.; Yazawa, R. Application of Surrogate Broodstock Technology in Aquaculture. *Fish. Sci.* **2019**, *85*, 429–437. [CrossRef]
51. Evans, T.G.; Kültz, D. The Cellular Stress Response in Fish Exposed to Salinity Fluctuations. *J. Exp. Zool. Part A Ecol. Integr. Physiol.* **2020**, *333*, 421–435. [CrossRef]
52. Takvam, M.; Wood, C.M.; Kryvi, H.; Nilsen, T.O. Ion Transporters and Osmoregulation in the Kidney of Teleost Fishes as a Function of Salinity. *Front. Physiol.* **2021**, *12*, 664588. [CrossRef]
53. Deng, Z.; Zhang, Z.; Zhao, R.; Sun, J.; Hao, P.; Zhang, L.; Wang, X.; Cui, Y.; Liu, F.; Wang, R. Effects of High-Salinity on the Expression of Aquaporins and Ion Transport-Related Genes in Chinese Shrimp (*Fenneropenaeus chinensis*). *Aquac. Rep.* **2023**, *30*, 101577. [CrossRef]
54. Wang, Z.; Chen, Y.; Wang, C.; Zhao, N.; Zhang, Z.; Deng, Z.; Cui, Y.; Wang, R.; Li, Y. Aquaporins in Pacific White Shrimp (*Litopenaeus vannamei*): Molecular Characterization, Expression Patterns, and Transcriptome Analysis in Response to Salinity Stress. *Front. Mar. Sci.* **2022**, *9*, 817868. [CrossRef]
55. Jaffer, Y.D.; Bhat, I.A.; Mir, I.N.; Bhat, R.A.H.; Sidiq, M.J.; Jana, P. Adaptation of Cultured Decapod Crustaceans to Changing Salinities: Physiological Responses, Molecular Mechanisms and Disease Implications. *Rev. Aquac.* **2024**, *16*, 1520–1543. [CrossRef]
56. Mohammed-Geba, K.; El-Nab, S.; Awad, E.; Nofal, A.I. DNA Barcoding Identifies a Unique Haplotype of Nile Tilapia *Oreochromis niloticus* Thriving in Egyptian Freshwater and Brackish Water Lakes. *Int. J. Ecotoxicol. Ecobiol.* **2017**, *2*, 172–177. [CrossRef]
57. Dawood, M.; Madkour, K.; Mugwanya, M.; Kimera, F.; Sewilam, H. Growth Performances, Body Composition, and Blood Variables of Nile Tilapia (*Oreochromis niloticus*) Juveniles Grown in a Recirculating Aquaculture System under Different Water Salinity Levels. *J. Appl. Aquac.* **2024**, *36*, 612–626. [CrossRef]
58. Losee, J.P.; Palm, D.; Claiborne, A.; Madel, G.; Persson, L.; Quinn, T.P.; Brodin, T.; Hellström, G. Anadromous Trout from Opposite Sides of the Globe: Biology, Ocean Ecology, and Management of Anadromous Brown and Cutthroat Trout. *Rev. Fish Biol. Fish.* **2024**, *34*, 461–490. [CrossRef]
59. Yancey, P.H.; Siebenaller, J.F. Co-Evolution of Proteins and Solutions: Protein Adaptation versus Cytoprotective Micromolecules and Their Roles in Marine Organisms. *J. Exp. Biol.* **2015**, *218*, 1880–1896. [CrossRef]
60. Glover, C.N.; Wood, C.M.; Goss, G.G. Drinking and Water Permeability in the Pacific Hagfish. *Eptatretus stoutii. J. Comp. Physiol. B* **2017**, *187*, 1127–1135. [CrossRef]
61. Herrera, I.; de Carvalho-Souza, G.F.; González-Ortegón, E. Physiological Responses of the Invasive Blue Crabs *Callinectes Sapidus* to Salinity Variations: Implications for Adaptability and Invasive Success. *Comp. Biochem. Physiol. Part A Mol. Integr. Physiol.* **2024**, *297*, 111709. [CrossRef]
62. Gutiérrez, J.S. Living in Environments with Contrasting Salinities: A Review of Physiological and Behavioural Responses in Waterbirds. *Ardeola* **2014**, *61*, 233–256. [CrossRef]
63. Scarlett, K.R.; Lovin, L.M.; Steele, W.B.; Kim, S.; Brooks, B.W. Identifying Behavioral Response Profiles of Two Common Larval Fish Models to a Salinity Gradient. *Arch. Environ. Contam. Toxicol.* **2022**, *83*, 180–192. [CrossRef]
64. Tietze, S.M.; Gerald, G.W. Trade-Offs between Salinity Preference and Antipredator Behaviour in the Euryhaline Sailfin Molly *Poecilia latipinna. J. Fish Biol.* **2016**, *88*, 1918–1931. [CrossRef]
65. Chow, S.C.; Ching, L.Y.; Wong, A.M.F.; Wong, C.K.C. Cloning and Regulation of Expression of the Na^{+}-Cl^{-}-Taurine Transporter in Gill Cells of Freshwater Japanese Eels. *J. Exp. Biol.* **2009**, *212*, 3205–3210. [CrossRef] [PubMed]

66. Edwards, S.L.; Marshall, W.S. Principles and Patterns of Osmoregulation and Euryhalinity in Fishes. *Fish Physiol.* **2012**, *32*, 1–44.
67. Breves, J.P.; Hasegawa, S.; Yoshioka, M.; Fox, B.K.; Davis, L.K.; Lerner, D.T.; Takei, Y.; Hirano, T.; Grau, E.G. Acute Salinity Challenges in Mozambique and Nile Tilapia: Differential Responses of Plasma Prolactin, Growth Hormone and Branchial Expression of Ion Transporters. *Gen. Comp. Endocrinol.* **2010**, *167*, 135–142. [CrossRef]
68. Yang, Y.; Ni, J.; Niu, D.; Zheng, G.; Li, Y. Physiological Response of the Razor Clam *Sinonovacula constricta* Exposed to Hyposalinity Stress. *Aquac. Fish.* **2024**, *9*, 663–673. [CrossRef]
69. Yang, Z.; Zhu, L.; Zhao, X.; Cheng, Y. Effects of Salinity Stress on Osmotic Pressure, Free Amino Acids, and Immune-Associated Parameters of the Juvenile Chinese Mitten Crab. *Eriocheir sinensis. Aquaculture* **2022**, *549*, 737776. [CrossRef]
70. Zikos, A.; Seale, A.P.; Lerner, D.T.; Grau, E.G.; Korsmeyer, K.E. Effects of Salinity on Metabolic Rate and Branchial Expression of Genes Involved in Ion Transport and Metabolism in Mozambique Tilapia (*Oreochromis mossambicus*). *Comp. Biochem. Physiol. Part A Mol. Integr. Physiol.* **2014**, *178*, 121–131. [CrossRef]
71. Scott, G.R.; Richards, J.G.; Forbush, B.; Isenring, P.; Schulte, P.M. Changes in Gene Expression in Gills of the Euryhaline Killifish *Fundulus heteroclitus* after Abrupt Salinity Transfer. *Am. J. Physiol.—Cell Physiol.* **2004**, *287*, C300–C309. [CrossRef]
72. Zhou, T.; Meng, Q.; Sun, R.; Xu, D.; Zhu, F.; Jia, C.; Zhou, S.; Chen, S.; Yang, Y. Structure and Gene Expression Changes of the Gill and Liver in Juvenile Black Porgy (*Acanthopagrus schlegelii*) under Different Salinities. *Comp. Biochem. Physiol. Part D Genom. Proteom.* **2024**, *50*, 101228. [CrossRef]
73. Anufriieva, E.V.; Shadrin, N.V. General Patterns of Salinity Influence on the Energy Balance of Aquatic Animals in Hypersaline Environment. *Biol. Bull. Rev.* **2023**, *13*, 420–430. [CrossRef]
74. Hasan-Nataj-Niazi, E.; Agh, N.; Noori, F.; Atashbar, B.; Van Stappen, G. Substitution of Microalgae by Bioflocs as a Food Source for the Brine Shrimp *Artemia Franciscana. Aquac. Res.* **2022**, *53*, 4374–4387. [CrossRef]
75. Ern, R.; Esbaugh, A.J. Effects of Salinity and Hypoxia-Induced Hyperventilation on Oxygen Consumption and Cost of Osmoregulation in the Estuarine Red Drum (*Sciaenops ocellatus*). *Comp. Biochem. Physiol. Part A Mol. Integr. Physiol.* **2018**, *222*, 52–59. [CrossRef] [PubMed]
76. Lee, C.E.; Moss, W.E.; Olson, N.; Chau, K.F.; Chang, Y.; Johnson, K.E. Feasting in Fresh Water: Impacts of Food Concentration on Freshwater Tolerance and the Evolution of Food × Salinity Response during the Expansion from Saline into Fresh Water Habitats. *Evol. Appl.* **2013**, *6*, 673–689. [CrossRef] [PubMed]
77. Waqas, W.; Yuan, Y.; Ali, S.; Zhang, M.; Shafiq, M.; Ali, W.; Chen, Y.; Xiang, Z.; Chen, R.; Ikhwanuddin, M.; et al. Toxic Effects of Heavy Metals on Crustaceans and Associated Health Risks in Humans: A Review. *Environ. Chem. Lett.* **2024**, *22*, 1391–1411. [CrossRef]
78. Bal, A.; Panda, F.; Pati, S.G.; Anwar, T.N.; Das, K.; Paital, B. Influence of Anthropogenic Activities on Redox Regulation and Oxidative Stress Responses in Different Phyla of Animals in Coastal Water via Changing in Salinity. *Water* **2022**, *14*, 4026. [CrossRef]
79. Lal, M.; Kumari, A.; Pooja; Sheokand, S. Reactive Oxygen Species, Reactive Nitrogen Species and Oxidative Metabolism under Waterlogging Stress. In *Reactive Oxygen, Nitrogen and Sulfur Species in Plants: Production, Metabolism, Signaling and Defense Mechanisms*; Wiley: Hoboken, NJ, USA, 2019; pp. 777–812.
80. Hoseinifar, S.H.; Yousefi, S.; Van Doan, H.; Ashouri, G.; Gioacchini, G.; Maradonna, F.; Carnevali, O. Oxidative Stress and Antioxidant Defense in Fish: The Implications of Probiotic, Prebiotic, and Synbiotics. *Rev. Fish. Sci. Aquac.* **2020**, *29*, 198–217. [CrossRef]
81. Paital, B.; Chainy, G.B.N. Antioxidant Defenses and Oxidative Stress Parameters in Tissues of Mud Crab (*Scylla serrata*) with Reference to Changing Salinity. *Comp. Biochem. Physiol.—C Toxicol. Pharmacol.* **2010**, *151*, 142–151. [CrossRef]
82. Shen, M.; Wang, Y.; Tang, Y.; Zhu, F.; Jiang, J.; Zhou, J.; Li, Q.; Meng, Q.; Zhang, Z. Effects of Different Salinity Reduction Intervals on Osmoregulation, Anti-Oxidation and Apoptosis of *Eriocheir sinensis* Megalopa. *Comp. Biochem. Physiol. Part A Mol. Integr. Physiol.* **2024**, *291*, 111593. [CrossRef]
83. Lee, M.-C.; Park, J.C.; Kim, D.-H.; Kang, S.; Shin, K.-H.; Park, H.G.; Han, J.; Lee, J.-S. Interrelationship of Salinity Shift with Oxidative Stress and Lipid Metabolism in the Monogonont Rotifer *Brachionus Koreanus. Comp. Biochem. Physiol. Part A Mol. Integr. Physiol.* **2017**, *214*, 79–84. [CrossRef]
84. Fuhrmann, M.; Delisle, L.; Petton, B.; Corporeau, C.; Pernet, F. Metabolism of the Pacific Oyster, *Crassostrea gigas*, Is Influenced by Salinity and Modulates Survival to the Ostreid Herpesvirus OsHV-1. *Biol. Open* **2018**, *7*, bio028134. [CrossRef]
85. Liu, Y.; Wang, W.N.; Wang, A.L.; Wang, J.M.; Sun, R.Y. Effects of Dietary Vitamin E Supplementation on Antioxidant Enzyme Activities in *Litopenaeus vannamei* (Boone, 1931) Exposed to Acute Salinity Changes. *Aquaculture* **2007**, *265*, 351–358. [CrossRef]
86. Wang, X.; Yao, Q.; Zhang, D.M.; Lei, X.Y.; Wang, S.; Wan, J.W.; Liu, H.J.; Chen, Y.K.; Zhao, Y.L.; Wang, G.Q.; et al. Effects of Acute Salinity Stress on Osmoregulation, Antioxidant Capacity and Physiological Metabolism of Female Chinese Mitten Crabs (*Eriocheir sinensis*). *Aquaculture* **2022**, *552*, 737989. [CrossRef]

87. Xiao, Y.; Zhang, Y.M.; Xu, W.B.; Chen, D.Y.; Li, B.W.; Cheng, Y.X.; Guo, X.L.; Dong, W.R.; Shu, M.A. The Effects of Salinities Stress on Histopathological Changes, Serum Biochemical Index, Non-Specific Immune and Transcriptome Analysis in Red Swamp Crayfish *Procambarus clarkii*. *Sci. Total Environ.* **2022**, *840*, 156502. [CrossRef] [PubMed]
88. Boonsanit, P.; Pairohakul, S. Effects of Salinity on Haemolymph Osmolality, Gill Na^+/K^+ ATPase and Antioxidant Enzyme Activities in the Male Mud Crab *Scylla olivacea* (Herbst, 1796). *Mar. Biol. Res.* **2021**, *17*, 86–97. [CrossRef]
89. Mozanzadeh, M.T.; Safari, O.; Oosooli, R.; Mehrjooyan, S.; Najafabadi, M.Z.; Hoseini, S.J.; Saghavi, H.; Monem, J. The Effect of Salinity on Growth Performance, Digestive and Antioxidant Enzymes, Humoral Immunity and Stress Indices in Two Euryhaline Fish Species: Yellowfin Seabream (*Acanthopagrus latus*) and Asian Seabass (*Lates calcarifer*). *Aquaculture* **2021**, *534*, 736329. [CrossRef]
90. Dawood, M.A.O.; Noreldin, A.E.; Sewilam, H. Long Term Salinity Disrupts the Hepatic Function, Intestinal Health, and Gills Antioxidative Status in Nile Tilapia Stressed with Hypoxia. *Ecotoxicol. Environ. Saf.* **2021**, *220*, 112412. [CrossRef]
91. Chang, C.H.; Mayer, M.; Rivera-Ingraham, G.; Blondeau-Bidet, E.; Wu, W.Y.; Lorin-Nebel, C.; Lee, T.H. Effects of Temperature and Salinity on Antioxidant Responses in Livers of Temperate (*Dicentrarchus labrax*) and Tropical (*Chanos chanos*) Marine Euryhaline Fish. *J. Therm. Biol.* **2021**, *99*, 103016. [CrossRef]
92. Dawood, M.A.O.; Alkafafy, M.; Sewilam, H. The Antioxidant Responses of Gills, Intestines and Livers and Blood Immunity of Common Carp (*Cyprinus carpio*) Exposed to Salinity and Temperature Stressors. *Fish Physiol. Biochem.* **2022**, *48*, 397–408. [CrossRef]
93. Park, M.S.; Shin, H.S.; Choi, C.Y.; Kim, N.N.; Park, D.W.; Kil, G.S.; Lee, J. Effect of Hypoosmotic and Thermal Stress on Gene Expression and the Activity of Antioxidant Enzymes in the Cinnamon Clownfish, *Amphiprion melanopus*. *Anim. Cells Syst.* **2011**, *15*, 219–225. [CrossRef]
94. Andreyeva, A.; Gostyukhina, O.; Gavruseva, T.; Sigacheva, T.; Tkachuk, A.; Podolskaya, M.; Chelebieva, E.; Kladchenko, E. Mediterranean Mussels (*Mytilus galloprovincialis*) Under Salinity Stress: Effects on Antioxidant Capacity and Gill Structure. *J. Exp. Zool. Part A Ecol. Integr. Physiol.* **2024**, *343*, 184–196. [CrossRef]
95. Bayir, A.; Sirkecioglu, A.N.; Bayir, M.; Haliloglu, H.I.; Kocaman, E.M.; Aras, N.M. Metabolic Responses to Prolonged Starvation, Food Restriction, and Refeeding in the Brown Trout, *Salmo trutta*: Oxidative Stress and Antioxidant Defenses. *Comp. Biochem. Physiol.—B Biochem. Mol. Biol.* **2011**, *159*, 191–196. [CrossRef]
96. Xu, Z.; Gan, L.; Li, T.; Xu, C.; Chen, K.; Wang, X.; Qin, J.G.; Chen, L.; Li, E. Transcriptome Profiling and Molecular Pathway Analysis of Genes in Association with Salinity Adaptation in Nile Tilapia *Oreochromis niloticus*. *PLoS ONE* **2015**, *10*, e0136506. [CrossRef] [PubMed]
97. Yang, Y.; Ma, Q.; Jin, S.; Huang, B.; Wang, Z.; Chen, G. Identification of Mapk Genes, and Their Expression Profiles in Response to Low Salinity Stress, in Cobia (*Rachycentron canadum*). *Comp. Biochem. Physiol. Part—B Biochem. Mol. Biol.* **2024**, *271*, 110950. [CrossRef] [PubMed]
98. Duan, Y.; Zhang, W.; Chen, X.; Wang, M.; Zhong, L.; Liu, J.; Bian, W.; Zhang, S. Genome-Wide Identification and Expression Analysis of Mitogen-Activated Protein Kinase (MAPK) Genes in Response to Salinity Stress in Channel Catfish (*Ictalurus punctatus*). *J. Fish Biol.* **2022**, *101*, 972–984. [CrossRef]
99. Chu, P.; Luo, S.; Wang, H.; Zhang, K.; Wen, X.; Yin, S.; Wang, T. Interactive Effects of Temperature and Salinity on the Apoptosis, Antioxidant Enzymes, and MAPK Signaling Pathway of Juvenile Pufferfish (*Takifugu fasciatus*). *Aquac. Rep.* **2023**, *29*, 101483. [CrossRef]
100. Jin, M.; Luo, J.; Zhu, T.; Fang, F.; Xie, S.; Lu, J.; Betancor, M.B.; Tocher, D.R.; Zhou, Q. Examination of Role of the AMP-Activated Protein Kinase (Ampk) Signaling Pathway during Low Salinity Adaptation in the Mud Crab, *Scylla paramamosain*, with Reference to Glucolipid Metabolism. *Aquaculture* **2024**, *593*, 741329. [CrossRef]
101. Cui, W.; Ma, A.; Wang, X. Response of the PI3K-AKT Signalling Pathway to Low Salinity and the Effect of Its Inhibition Mediated by Wortmannin on Ion Channels in Turbot *Scophthalmus maximus*. *Aquac. Res.* **2020**, *51*, 2676–2686. [CrossRef]
102. Xu, C.; Li, E.; Xu, Z.; Wang, S.; Chen, K.; Wang, X.; Li, T.; Qin, J.G.; Chen, L. Molecular Characterization and Expression of AMP-Activated Protein Kinase in Response to Low-Salinity Stress in the Pacific White Shrimp *Litopenaeus vannamei*. *Comp. Biochem. Physiol. Part—B Biochem. Mol. Biol.* **2016**, *198*, 79–90. [CrossRef]
103. Farhadi, A.; Liu, Y.; Xu, C.; Wang, X.; Li, E. The Role of the Renin-Angiotensin System (RAS) in Salinity Adaptation in Pacific White Shrimp (*Litopenaeus vannamei*). *Front. Endocrinol.* **2022**, *13*, 1089419. [CrossRef]
104. Wang, H.; Tang, L.; Wei, H.; Lu, J.; Mu, C.; Wang, C. Transcriptomic Analysis of Adaptive Mechanisms in Response to Sudden Salinity Drop in the Mud Crab, *Scylla paramamosain*. *BMC Genom.* **2018**, *19*, 421. [CrossRef]
105. Li, Y.D.; Si, M.R.; Jiang, S.G.; Yang, Q.B.; Jiang, S.; Yang, L.S.; Huang, J.H.; Chen, X.; Zhou, F.L.; Li, E.C. Transcriptome and Molecular Regulatory Mechanisms Analysis of Gills in the Black Tiger Shrimp *Penaeus monodon* under Chronic Low-Salinity Stress. *Front. Physiol.* **2023**, *14*, 1118341. [CrossRef]

106. Luo, Z.; Huang, W.; Wang, G.; Sun, H.; Chen, X.; Luo, P.; Liu, J.; Hu, C.; Li, H.; Shu, H. Identification and Characterization of P38MAPK in Response to Acute Cold Stress in the Gill of Pacific White Shrimp (*Litopenaeus vannamei*). *Aquac. Rep.* **2020**, *17*, 100365. [CrossRef]
107. Zheng, W.; Xu, X.W.; Zechen, E.; Liu, Y.; Chen, S. Genome-Wide Identification of the MAPK Gene Family in Turbot and Its Involvement in Abiotic and Biotic Stress Responses. *Front. Mar. Sci.* **2022**, *9*, 1005401. [CrossRef]
108. Lee, K.; Eun, H.S.; Ho, S.K.; Jae, H.Y.; Hay, J.H.; Mi, S.J.; Sang, M.L.; Kyung, E.K.; Min, C.K.; Moo, J.C.; et al. Regulation of MAPK Phosphatase 1 (AtMKP1) by Calmodulin in Arabidopsis. *J. Biol. Chem.* **2008**, *283*, 23581–23588. [CrossRef] [PubMed]
109. Tian, Y.; Wen, H.; Qi, X.; Zhang, X.; Li, Y. Identification of Mapk Gene Family in *Lateolabrax maculatus* and Their Expression Profiles in Response to Hypoxia and Salinity Challenges. *Gene* **2019**, *684*, 20–29. [CrossRef]
110. Tao, S.; Li, X.; Wang, J.; Bai, Y.; Wang, J.; Yang, Y.; Zhao, Z. Examination of the Relationship of Carbonate Alkalinity Stress and Ammonia Metabolism Disorder-Mediated Apoptosis in the Chinese Mitten Crab, *Eriocheir sinensis*: Potential Involvement of the ROS/MAPK Signaling Pathway. *Aquaculture* **2024**, *579*, 740179. [CrossRef]
111. Gao, G.; Liang, G.; Wei, H.; Li, X.; Chen, Y.; Wang, H.; Wang, C.; Mu, C.; Liu, S. Involvement of the JNK Signaling Pathway in Regulating Yolk Accumulation in the Swimming Crab, *Portunus trituberculatus*. *Aquaculture* **2022**, *551*, 737890. [CrossRef]
112. Yu, J.S.L.; Cui, W. Proliferation, Survival and Metabolism: The Role of PI3K/AKT/MTOR Signalling in Pluripotency and Cell Fate Determination. *Development* **2016**, *143*, 3050–3060. [CrossRef]
113. Le, O.T.T.; Nguyen, T.T.N.; Lee, S.Y. Phosphoinositide Turnover in Toll-like Receptor Signaling and Trafficking. *BMB Rep.* **2014**, *47*, 361–368. [CrossRef]
114. Hu, Y.-C.; Chu, K.-F.; Yang, W.-K.; Lee, T.-H. Na+, K+-ATPase B1 Subunit Associates with A1 Subunit Modulating a "Higher-NKA-in-Hyposmotic Media" Response in Gills of Euryhaline Milkfish, *Chanos chanos*. *J. Comp. Physiol. B* **2017**, *187*, 995–1007. [CrossRef]
115. Liu, Y.; Li, E.; Xu, C.; Su, Y.; Qin, J.G.; Chen, L.; Wang, X. Brain Transcriptome Profiling Analysis of Nile Tilapia (*Oreochromis niloticus*) under Long-Term Hypersaline Stress. *Front. Physiol.* **2018**, *9*, 219. [CrossRef]
116. Hardie, D.G.; Ross, F.A.; Hawley, S.A. AMPK: A Nutrient and Energy Sensor That Maintains Energy Homeostasis. *Nat. Rev. Mol. Cell Biol.* **2012**, *13*, 251–262. [CrossRef] [PubMed]
117. Chen, K.; Li, E.; Li, T.; Xu, C.; Wang, X.; Lin, H.; Qin, J.G.; Chen, L. Transcriptome and Molecular Pathway Analysis of the Hepatopancreas in the Pacific White Shrimp *Litopenaeus vannamei* under Chronic Low-Salinity Stress. *PLoS ONE* **2015**, *10*, e0131503. [CrossRef] [PubMed]
118. Lu, Y.; Nie, M.; Wang, L.; Xiong, Y.; Wang, F.; Wang, L.; Xiao, P.; Wu, Z.; Liu, Y.; You, F. Energy Response and Modulation of AMPK Pathway of the Olive Flounder *Paralichthys olivaceus* in Low-Temperature Challenged. *Aquaculture* **2018**, *484*, 205–213. [CrossRef]
119. Wu, P.; Zhang, L.; Jiang, W.; Liu, Y.; Jiang, J.; Kuang, S.; Li, S.; Tang, L.; Tang, W.; Zhou, X.; et al. Dietary Vitamin A Improved the Flesh Quality of Grass Carp (*Ctenopharyngodon idella*) in Relation to the Enhanced Antioxidant Capacity through Nrf2/Keap 1a Signaling Pathway. *Antioxidants* **2022**, *11*, 148. [CrossRef]
120. Misra, J.R.; Irvine, K.D. The Hippo Signaling Network and Its Biological Functions. *Annu. Rev. Genet.* **2018**, *52*, 65–87. [CrossRef]
121. Yang, L.; Wang, Z.-A.; Geng, R.; Deng, H.; Niu, S.; Zuo, H.; Weng, S.; He, J.; Xu, X. White Spot Syndrome Virus (WSSV) Inhibits Hippo Signaling and Activates Yki To Promote Its Infection in *Penaeus vannamei*. *Microbiol. Spectr.* **2023**, *11*, e0236322. [CrossRef]
122. Liu, S.; Guo, J.; Cheng, X.; Li, W.; Lyu, S.; Chen, X.; Li, Q.; Wang, H. Molecular Evolution of Transforming Growth Factor-β (TGF-β) Gene Family and the Functional Characterization of Lamprey TGF-B2. *Front. Immunol.* **2022**, *13*, 836226. [CrossRef]
123. Riley, S.E.; Feng, Y.; Hansen, C.G. Hippo-Yap/Taz Signalling in Zebrafish Regeneration. *npj Regen. Med.* **2022**, *7*, 9. [CrossRef]
124. Huang, D.; Li, X.; Sun, L.; Huang, P.; Ying, H.; Wang, H.; Wu, J.; Song, H. Regulation of Hippo Signalling by P38 Signalling. *J. Mol. Cell Biol.* **2016**, *8*, 328–337. [CrossRef]
125. Meng, Z.; Moroishi, T.; Guan, K.-L. Mechanisms of Hippo Pathway Regulation. *Genes Dev.* **2016**, *30*, 1–17. [CrossRef]
126. Sun, Y.; Zhang, X.; Wang, G.; Lin, S.; Zeng, X.; Wang, Y.; Zhang, Z. PI3K-AKT Signaling Pathway Is Involved in Hypoxia/Thermal-Induced Immunosuppression of Small Abalone *Haliotis diversicolor*. *Fish Shellfish Immunol.* **2016**, *59*, 492–508. [CrossRef] [PubMed]
127. Li, Y.; Zhou, F.; Fan, H.; Jiang, S.; Yang, Q.; Huang, J.; Yang, L.; Chen, X.; Zhang, W.; Jiang, S. Molecular Technology for Isolation and Characterization of Mitogen-Activated Protein Kinase Kinase 4 from *Penaeus monodon*, and the Response to Bacterial Infection and Low-Salinity Challenge. *J. Mar. Sci. Eng.* **2022**, *10*, 1642. [CrossRef]
128. Li, X.; Shen, Y.; Bao, Y.; Wu, Z.; Yang, B.; Jiao, L.; Zhang, C.; Tocher, D.R.; Zhou, Q.; Jin, M. Physiological Responses and Adaptive Strategies to Acute Low-Salinity Environmental Stress of the Euryhaline Marine Fish Black Seabream (*Acanthopagrus schlegelii*). *Aquaculture* **2022**, *554*, 738117. [CrossRef]
129. Shang, X.; Xu, W.; Zhang, Y.; Sun, Q.; Li, Z.; Geng, L.; Teng, X. Transcriptome Analysis Revealed the Mechanism of *Luciobarbus capito* (*L. capito*) Adapting High Salinity: Antioxidant Capacity, Heat Shock Proteins, Immunity. *Mar. Pollut. Bull.* **2023**, *192*, 115017. [CrossRef]

130. Accili, D.; Arden, K.C. FoxOs at the Crossroads of Cellular Metabolism, Differentiation, and Transformation. *Cell* **2004**, *117*, 421–426. [CrossRef]
131. Birrer, S.C.; Reusch, T.B.H.; Roth, O. Salinity Change Impairs Pipefish Immune Defence. *Fish Shellfish Immunol.* **2012**, *33*, 1238–1248. [CrossRef]
132. Ching, B.; Chen, X.L.; Yong, J.H.A.; Wilson, J.M.; Hiong, K.C.; Sim, E.W.L.; Wong, W.P.; Lam, S.H.; Chew, S.F.; Ip, Y.K. Increases in Apoptosis, Caspase Activity and Expression of P53 and Bax, and the Transition between Two Types of Mitochondrion-Rich Cells, in the Gills of the Climbing Perch, *Anabas testudineus*, during a Progressive Acclimation from Freshwater to Seawater. *Front. Physiol.* **2013**, *4*, 135. [CrossRef]
133. Zhao, R.; Sun, L.; Zhang, L.; Hao, P.; Wang, X.; Wang, B.; Song, G.; Cui, Y.; Liu, F.; Wang, R. RNAi-Mediated Knockdown of the Aquaporin 4 Gene Impairs Salinity Tolerance and Delays the Molting Process in Pacific White Shrimp, *Litopenaeus vannamei*. *Aquac. Rep.* **2024**, *35*, 101974. [CrossRef]
134. Zhou, J.; Li, N.; Wang, H.; Mu, C.; Wang, C. Transcriptome Analysis Reveals Hepatopancreatic Genes and Pathways Associated with Metabolism Responding to Water Salinities during the Overwintering of Mud Crab *Scylla paramamosain*. *Aquac. Res.* **2021**, *52*, 3212–3225. [CrossRef]
135. Esmaeili, N.; Ma, H.; Kadri, S.; Tocher, D.R. Protein and Lipid Nutrition in Crabs. *Rev. Aquac.* **2024**, *16*, 1499–1519. [CrossRef]
136. Whitehead, A.; Roach, J.L.; Zhang, S.; Galvez, F. Salinity- and Population-Dependent Genome Regulatory Response during Osmotic Acclimation in the Killifish (*Fundulus heteroclitus*) Gill. *J. Exp. Biol.* **2012**, *215*, 1293–1305. [CrossRef] [PubMed]
137. Makvandi-Nejad, S.; Moghadam, H. Genetics and Epigenetics in Aquaculture Breeding. In *Epigenetics in Aquaculture*; John Wiley & Sons: Hoboken, NJ, USA, 2023; pp. 439–449.
138. Guo, J.; Lv, J.; Sun, D.; Liu, P.; Gao, B. Elucidating the Role of DNA Methylation in Low-Salinity Adaptation of the Swimming Crab Portunus Trituberculatus. *Aquaculture* **2024**, *595*, 741707. [CrossRef]
139. Zhang, X.; Li, Q.; Kong, L.; Yu, H. DNA Methylation Changes Detected by Methylation-Sensitive Amplified Polymorphism in the Pacific Oyster (*Crassostrea gigas*) in Response to Salinity Stress. *Genes Genom.* **2017**, *39*, 1173–1181. [CrossRef]
140. Jeremias, G.; Barbosa, J.; Marques, S.M.; De Schamphelaere, K.A.C.; Van Nieuwerburgh, F.; Deforce, D.; Gonçalves, F.J.M.; Pereira, J.L.; Asselman, J. Transgenerational Inheritance of DNA Hypomethylation in *Daphnia magna* in Response to Salinity Stress. *Environ. Sci. Technol.* **2018**, *52*, 10114–10123. [CrossRef]
141. Kim, M.-S.; Kim, D.-H.; Lee, J.-S. A Review of Environmental Epigenetics in Aquatic Invertebrates. *Mar. Pollut. Bull.* **2024**, *208*, 117011. [CrossRef]
142. Pierron, F.; Lorioux, S.; Héroin, D.; Daffe, G.; Etcheverria, B.; Cachot, J.; Morin, B.; Dufour, S.; Gonzalez, P. Transgenerational Epigenetic Sex Determination: Environment Experienced by Female Fish Affects Offspring Sex Ratio. *Environ. Pollut.* **2021**, *277*, 116864. [CrossRef]
143. Heckwolf, M.J.; Meyer, B.S.; Döring, T.; Eizaguirre, C.; Reusch, T.B.H. Transgenerational Plasticity and Selection Shape the Adaptive Potential of Sticklebacks to Salinity Change. *Evol. Appl.* **2018**, *11*, 1873–1885. [CrossRef]
144. Yang, J.; Liu, M.; Zhou, T.; Li, Q.; Lin, Z. Genome-Wide Methylome and Transcriptome Dynamics Provide Insights into Epigenetic Regulation of Kidney Functioning of Large Yellow Croaker (*Larimichthys crocea*) during Low-Salinity Adaption. *Aquaculture* **2023**, *571*, 739410. [CrossRef]
145. Qiao, R.; Sheng, C.; Lu, Y.; Zhang, Y.; Ren, H.; Lemos, B. Microplastics Induce Intestinal Inflammation, Oxidative Stress, and Disorders of Metabolome and Microbiome in Zebrafish. *Sci. Total Environ.* **2019**, *662*, 246–253. [CrossRef]
146. Zhang, M.; Sun, Y.; Liu, Y.; Qiao, F.; Chen, L.; Liu, W.T.; Du, Z.; Li, E. Response of Gut Microbiota to Salinity Change in Two Euryhaline Aquatic Animals with Reverse Salinity Preference. *Aquaculture* **2016**, *454*, 72–80. [CrossRef]
147. Schmidt, V.T.; Smith, K.F.; Melvin, D.W.; Amaral-Zettler, L.A. Community Assembly of a Euryhaline Fish Microbiome during Salinity Acclimation. *Mol. Ecol.* **2015**, *24*, 2537–2550. [CrossRef] [PubMed]
148. Zhu, T.; Li, X.; Wu, X.; Yang, D. Temperature Acclimation Alters the Thermal Tolerance and Intestinal Heat Stress Response in a Tibetan Fish *Oxygymnocypris stewarti*. *Front. Microbiol.* **2022**, *13*, 898145. [CrossRef] [PubMed]
149. Tian, L.; Tan, P.; Yang, L.; Zhu, W.; Xu, D. Effects of Salinity on the Growth, Plasma Ion Concentrations, Osmoregulation, Non-Specific Immunity, and Intestinal Microbiota of the Yellow Drum (*Nibea albiflora*). *Aquaculture* **2020**, *528*, 735470. [CrossRef]
150. Liu, Y.; Gao, J.; Nie, Z.; Wang, J.; Sun, Y.; Xu, G. Integration of Metagenome and Metabolome Analysis Reveals the Correlation of Gut Microbiota, Oxidative Stress, and Inflammation in *Coilia nasus* under Air Exposure Stress and Salinity Mitigation. *Comp. Biochem. Physiol. Part D Genom. Proteom.* **2024**, *49*, 101175. [CrossRef]
151. Sun, S.; Gong, C.; Deng, C.; Yu, H.; Zheng, D.; Wang, L.; Sun, J.; Song, F.; Luo, J. Effects of Salinity Stress on the Growth Performance, Health Status, and Intestinal Microbiota of Juvenile *Micropterus salmoides*. *Aquaculture* **2023**, *576*, 739888. [CrossRef]
152. Dehler, C.E.; Secombes, C.J.; Martin, S.A.M. Seawater Transfer Alters the Intestinal Microbiota Profiles of Atlantic Salmon (*Salmo salar* L.). *Sci. Rep.* **2017**, *7*, 13877. [CrossRef]
153. Kim, J.A.; Park, Y.-S.; Kim, J.-H.; Choi, C.Y. Hyposalinity Elicits Physiological Responses and Alters Intestinal Microbiota in Korean Rockfish *Sebastes schlegelii*. *Fish Physiol. Biochem.* **2024**, *50*, 2315–2326. [CrossRef]

154. Lai, K.P.; Boncan, D.A.T.; Qin, X.; Chan, T.F.; Tse, W.K.F. Roles and Occurrences of Microbiota in the Osmoregulatory Organs, Gills and Gut, in Marine Medaka upon Hypotonic Stress. *Comp. Biochem. Physiol. Part D Genom. Proteom.* **2024**, *52*, 101285. [CrossRef]
155. Mkulo, E.M.; Iddrisu, L.; Zhong, J.; Zheng, A.; Jin, M.; Wang, L.; Mpwaga, A.Y.; Zhang, H.; Tang, B.; Zhou, H. Exploring Salinity Adaptation in Teleost Fish, Focusing on Omics Perspectives on Osmoregulation and Gut Microbiota. *Front. Mar. Sci.* **2025**, *12*, 1559871. [CrossRef]
156. Amparyup, P.; Charoensapsri, W.; Tassanakajon, A. Prophenoloxidase System and Its Role in Shrimp Immune Responses against Major Pathogens. *Fish Shellfish Immunol.* **2013**, *34*, 990–1001. [CrossRef]
157. Johansson, M.E.V.; Hansson, G.C. Immunological Aspects of Intestinal Mucus and Mucins. *Nat. Rev. Immunol.* **2016**, *16*, 639–649. [CrossRef] [PubMed]
158. Lin, G.; Zheng, M.; Li, S.; Xie, J.; Fang, W.; Gao, D.; Huang, J.; Lu, J. Response of Gut Microbiota and Immune Function to Hypoosmotic Stress in the Yellowfin Seabream (*Acanthopagrus latus*). *Sci. Total Environ.* **2020**, *745*, 140976. [CrossRef] [PubMed]
159. Patil, P.K.; Antony, L.; Avunje, S.; Viswanathan, B.; Lalitha, N.; Jangam, A.K.; Kumar, D.; Solanki, H.G.; Reddy, M.A.; Alavandi, S.V. Bioaugmentation with Nitrifying and Denitrifying Microbial Consortia for Mitigation of Nitrogenous Metabolites in Shrimp Ponds. *Aquaculture* **2021**, *541*, 736819. [CrossRef]
160. Champion, C.; Broadhurst, M.K.; Ewere, E.E.; Benkendorff, K.; Butcherine, P.; Wolfe, K.; Coleman, M.A. Resilience to the Interactive Effects of Climate Change and Discard Stress in the Commercially Important Blue Swimmer Crab (*Portunus armatus*). *Mar. Environ. Res.* **2020**, *159*, 105009. [CrossRef]
161. Pedrosa, J.A.M.; Cocchiararo, B.; Bordalo, M.D.; Rodrigues, A.C.M.; Soares, A.M.V.M.; Barata, C.; Nowak, C.; Pestana, J.L.T. The Role of Genetic Diversity and Past-History Selection Pressures in the Susceptibility of *Chironomus riparius* Populations to Environmental Stress. *Sci. Total Environ.* **2017**, *576*, 807–816. [CrossRef]
162. Asem, A.; Eimanifar, A.; Van Stappen, G.; Sun, S.C. The Impact of One-Decade Ecological Disturbance on Genetic Changes: A Study on the Brine Shrimp *Artemia urmiana* from Urmia Lake, Iran. *PeerJ* **2019**, *7*, e7190. [CrossRef]
163. Pauls, S.U.; Nowak, C.; Bálint, M.; Pfenninger, M. The Impact of Global Climate Change on Genetic Diversity within Populations and Species. *Mol. Ecol.* **2013**, *22*, 925–946. [CrossRef]
164. Laurenzano, C.; Costa, T.M.; Schubart, C.D. Contrasting Patterns of Clinal Genetic Diversity and Potential Colonization Pathways in Two Species of Western Atlantic Fiddler Crabs. *PLoS ONE* **2016**, *11*, e0166518. [CrossRef]
165. Peres, P.A.; Mantelatto, F.L. Salinity Tolerance Explains the Contrasting Phylogeographic Patterns of Two Swimming Crabs Species along the Tropical Western Atlantic. *Evol. Ecol.* **2020**, *34*, 589–609. [CrossRef]
166. Geletu, T.T.; Zhao, J. Genetic Resources of Nile Tilapia (*Oreochromis niloticus linnaeus*, 1758) in Its Native Range and Aquaculture. *Hydrobiologia* **2023**, *850*, 2425–2445. [CrossRef]
167. Sunde, J.; Tamario, C.; Tibblin, P.; Larsson, P.; Forsman, A. Erratum: Author Correction: Variation in Salinity Tolerance between and within Anadromous Subpopulations of Pike (*Esox lucius*). *Sci. Rep.* **2018**, *8*, 6813.
168. Talukdar, A.; Dharmendra Deo, A.; Prasad Sahu, N.; Sardar, P.; Aklakur, M.; Harikrishna, V.; Prakash, S.; Shamna, N.; Jana, P. Effects of Different Levels of Dietary Protein on the Growth Performance, Nutrient Utilization, Digestive Enzymes and Physiological Status of White Shrimp, *Litopenaeus vannamei* Juveniles Reared in Inland Saline Water. *Aquac. Nutr.* **2021**, *27*, 77–90. [CrossRef]
169. Jin, Y.; Liu, F.J.; Liu, Y.J.; Tian, L.X.; Zhang, Z.H. Dietary Tryptophan Requirements of Juvenile Pacific White Shrimp, *Litopenaeus vannamei* (Boone) Reared in Low-Salinity Water. *Aquac. Int.* **2017**, *25*, 955–968. [CrossRef]
170. Xie, S.; Wei, D.; Fang, W.; Wan, M.; Guo, T.; Liu, Y.; Yin, P.; Tian, L.; Niu, J. Optimal Dietary Lipid Requirement of Postlarval White Shrimp, *Litopenaeus vannamei* in Relation to Growth Performance, Stress Tolerance and Immune Response. *Aquac. Nutr.* **2019**, *25*, 1231–1240. [CrossRef]
171. Yang, Q.; Zhang, W.; Tan, B.; Wang, F.; Chi, S.; Dong, X.; Zhang, H.; Liu, H.; Zhang, S. Effects of Dietary N-3HUFA on Juvenile White Shrimp, *Litopenaeus vannamei*: Growth, Feed Utilization, Antioxidant Enzymes Activities and Fatty Acid Compositions. *Aquac. Res.* **2019**, *50*, 882–894. [CrossRef]
172. Wen, M.; Liu, Y.J.; Tian, L.X.; Wang, S. Vitamin D3 Requirement in Practical Diet of White Shrimp, *Litopenaeus vannamei*, at Low Salinity Rearing Conditions. *J. World Aquac. Soc.* **2015**, *46*, 531–538. [CrossRef]
173. Ebadi, H.; Zakeri, M.; Mousavi, S.M.; Yavari, V.; Souri, M. The Interaction Effects of Dietary Lipid, Vitamin E and Vitamin C on Growth Performance, Feed Utilization, Muscle Proximate Composition and Antioxidant Enzyme Activity of White Leg Shrimp (*Litopenaeus vannamei*). *Aquac. Res.* **2021**, *52*, 2048–2060. [CrossRef]
174. Ciji, A.; Akhtar, M.S. Stress Management in Aquaculture: A Review of Dietary Interventions. *Rev. Aquac.* **2021**, *13*, 2190–2247. [CrossRef]
175. Zhang, J.; Liu, Y.J.; Tian, L.X.; Yang, H.J.; Liang, G.Y.; Yue, Y.R.; Xu, D.H. Effects of Dietary Astaxanthin on Growth, Antioxidant Capacity and Gene Expression in Pacific White Shrimp *Litopenaeus vannamei*. *Aquac. Nutr.* **2013**, *19*, 917–927. [CrossRef]
176. Zhang, Y.; Gao, W.; Yuan, Y.; Cui, W.; Xiang, Z.; Ye, S.; Ikhwanuddin, M.; Ma, H. Impact and Accumulation of Calcium on Soft-Shell Mud Crab *Scylla paramamosain* in Recirculating Aquaculture System. *Aquaculture* **2024**, *593*, 741323. [CrossRef]

177. Hossain, M.S.; Koshio, S.; Ishikawa, M.; Yokoyama, S.; Sony, N.M. Dietary Nucleotide Administration Influences Growth, Immune Responses and Oxidative Stress Resistance of Juvenile Red Sea Bream (*Pagrus major*). *Aquaculture* **2016**, *455*, 41–49. [CrossRef]
178. Yuan, X.; Yang, H.; Wang, L.; Zhou, Y.; Gabr, H.R. Effects of Salinity on Energy Budget in Pond-Cultured Sea Cucumber *Apostichopus japonicus* (Selenka) (Echinodermata: Holothuroidea). *Aquaculture* **2010**, *306*, 348–351. [CrossRef]
179. Lund, I.; Skov, P.V.; Hansen, B.W. Dietary Supplementation of Essential Fatty Acids in Larval Pikeperch (*Sander Lucioperca*); Short and Long Term Effects on Stress Tolerance and Metabolic Physiology. *Comp. Biochem. Physiol.—A Mol. Integr. Physiol.* **2012**, *162*, 340–348. [CrossRef]
180. Kamalam, B.S.; Medale, F.; Panserat, S. Utilisation of Dietary Carbohydrates in Farmed Fishes: New Insights on Influencing Factors, Biological Limitations and Future Strategies. *Aquaculture* **2017**, *467*, 3–27. [CrossRef]
181. Yousefi, M.; Paktinat, M.; Mahmoudi, N.; Pérez-Jiménez, A.; Hoseini, S.M. Serum Biochemical and Non-Specific Immune Responses of Rainbow Trout (*Oncorhynchus mykiss*) to Dietary Nucleotide and Chronic Stress. *Fish Physiol. Biochem.* **2016**, *42*, 1417–1425. [CrossRef]
182. Palermo, F.A.; Cardinaletti, G.; Cocci, P.; Tibaldi, E.; Polzonetti-Magni, A.; Mosconi, G. Effects of Dietary Nucleotides on Acute Stress Response and Cannabinoid Receptor 1 MRNAs in *Sole, Solea solea*. *Comp. Biochem. Physiol. Part A Mol. Integr. Physiol.* **2013**, *164*, 477–482. [CrossRef]
183. Fu, M.; Collins, S.A.; Anderson, D.M. Maxi-Gen^{TM} Plus: A Nucleotide-Containing Product That Reduces Stress Indicators and Improves Growth Performance during Smoltification in Atlantic Salmon (*Salmo salar*). *Aquaculture* **2017**, *473*, 20–30. [CrossRef]
184. Yu, X.; Setyawan, P.; Bastiaansen, J.W.M.; Liu, L.; Imron, I.; Groenen, M.A.M.; Komen, H.; Megens, H.J. Genomic Analysis of a Nile Tilapia Strain Selected for Salinity Tolerance Shows Signatures of Selection and Hybridization with Blue Tilapia (*Oreochromis Aureus*). *Aquaculture* **2022**, *560*, 7385277. [CrossRef]
185. Minh, H.D.; Thuy, Y.D.; Thanh, L.P.; Minh, T.B.; Nam, S.V.; Thanh, H.D.T.; Bich, H.B.T.; Ngoc, T.N.T.; Quang, H.D.; Kestemont, P. Selective Breeding of Saline-Tolerant Striped Catfish (*Pangasianodon hypophthalmus*) for Sustainable Catfish Farming in Climate Vulnerable Mekong Delta, Vietnam. *Aquac. Rep.* **2022**, *25*, 101263. [CrossRef]
186. Luo, Z.; Yu, Y.; Zhang, Q.; Bao, Z.; Xiang, J.; Li, F. Comparative Transcriptome Analysis Reveals the Adaptation Mechanism to High Salinity in *Litopenaeus vannamei*. *Front. Mar. Sci.* **2022**, *9*, 864338. [CrossRef]
187. Fang, J.-F.; Li, Q. The Effects of Inbreeding on Stress Resistance of the Pacific Oyster *Crassostrea gigas* at Different Temperatures and Salinities. *Mar. Biol. Res.* **2023**, *19*, 249–260. [CrossRef]
188. Tan, K.; Zhang, H.; Zheng, H. Selective Breeding of Edible Bivalves and Its Implication of Global Climate Change. *Rev. Aquac.* **2020**, *12*, 2559–2572. [CrossRef]

fishes

Article

YOLOv8-BaitScan: A Lightweight and Robust Framework for Accurate Bait Detection and Counting in Aquaculture

Jian Li [1,2], Zehao Zhang [3,4], Yanan Wei [1] and Tan Wang [3,4,*]

1 Anhui Provincial Key Laboratory of Embryo Development and Reproductive Regulation, Fuyang 236037, China; 202201008@fynu.edu.cn (J.L.); weiyn_1991@163.com (Y.W.)
2 School of Biology and Food Engineering, Fuyang Normal University, Fuyang 236037, China
3 School of Information and Artificial Intelligence, Anhui Agricultural University, Hefei 230036, China
4 Key Laboratory of Agricultural Sensors, Ministry of Agriculture and Rural Affairs, Hefei 230036, China
* Correspondence: wang_tan@ahau.edu.cn

Abstract: Excessive bait wastage is a major issue in aquaculture, leading to higher farming costs, economic losses, and water pollution caused by bacterial growth from unremoved residual bait. To address this problem, we propose a bait residue detection and counting model named YOLOv8-BaitScan, based on an improved YOLO architecture. The key innovations are as follows: (1) By incorporating the channel prior convolutional attention (CPCA) into the final layer of the backbone, the model efficiently extracts spatial relationships and dynamically allocates weights across the channel and spatial dimensions. (2) The minimum points distance intersection over union (MPDIoU) loss function improves the model's localization accuracy for bait bounding boxes. (3) The structure of the Neck network is optimized by adding a tiny-target detection layer, which improves the recall rate for small, distant bait targets and significantly reduces the miss rate. (4) We design the lightweight detection head named Detect-Efficient, incorporating the GhostConv and C2f-GDC module into the network to effectively reduce the overall number of parameters and computational cost of the model. The experimental results show that YOLOv8-BaitScan achieves strong performance across key metrics: The recall rate increased from 60.8% to 94.4%, mAP@50 rose from 80.1% to 97.1%, and the model's number of parameters and computational load were reduced by 55.7% and 54.3%, respectively. The model significantly improves the accuracy and real-time detection capabilities for underwater bait and is more suitable for real-world aquaculture applications, providing technical support to achieve both economic and ecological benefits.

Keywords: intensive aquaculture; bait detection; computer vision; YOLOv8n

Key Contribution: We propose a bait detection algorithm based on improved YOLOv8n, which achieves high detection accuracy while maintaining being lightweight and provides technical support for intelligent feeding.

1. Introduction

In aquaculture, low bait utilization has led to rising farming costs and water pollution, posing serious challenges for the industry [1]. Statistics show that bait costs account for 40–60% of the total aquaculture expenses [2]. Therefore, improving bait utilization efficiency has become a pressing issue. In practice, dynamically adjusting feeding strategies based on the amount of residual bait is key to improving bait utilization and reducing water pollution risks [3]. However, there are still many challenges in accurately identifying and counting bait residues in aquaculture ponds. In complex underwater environments, bait detection is hindered by challenges such as bait aggregation, overlapping targets, missed detections of small or distant objects, and background noise caused by fish movement. These factors significantly affect detection accuracy and pose major obstacles to reliable bait residue recognition. In previous research work, many scholars have attempted to use acoustic methods to detect residual bait underwater. Llorens et al. utilized ultrasonic echo methods to enable the quantification of residual bait [4]. However, these acoustic systems rely on particle-specific acoustic signatures [5], and sonar-based methods often produce monochromatic, low-resolution images that limit detection performance [6]. In addition, acoustic equipment is usually expensive, which further limits its widespread use in practical production.

With the rapid advancement of machine vision, many researchers have applied it to bait detection. Foster et al. [7] proposed an image analysis algorithm to achieve automatic target identification and tracking. Parsonage et al. [8] developed a computer vision system to provide real-time feedback on bait particle consumption. Atoum et al. [9] achieved continuous detection of bait particles in video frames by applying a correlation filter to identify residual bait within local regions. An SVM classifier was used to suppress detection errors, providing a foundation for intelligent bait-feeding control. Li et al. [10] proposed an adaptive threshold segmentation method for detecting residual bait particles in underwater images. However, traditional machine vision methods struggle with complex backgrounds and small targets, limiting their effectiveness in practical applications.

Deep learning can automatically extract high-dimensional features from large datasets, resulting in significantly higher accuracy compared to traditional machine learning methods [11]. Currently, deep learning techniques have been widely used in aquaculture [12]: Rauf et al. [13] used deep convolutional neural networks to automatically identify fish based on visual features. Fernandes et al. [14] applied image segmentation techniques to predict weight and carcass traits of Nile tilapia. Labao and Naval [15] proposed a fish detection system based on deep network architecture for robust detection and counting of fish targets in a variety of biological backgrounds and lighting conditions. Zhao et al. [16] applied a particle advection method to extract motion features from the entire fish population, eliminating the need for tracking and foreground segmentation. They also used augmented motion-captured images and recurrent neural networks to accurately identify, localize, and classify abnormal fish behavior in intensive aquaculture environments. Xu et al. [17] proposed a fish species identification model based on SE-ResNet152 with class-balanced focal loss, addressing the overfitting caused by sample imbalance. However, these algorithms still fall short of meeting real-time detection requirements in practical aquaculture settings [18].

The YOLO family of algorithms offers a promising solution for real-time object detection. As representative single-stage detectors, they are known for their high accuracy and fast inference speed and have been widely applied in aquaculture tasks. Cai et al. [19] proposed an enhanced fish detection and counting method by using the YOLOv3 and MobileNet architectures. The method improves detection accuracy by incorporating a fixed selection strategy into the original YOLOv3 framework. Hu et al. [20] improved average model accuracy by 27.21% and reduced computational cost by approximately 30% by applying dense concatenation and de-redundancy techniques to the feature extraction module of the YOLOv4 backbone. These enhancements make the algorithm more suitable for practical aquaculture environments. In addition, the YOLO algorithm also shows strong advantages in small target detection tasks in underwater environments [21]. Even in low light environments and turty water environments, the YOLO algorithm can show excellent algorithm adaptability and complete the detection task well [22].

The complexity of underwater imaging and the small size of bait particles [23] make the YOLO algorithm prone to positional prediction errors. Moreover, bait particles often appear in clusters or overlap due to occlusion, while distant targets occupy only a few pixels in the image, making them especially prone to missed detection. Background noise from swimming fish and other dynamic elements further complicates the detection process [24].

Therefore, this paper proposes an improved YOLOv8n-based method for more efficient and accurate detection and counting of bait residues. The approach is designed to address key challenges such as target aggregation, occlusion, missed detections of small distant baits, and interference from complex underwater backgrounds. By removing redundant operations, the model is streamlined for higher efficiency and better suitability in real-world aquaculture environments.

The remainder of this paper is organized as follows: Section 2 presents the dataset and algorithmic improvements; Section 3 analyzes the results of ablation experiments; Section 4 discusses the experimental findings in depth; and Section 5 concludes this study.

2. Materials and Methods

2.1. Data Collection and Processing

The data for this experiment were obtained from the publicly available dataset [25] and were further processed. The size of the experimental platform was 3 m × 3 m × 1.2 m, and the water depth was 0.7 m. In order to ensure the stability and controllability of the experimental environment, the dissolved oxygen content of the pool was 7.0 ± 1.0 mg/L, and the water temperature was 13–16 °C, creating a suitable breeding environment for fish growth and behavior observation. Feeding was performed every 24 h at 14:00 Beijing time during the data collection period to establish a stable eating behavior pattern. Each feeding was carried out in batches, and the single feeding amount was 0.15–0.20% of the fish's body weight, and the total feeding amount was 1.5–2.0% of the fish's body weight. Data were collected on the remaining baits in the culture tank 25 min after the fish had eaten. Data were captured with an underwater HD industrial camera positioned at the edge of the circulating aquaculture unit 45 cm from the bottom at a 45-degree angle to the bottom. Due to the limited number of images in the initial dataset, we expanded the dataset through random flipping, cropping, and other augmentation techniques to improve model training efficacy, data augmentation examples on pepper imagesas shown in Figure 1. The augmented dataset, which contains 924 images, was randomly divided into training, validation, and test sets in an 8:1:1 ratio.

Figure 1. Data augmentation examples on pepper images: (a) original image, (b) random flipping, (c) cropping, and (d) other augmentation techniques.

The expanded dataset was annotated using LabelImg with the following guidelines: Baits were annotated if 50% or more of their volume was visible. In cases of overlapping baits, each was annotated independently. This process yielded 22,444 labeled samples, which provided high-quality annotated data for subsequent model training.

2.2. Proposed YOLOv8-BaitScan

YOLOv8n was selected as the baseline network for our experiments. However, the lightweight architecture of YOLOv8n limits its feature extraction capability, leading to reduced detection accuracy for baits. To address this limitation, we propose YOLOv8-BaitScan, an improved network based on YOLOv8n, designed to achieve higher bait counting accuracy and efficiency. Figure 2 shows the comparison between the improved YOLOv8-BaitScan and the original structure of YOLOv8. The specific improvements are as follows: (1) A channel prior convolutional attention (CPCA) mechanism is integrated into the final layer of the backbone network. This module enhances feature extraction by dynamically weighting channel and spatial dimensions while preserving prior channel information. (2) Adopt the *MPDIoU* loss function to strengthen the accuracy of bounding box regression and significantly improve the network's localization ability for tiny bait targets. (3) A multi-scale sequence feature fusion module is introduced to improve multi-scale information extraction, combining global semantic data from various scales to capture detailed spatial information of small targets. Additionally, a dedicated detection layer for tiny targets is added to the head, while the large-target detection layer is removed to minimize computational redundancy without compromising performance. (4) To achieve real-time monitoring, the redundancy in the number of model parameters and computation is reduced by introducing the GhostConv and C2f-GDC modules, and the lighter detection head Detect-Efficient is designed to significantly improve the real-time detection performance of the model.

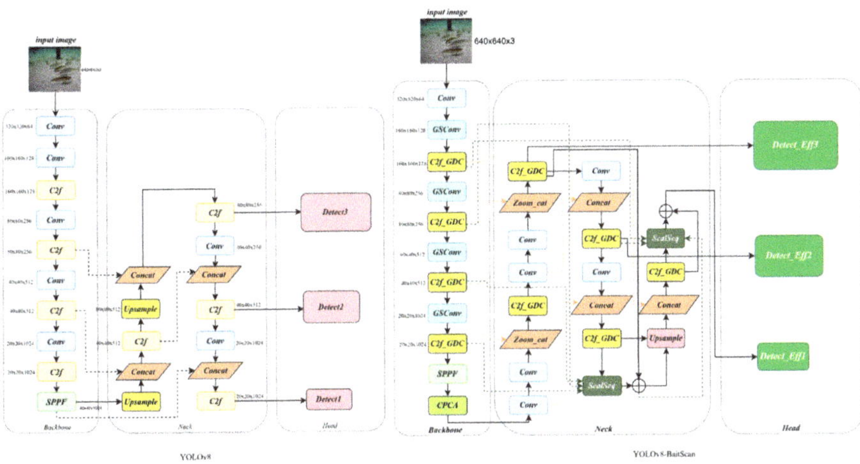

Figure 2. Structures comparison of YOLOv8 and YOLOv8-BaitScan structures.

2.2.1. CPCA Attention Mechanisms

The attention mechanism is a computational method inspired by the human visual system. It enables models to prioritize task-critical regions or channels during image processing, thereby enhancing performance in vision-related tasks. The mechanism fundamentally operates through weight assignment. For a given input feature map, the mechanism dynamically computes attention weights across spatial locations and channels. This process emphasizes salient regions or channels while suppressing redundant or irrelevant features.

CPCA (channel prior convolutional attention) [26] is an efficient attention mechanism that dynamically assigns weights across channel and spatial dimensions, enhancing the network's feature representation capability. This mechanism extracts spatial relationships via a multi-scale deep convolutional module while preserving a priori channel information, enabling the network to focus on channels and salient regions containing critical information.

The CPCA architecture comprises a channel attention module and a spatial attention module, as illustrated in Figure 3. Specifically, the input feature maps are first processed by the channel attention module: The channel attention module aggregates the spatial information of the input features through average pooling and maximum pooling operations to generate feature maps representing the global information of the channels, as in Equation (1):

$$CA(F) = \sigma(MLP(AvgPool(F)) + MLP(MaxPool(F))) \quad (1)$$

where F denotes the input feature map, σ denotes the sigmoid activation function, and *AvgPool* and *MaxPool* denote the average pooling and maximum pooling, respectively.

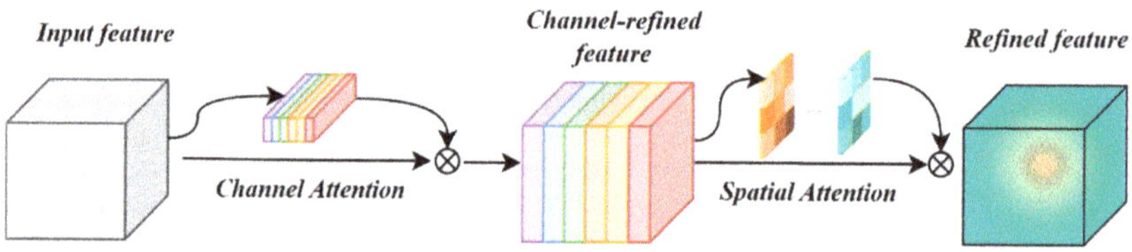

Figure 3. The structure of CPCA attention mechanism.

Subsequently, the spatial information is further processed by a shared *MLP* and combined with the original features to generate the channel attention map. After multiplying the channel attention map with the input features element by element, the channel priori information incorporating the channel weights is obtained. Next, the channel prior information is fed into the spatial attention module. This module extracts spatial relations using deep convolution, generates the spatial attention map, and fuses the channel and spatial attention through the channel blending operation, as in Equation (2):

$$SA(F) = Conv1x1\left(\sum_{i=0}^{3} Branch_i(DwConv(F))\right) \quad (2)$$

where *DwConv* denotes deep convolution, and $Branch_i$ denotes the ith branch with different convolution kernel sizes for capturing multi-scale information. Eventually, the fused spatial attention graph is multiplied element by element with the channel a priori information again to generate a more refined feature representation as output. This process significantly improves the network's ability to perceive important regions and key information and enhances the differentiation and robustness of the feature representation.

In this study, the CPCA mechanism is introduced into the bait small target detection task, making full use of its ability to focus on the key regions and extract key information. By guiding the network to focus on the region containing the bait target to be detected, CPCA effectively improves the detection accuracy of small targets. Aiming at the problem that the model does not pay enough attention to the actual bait area, which leads to insufficient detection accuracy, the CPCA mechanism can dynamically adjust the model's attention to different regions to achieve more attention to the area with bait samples so as to obtain more delicate features and finally improve the detection accuracy.

2.2.2. MPDIoU Loss Function

Bounding box regression (BBR) [27] is widely used in object detection for precise target localization. A well-designed loss function plays a critical role in improving regression accuracy.

MPDIoU loss [28], a novel loss function based on the minimum point distance (MPD), is proposed to optimize bounding box regression. Its geometric parameters are illustrated in Figure 4.

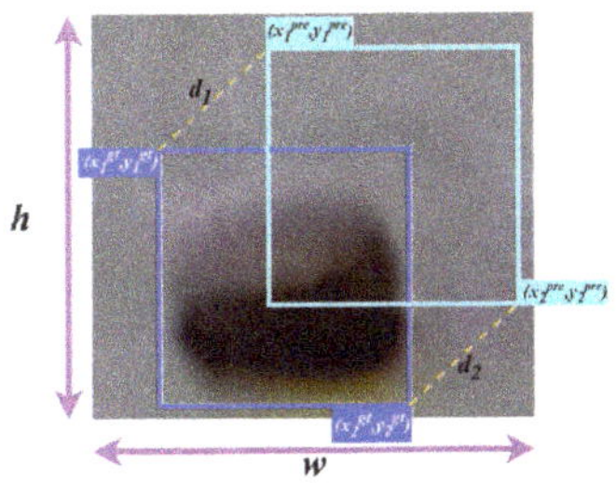

Figure 4. *MPDIoU* related parameters.

The calculation formula is shown in Equations (3) and (4):

$$MPDIoU = \frac{A \cap B}{A \cup B} - \frac{d_1^2}{w^2 + h^2} - \frac{d_2^2}{w^2 + h^2} \quad (3)$$

$$Loss(MPDIoU) = 1 - MPDIoU \quad (4)$$

where (x_i^{gt}, y_i^{gt}) denotes the actual labeled bounding box coordinates, and (x_i^{pre}, y_i^{pre}) denotes the predicted bounding box coordinates. The distances are calculated as follows: $d_1^2 = (x_1^{pre} - x_1^{gt})^2 + (y_1^{pre} - y_1^{gt})^2$, $d_2^2 = (x^{2pre} - x_2^{gt})^2 + (y_2^{pre} - y_2^{gt})^2$. *MPDIoU* evaluates the overlap degree by measuring the distance between the predicted and actual bounding boxes.

MPDIoU, as a new metric, improves the accuracy and efficiency of bounding box regression by comparing the similarity between the predicted bounding box and the actual labeled bounding box in the bounding box regression process. It focuses on optimizing the detection performance of small and complex boundary targets in the target detection and segmentation task and overcomes the shortcomings of traditional IoU loss in small target processing and boundary matching. Therefore, the *MPDIoU* loss is introduced in this study to enhance the model's localization accuracy for bait targets and solve the problem of predicting bounding box duplication in the case of overlapping bait particles. In addition, in terms of model training, the mathematical calculation formula of *MPDIoU* loss is easier, which can enhance the convergence speed of the model to some extent and shorten the model training time during the training process.

2.2.3. Neck Network Optimization and Replacement of Small Target Detection Layer

Given the characteristics of the small-target bait detection dataset, where some bait targets are small in size and distributed over long distances, the traditional P3, P4, and P5 detection layers have a low recall rate for long-distance small-target baits. To solve this issue, the detection head configuration was optimized through extensive experiments. The redundant P5 large-target detection layer, which was not contributing to small-target detection, was removed, and a P2 detection layer was added to significantly enhance recall for long-distance small-target baits. Additionally, to further enhance the ScalSeq [29] module's ability to fuse multi-scale features, the structure of the Neck network was optimized by adding a new ScalSeq layer to the original one, enabling deeper fusion of feature information from the P2, P3, and P4 layers, as shown in Figure 5. With this improvement, the model can capture the detailed features of tiny targets more efficiently and shows excellent detection performance, especially at long distances and in complex backgrounds. The optimized Neck network not only improves the detection accuracy of the overall model but also significantly improves the efficiency and robustness of the bait target counting task, which provides new technical ideas and practical tools for the field of small target detection.

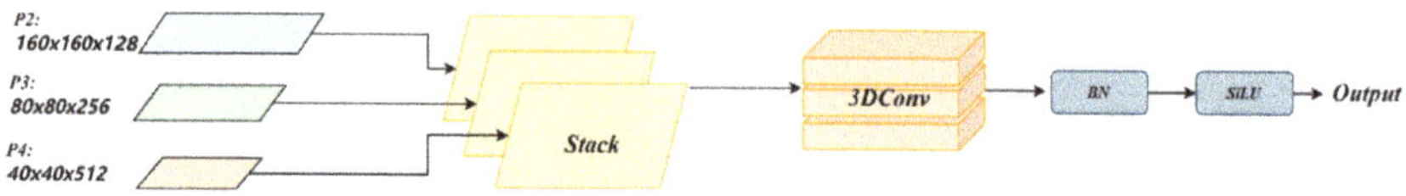

Figure 5. Structure of multi-scale sequence feature fusion module.

The size of the newly added P2 detection layer is 160 × 160, which is able to retain more spatial details of the target, and its receptive field is smaller, which is more suitable for perceiving local regional details without ignoring small target information. The traditional deep detection layers P5 and P6 often blur or disappear the small target information because of multiple downsampling, while the P2 layer is located in the shallow layer of the model and can retain the complete key features such as the edge and texture of the small target. Due to the small size and irregular shape of bait particles in intensive aquaculture, the introduction of P2 layer can make the model detect these small target bait more accurately and then significantly improve the accuracy of model detection.

2.2.4. GhostConv and C2f-GDC Modules and Detect-Efficient

To balance detection accuracy with model parameters and computation, this study introduces the GhostConv [30] and C2f-GDC modules and designs a lightweight detection head, Detect-Efficient. This approach significantly reduces both computational cost and the number of model parameters while maintaining detection accuracy, providing a more efficient solution for practical use.

As shown in Figure 6, the structure of GhostConv consists of three main steps: standard convolution, Ghost feature generation, and feature map stitching. First, GhostConv performs preliminary feature extraction on the input image by standard convolution operation and, at the same time, compresses the number of channels to reduce the computational complexity. Then, the Ghost module is used to generate more feature maps by simple linear operations on the preliminary generated feature maps, thus enriching the feature representation. Finally, all generated feature maps are stitched together to form the final output feature maps. In this way, GhostConv significantly reduces the number of parameters and computational complexity, while preserving detailed feature information.

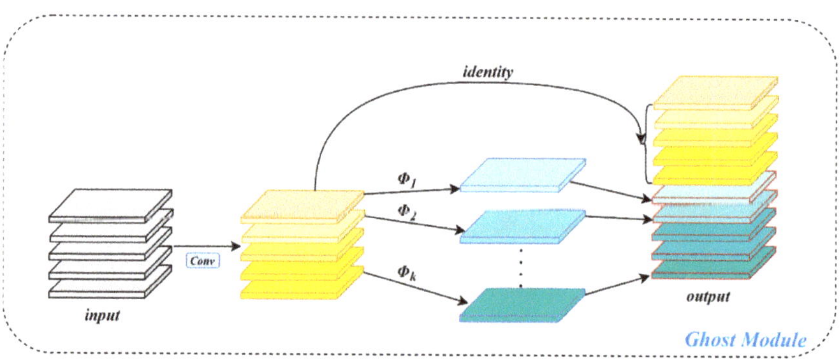

Figure 6. GhostConv module structure.

In this study, we developed the C2f-GDC module by combining DynamicConv [31] with GhostBottleNeck and incorporating it into the C2f module, as shown in Figure 7.

The structure of DynamicConv adaptively adjusts the weight of the convolution kernel according to the features of the input image. This alleviates the computational burden caused by a large number of stacked convolutional layers. In the C2f module, we replace the original BottleNeck structure with GhostBottleNeck. GhostBottleNeck consists of two stacked Ghost Modules. The first Ghost Module acts as an extension layer, increasing the number of channels to enrich feature representation. The second Ghost Module compresses the number of channels to align the output with the shortcut paths. Finally, a shortcut connection links the inputs and outputs, ensuring efficient feature delivery.

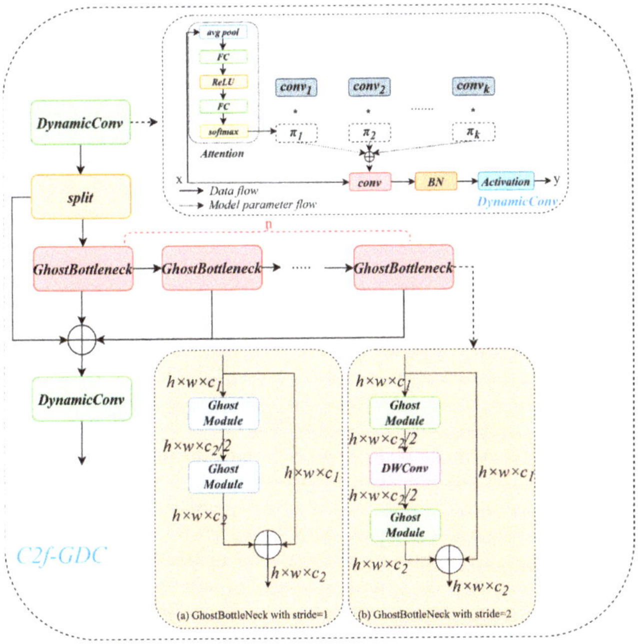

Figure 7. Structure of C2f-GDC.

The detection head is a critical component of the YOLO family of algorithms. Located at the end of the network, it is primarily responsible for target localization and classification. However, the detection head contributes significantly to the overall computational complexity of the YOLOv8 network. To better balance computation and detection accuracy, this study proposes a more lightweight detection head named Detect-Efficient, with a structural comparison to the traditional detection head shown in Figure 8. The traditional YOLOv8 detection head typically extracts information sequentially through two 3×3 convolutions and one 1×1 convolution, followed by separate calculations for Bbox-loss and Cls-loss. However, this structure suffers from parameter and computational redundancy in the information extraction process.

To address this issue, we propose an improvement strategy: Before calculating Bbox-loss and Cls-loss independently, we allow both tasks to share the features extracted by the convolution layers. Additionally, we replace the original convolution layers with PConv [32] (see Figure 9), which significantly reduces both parameters and computational load. This optimization greatly improves the efficiency of the Detect-Efficient head.

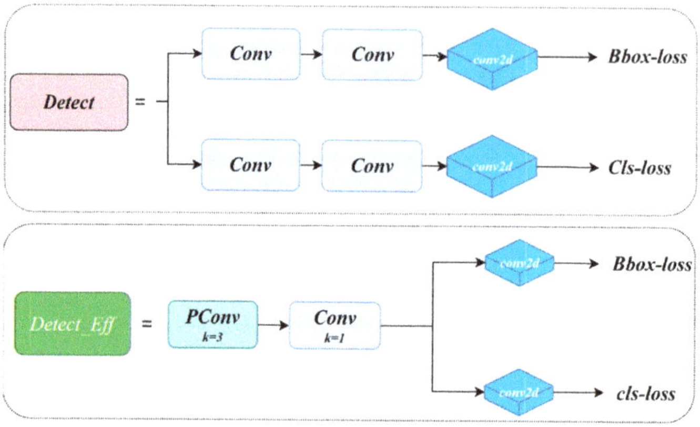

Figure 8. Comparison between Detect-Efficient and traditional Detect structure.

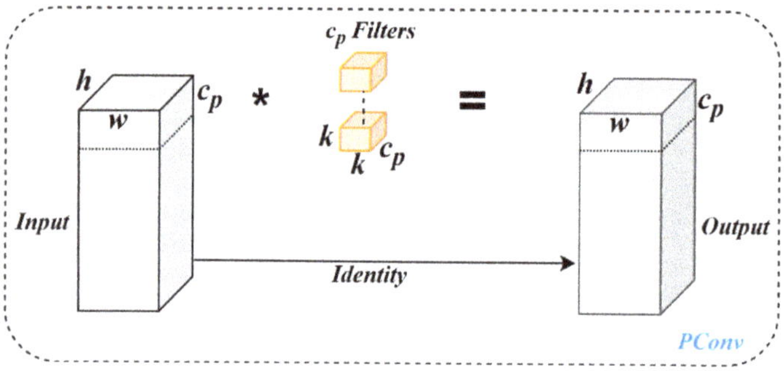

Figure 9. PConv structure.

3. Results

3.1. Assessment of Indicators

Precision, recall, and F1-score are commonly used metrics to evaluate the correctness of object detection, with their calculation formulas shown in Equations (5)–(7). Precision is used to indicate the percentage of correctly predicted baits in the prediction results, recall indicates the percentage of correctly predicted baits to the total number of all baits, and the F1-score value, as a composite metric, is the reconciled average of precision and recall. Param and GFLOPS are model parametric and computational metrics, respectively, typically used to assess the computational complexity of the model. These metrics will be employed to evaluate the model's performance before and after optimization in the subsequent sections of this study.

$$Precison = \frac{TP}{TP + FP} \tag{5}$$

$$Recall = \frac{TP}{TP + FN} \tag{6}$$

$$F1 - score = \frac{2^{*}Precision^{*}Recall}{Precision + Recall} \tag{7}$$

$$AP = \sum_{n=1}^{N} (R_n - R_{n-1}) \cdot P_n \tag{8}$$

$$\text{mAP} = \frac{\sum_1^N \int_0^1 P(R)dR}{N} \tag{9}$$

In this context, *TP* (True Positive) denotes correctly detected bait samples, *FN* (False Negative) denotes samples incorrectly identified as background, *TN* (True Negative) denotes samples correctly identified as background, and *FP* (False Positive) denotes samples incorrectly identified as bait. *AP* represents the area under the precision–recall (PR) curve and is used to measure the average accuracy of the model for a single category, with its value ranging from 0 to 1, as shown in Equation (8), where Pn is the precision value of the NTH point, Rn is the recall value of the NTH point, and *N* is the number of samples on the P-R curve. mAP denotes the average accuracy of multiple categories (single-category detection in this study) and is calculated as shown in Equation (9). mAP@50 and mAP@50:95 are the two metrics commonly used in mAP. mAP@50 indicates the average accuracy of the model at an IoU threshold of 0.5, while mAP@50:95 represents the average of mAP values calculated at IoU thresholds ranging from 0.5 to 0.95 (i.e., mAP@50, mAP@55, . . ., mAP@90, and mAP@95).

3.2. Experimental Parameter Setting

The experimental platform is shown in Table 1. The experimental hyperparameter settings are shown in Table 2. The experimental hyperparameters refer to the optimal hyperparameter settings given by the official authorities and are adjusted appropriately according to the specific experimental requirements.

Table 1. Parameters of the experimental platform.

Configuration	Parameter
CPU	Intel(R)Core(TM)i9-14900K
GPU	NVIDIA GeForce RTX4090 24G
Operating System	Win11
Frameworks	Pytorch 2.2.0
Language	Python 3.10
CUDA	12.1

Table 2. Hyperparameter settings.

Parameter Name	Set-Up
Image size	640 × 640
Batch size	16
Learning rate	0.01
Optimizer	SGD
Momentum factor	0.937
Epochs	300

3.3. Ablation Experiment

In order to verify that the improved algorithm has superior detection accuracy as well as speed for the bait counting task, the baseline model YOLOv8n is compared with the seven improved models based on YOLOv8n, and the results are shown in Table 3. The experimental environment and parameter settings were kept consistent during this experiment to ensure the credibility of the experimental results. In addition, pre-trained weights were loaded in the training process to ensure the reliability of the experimental results.

Table 3. The comparative results of ablation experiments.

Model	P(%)	R(%)	mAP@50 (%)	mAP@50:95 (%)	F1(%)	Param (M)	GFLOPS
YOLOv8n	94.4	60.8	80.1	46.0	75.0	3.00	8.1
YOLOv8n+①	88.5	69.4	81.8	46.4	76.0	3.13	8.3
YOLOv8n+②	91.5	67.1	81.7	47.2	77.0	3.00	8.1
YOLOv8n+③	94.9	93.7	97.6	59.4	94.0	2.32	11.8
YOLOv8n+④	93.0	62.3	79.6	45.3	75.0	2.81	7.7
YOLOv8n+⑤	93.3	61.7	79.6	45.8	75.0	2.58	6.7
YOLOv8n+⑥	93.1	63.8	81.1	47.4	76.0	2.41	5.5
YOLOv8-BaitScan	93.0	94.4	97.1	56.6	93.0	1.33	3.7

①: CPCA Attention Module, ②: MPDIOU loss, ③: Neck improved and replaced detection layer, ④: GhostConv module, ⑤: C2f-GDC, and ⑥: Detect-Efficient.

Applying the CPCA attention mechanism to the Backbone enhances the network's focus on small bait regions, thereby improving detection precision. The MPDIoU loss further boosts detection performance by increasing localization accuracy and addressing bounding box duplication caused by overlapping bait particles. The improved Neck network and newly designed detection head significantly increase the recall of small, long-distance bait targets by effectively fusing multi-scale feature sequences. GhostConv and C2f-GDC reduce the model's parameter count and computational cost with only a minimal impact on overall accuracy. Detect-Efficient, the proposed lightweight detection head, further reduces complexity by introducing PConv and enabling feature sharing across convolution layers. Additionally, with PConv, the convolutional kernel dynamically adjusts its range by ignoring missing or invalid data points, which leads to a notable improvement in detection accuracy.

Ultimately, the YOLOv8-BaitScan proposed in this study achieves significant performance improvements compared to the baseline YOLOv8n. Specifically, it improves recall by 33.6%, mAP@50 by 17%, mAP@50:95 by 10.6%, and F1-score by 18%. At the same time, the number of parameters and GFLOPS are reduced by 55.7% and 54.3%, respectively.

3.3.1. Analysis of the Effect of Introducing the CPCA Attention Mechanism

After integrating the CPCA module, all evaluation metrics improved except for precision(see Figure 10). Specifically, recall increased by 8.6%, reaching 69.4%; mAP@50 increased by 1.7% to 81.8%; mAP@50:95 improved by 0.4%; and the F1-score rose by 1%. These results demonstrate that the CPCA attention mechanism effectively enhances the overall detection performance of the model.

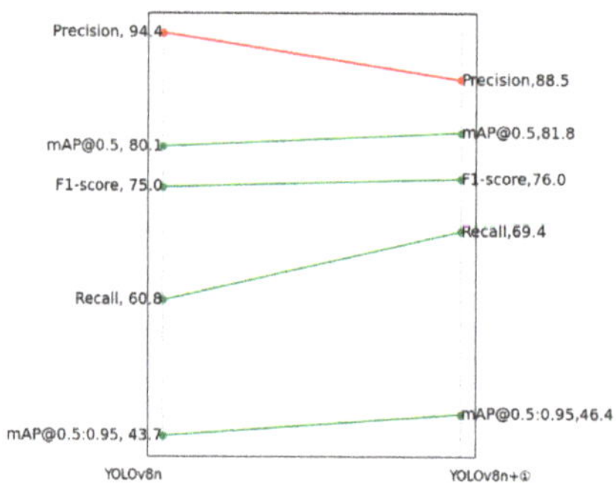

Figure 10. The effect comparison of CPCA is introduced.

3.3.2. Analysis of the Effect of Replacing the MPDIoU Loss Function

The introduction of the *MPDIoU* loss function improves the model's localization accuracy for bait targets, thereby enhancing overall detection performance. As shown in Figure 11, comparing the effect of using the original loss function and *MPDIoU* loss, it can be seen from the results that the *MPDIoU* loss function is more suitable for the current bait residue detection task, and its effect is better than that of the original loss function in both the mAP@50 and mAP@50:95 metrics, which enhances the detection performance of the model. In addition to performance gains, *MPDIoU* has a more concise mathematical formulation, which accelerates the training process and allows the model to converge faster. As illustrated in Figure 11, the model using *MPDIoU* shows a noticeable increase in mAP starting from epoch 20, whereas the model using the original loss function does not show a similar rise until epoch 35.

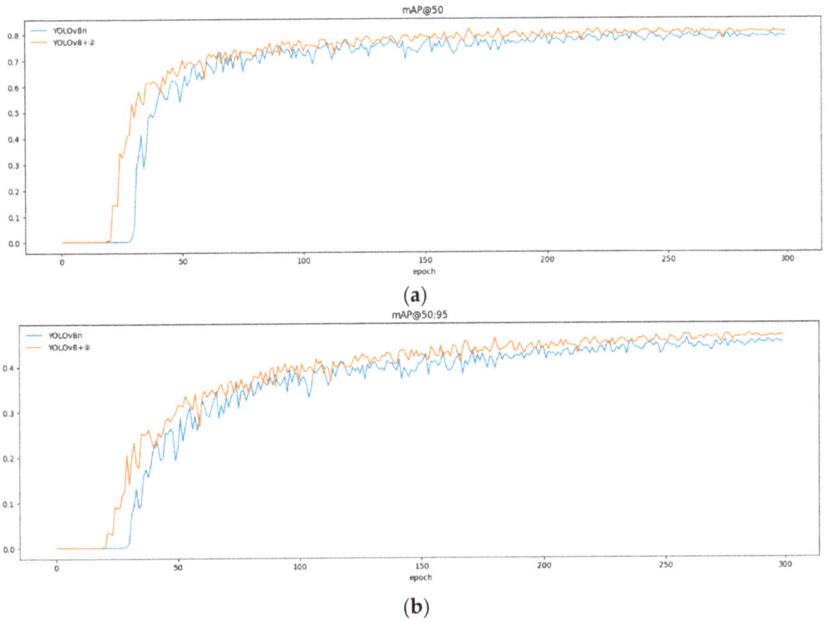

Figure 11. (**a**) mAP@50 comparison and (**b**) mAP@50:95 comparison.

3.3.3. Analysis of the Effect of Neck Network Optimization and Replacement of Small Target Detection Layers

By redesigning the Neck part of the network, we enable the network to capture more feature information and reduce information loss during feature fusion. The bait detection accuracy is further improved by combining high-level information from deep feature maps with detailed information from shallow feature maps through the ScalSeq module, resulting in a multi-scale feature representation. In addition, by eliminating the redundant P5 large-target detection layer and adding the P2 small-target detection layer, the recall of long-range small baits is significantly improved, and thus the detection accuracy is significantly improved, with the recall increased to 93.7% and the mAP@50 increased to 97.6%. Figure 12 shows the comparison of the indexes between the original model and the Neck partially improved and replaced one detection layer. As can be seen from the results, all the indexes are significantly improved after the introduction of the Neck partially improved and the replacement of the P2 detection layer, but then it also brings a substantial increase in the amount of computation, and the GFLOPS increases from 8.1 to 11.8, which is a 31% increase in computation, and increases the burden of the model inference.

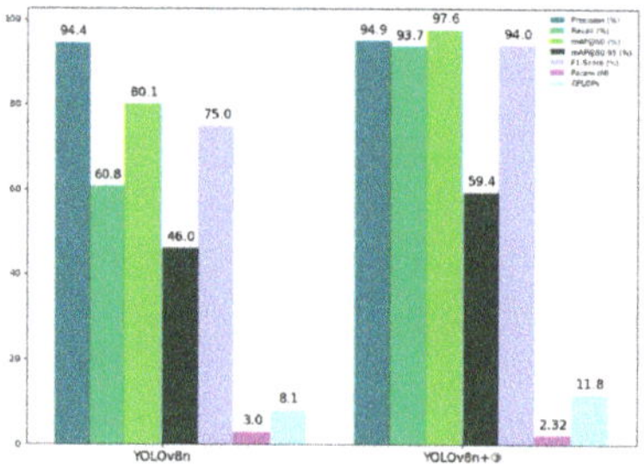

Figure 12. The results of comparison after optimizing the Neck structure.

3.3.4. Analysis of the Effect of Introducing GhostConv, C2f-GDC Module, and Lightweight Detector Heads

Figure 13a shows the comparison between the number of model parameters and computation amount and the original model after the introduction of GhostConv, C2f-GDC, and Detect-Efficient, and Figure 13b shows the comparison between the model detection accuracy and the original model after the introduction of GhostConv, C2f-GDC, and Detect-Efficient. It can be seen that all the above methods can reduce the overall computation and number of parameters of the model to different degrees. Meanwhile, GhostConv and C2f-GDC reduce the computation while keeping the mAP@50 down by only 0.5%, whereas Detect-Efficient significantly reduces the computation and number of parameters while increasing the mAP@50 by 1.0%, which proves that Detect-Efficient detection head is more suitable for bait residue detection tasks.

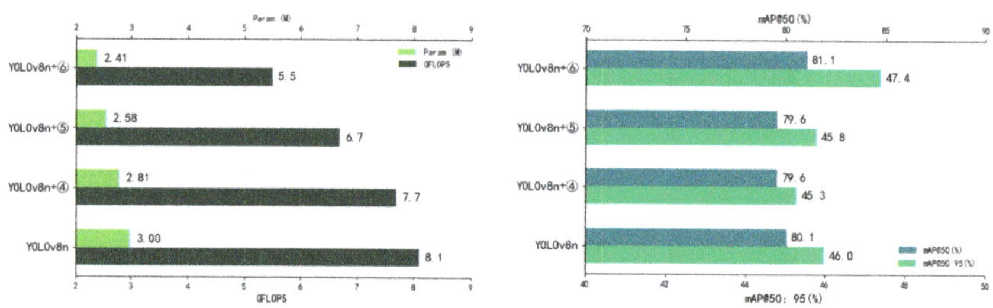

(a) (b)

Figure 13. (**a**) Comparison of computational load before and after improvement. (**b**) Comparison of detection accuracy before and after improvement.

3.4. Comparison of Different Attention Mechanisms

In order to verify the efficiency of the CPCA attention mechanism in the bait detection task, this paper compares the common attention mechanisms: MLCA, DA, CAFM, and MPCA. Table 4 shows a comparison of the experimental results of different attention mechanisms.

Among the evaluated attention mechanisms, the integration of MLCA and DA resulted in a decrease in mAP@50, indicating that these mechanisms are not well-suited for bait detection tasks. In contrast, the incorporation of CAFM and MPCA led to modest improvements in mAP@50, with increases of only 0.3% and 0.7%, respectively. Notably, the CPCA attention mechanism demonstrated a more substantial enhancement, achieving a 1.7% increase in mAP@50. Based on these experimental results, this study adopts the

CPCA attention mechanism within the backbone network of YOLOv8n to improve the detection performance for bait targets.

Table 4. The comparative experimental results of different attention mechanisms.

Model	P(%)	R(%)	mAP@50(%)
YOLOv8n	94.4	60.8	80.1
YOLOv8n-MLCA	92.5	62.0	79.5
YOLOv8n-DA	93.5	62.8	80.0
YOLOv8n-CAFM	93.1	63.1	80.4
YOLOv8n-MPCA	94.0	62.3	80.8
YOLOv8n-CPCA	88.5	69.4	81.8

3.5. Comparison of Different Loss Functions

To investigate the impact of different loss functions on the performance of the YOLOv8-BaitScan model and to identify the most suitable loss function, the original CIoU loss was individually replaced with GIoU, SIoU, DIoU, and *MPDIoU*. The performance differences in bait target detection were then systematically evaluated. The experimental results are summarized in Table 5.

Table 5. The comparative experimental results of different loss functions.

Model	P(%)	R(%)	mAP@50(%)
CIOU	94.4	60.8	80.1
GIoU	93.9	61.1	79.9
SIoU	94.1	62.2	80.6
DIOU	94.4	63.2	81.2
MPDIoU	91.5	67.1	81.7

As presented in Table 5, the *MPDIoU* loss function demonstrates notable advantages. Compared to the original CIoU loss function, *MPDIoU* yields a 6.3% increase in recall and a 1.6% improvement in mAP@50. In contrast, the use of GIoU not only fails to enhance detection performance but also results in a 0.2% decrease in mAP@50. Although SIoU contributes a 0.5% increase in mAP@50, its impact remains relatively limited when compared to *MPDIoU*. DIoU achieves a 1.1% improvement in mAP@50; however, its performance gain is still inferior to that of *MPDIoU*.

3.6. Comparison of Different Models

To highlight the advantages of the proposed YOLOv8-BaitScan model, we conducted a comparative evaluation against Faster R-CNN, YOLOv7-tiny, YOLOv8n, and YOLOv10n. The experimental results are summarized in Table 6.

Table 6. The comparative experimental results of different models.

Model	P(%)	R(%)	mAP@50 (%)	mAP@50:95 (%)	F1(%)	Param(M)	GFLOPS
Faster R-CNN	83.9	76.8	83.8	43.7	78.0	41.3	251.4
YOLOv7-tiny	79.1	66.6	72.8	28.0	72.0	6.00	13.0
YOLOv8n	94.4	60.8	80.1	46.0	75.0	3.00	8.1
YOLOv10n	89.1	62.1	71.7	38.2	73.0	2.26	6.5
YOLOv8-BaitScan	93.0	94.4	97.1	56.6	93.0	1.33	3.7

As shown in Table 6, YOLOv8-BaitScan achieves a mAP@50 of 97.1%, significantly outperforming the other detection models. Furthermore, it requires the least computational resources, with only 1.33 million parameters and a computational cost of 3.7 GFLOPS. Although YOLOv10n is also a lightweight model, its detection accuracy is substantially lower, reaching only 71.7% mAP@50. In terms of the F1-score, YOLOv8-BaitScan attains an outstanding 93.0%, which is 15%, 21%, 18%, and 20% higher than Faster R-CNN, YOLOv7-tiny, YOLOv8n, and YOLOv10n, respectively.

3.7. Visualization

To further validate the superiority of the YOLOv8-BaitScan model, we conducted a visual comparison of the detection results against ground truth annotations, aiming to comprehensively highlight the improvements introduced by the proposed algorithm.

Figure 14 shows the detection performance of YOLOv8-BaitScan in different difficult recognition scenarios. (a) shows the detection results in the case of overlapping baits. The results show that YOLOv8-BaitScan can better identify overlapping baits, and (b) shows the detection results of long-distance bait. The optimized model greatly reduces the probability of missed detection of long-distance bait, and (c) shows the detection results under a large number of bait aggregation scenes. The results show that the optimized model has a better recognition effect for bait in aggregation scenes, and (d) shows the detection results in the case of background interference. The results show that YOLOv8-BaitScan can better complete the bait detection task in the case of background interference caused by fish swimming. The enhanced YOLOv8-BaitScan exhibits a marked improvement in prediction performance compared to the original YOLOv8n. It notably enhances the recall of long-range bait targets and achieves greater accuracy in bounding box localization. These advantages are particularly evident in challenging scenarios involving dense clustering, overlapping targets, and complex background interference. The results confirm the effectiveness and robustness of the proposed method.

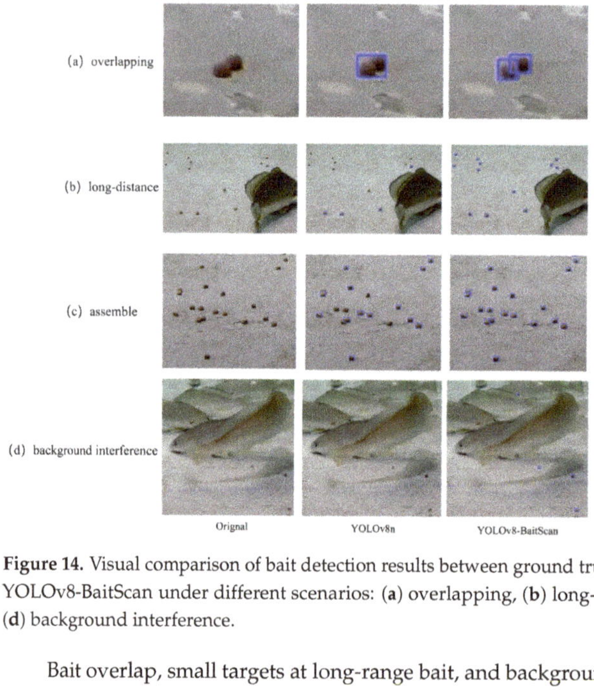

Figure 14. Visual comparison of bait detection results between ground truth, the baseline model, and YOLOv8-BaitScan under different scenarios: (**a**) overlapping, (**b**) long-distance, (**c**) assemble, and (**d**) background interference.

Bait overlap, small targets at long-range bait, and background noise interference are great challenges in the bait residue counting task. Figure 15 shows the heat map of the YOLOv8n algorithm and the YOLOv8-BaitScan proposed in this paper in the bait prediction results in different cases, which can more intuitively show the degree of attention of the algorithm to different regions in the picture before and after the improvement. Compared with the original YOLOv8n model, YOLOv8-BaitScan reduces the probability of bait target miss detection by applying more attention weights to the long-distance small-target bait and adding a new tiny-target detection layer, which makes the model more sensitive to the long-distance small-target bait, and thus reduces the probability of bait target miss detection to a great extent. In addition, the use of the *MPDIoU* loss function enhances the localization precision of the predicted bounding boxes, particularly in cases involving dense aggregation and overlap. These improvements collectively contribute to the superior performance and robustness of YOLOv8-BaitScan in practical bait residue counting tasks.

Figure 15. Heat maps comparison of bait detection results.

4. Discussion

This study proposed YOLOv8-BaitScan to optimize the structure of the YOLOv8 network for bait detection tasks, aiming to achieve accurate and efficient detection of the number of bait in intensive aquaculture ponds and provide technical support for precision breeding. This model introduces the CPCA attention mechanism in the backbone part of the network, replaces the *MPDIoU* loss function, and optimizes the configuration of the detection layer. Finally, the computational load required by the model is reduced by the lightweight structure. Experimental results show that YOLOv8-BaitScan performs well in actual intensive aquaculture scenarios, and all indicators are significantly improved compared with the original benchmark model.

However, although the proposed YOLOv8-BaitScan performs well in the bait detection task, there are still some limitations of this model that need to be further solved to obtain better adaptability. Firstly, the required computational load will be limited on marginal computing power devices. Based on the current findings, there are several avenues for improving YOLOv8-BaitScan. One key area for future work is addressing the model's computational load on resource-constrained devices. While the lightweight structure of YOLOv8-BaitScan offers some reduction in computational demands, further optimization techniques, such as pruning or quantization, could be explored to enable the model to run more efficiently on edge devices with limited computational power. Additionally, exploring hardware acceleration methods, such as the use of GPUs or specialized AI chips, may enhance real-time performance in practical deployment scenarios. Secondly, the baits fed in the process of intensive culture are diverse, so the performance of the model on different types of baits has not been further tested and verified, and the data set needs to be further expanded. Furthermore, the incorporation of domain-specific knowledge related to bait behavior and interaction with the surrounding environment could enhance the model's performance. For instance, integrating temporal information, such as bait movement patterns or feeding frequency, could provide a more dynamic and accurate detection system. The combination of visual and temporal data might lead to more reliable predictions, especially in dynamic aquaculture settings. In addition, how to make the model automatically recognize other background interference such as algae as background is our further work. The interference such as algae was used as the background to make the model ignore its characteristics so as to realize the anti-interference of the model to the complex background.

In summary, while YOLOv8-BaitScan demonstrates promising results in bait detection for intensive aquaculture, there is still room for improvement in terms of computational efficiency, bait diversity handling, and model adaptability to real-world conditions. Future research should focus on these aspects to further enhance the model's practical utility in precision aquaculture.

5. Conclusions

In this study, we introduce YOLOv8-BaitScan, an enhanced method for bait residue counting based on an improved version of YOLOv8n. This method addresses the challenges of detecting bait residues in complex aquaculture environments, where issues such as bait aggregation, target occlusion, long-distance detection of small targets, and background interference are common.

To assess the performance of the proposed algorithm, we conducted a series of ablation experiments. The results demonstrate that YOLOv8-BaitScan outperforms the YOLOv8n algorithm in several key metrics. Specifically, mAP@50 increased from 80.1% to 97.1%, and the F1-score improved from 75% to 93.0%. Additionally, to meet the real-time requirements of practical applications, YOLOv8-BaitScan is designed to be lightweight, reducing the model size to 1.33M parameters (a 55.7% reduction) and the computational complexity to 3.7 GFLOPS (a 54.3% reduction). These optimizations lead to significant improvements in the model's efficiency and inference speed.

In conclusion, YOLOv8-BaitScan achieves a balance between high detection accuracy and computational efficiency, making it a promising solution for bait residue counting in aquaculture. It offers strong technical support for the development of scientific and effective feeding strategies, with potential applications in real-time monitoring systems for aquaculture.

Author Contributions: Conceptualization, T.W. and J.L.; methodology, J.L., Z.Z., and Y.W.; software, Z.Z.; validation, J.L.; formal analysis, Y.W.; investigation, Z.Z.; resources, J.L. and T.W.; data curation, J.L.; writing—original draft preparation, J.L. and Z.Z.; writing—review and editing, T.W.; supervision, T.W.; project administration, J.L.; funding acquisition, J.L. All authors have read and agreed to the published version of the manuscript.

Funding: This research was funded by Biological and Medical Sciences of Applied Summit Nurturing Disciplines in Anhui Province (Anhui Education Secretary Department [2023]13), the Key projects of natural science research projects in Anhui universities (Grant Number. 2022AH051345 and 2024AH050462), the Scientific research project of Fuyang Normal University (Project Number: 2023KYQD0013), the School-level Undergraduate Teaching Project of Fuyang Normal University (Project Number: 2024YLKC0018; 2024KCSZSF07), and the Technical Service Contract Between Shenzhen Mairuide Technology Co., Ltd. and Fuyang Normal University (Project Code: HX2023021000).

Institutional Review Board Statement: This study was conducted on picture data only, and no experiments were conducted directly on animals.

Informed Consent Statement: Not applicable.

Data Availability Statement: The code can be requested at the following link: https://github.com/ABCDZzz22/YOLOv8-BaitScan (accessed on 3 June 2025). The data are from public datasets, which are introduced in Section 2.1.

Acknowledgments: The authors thank all the editors and reviewers for their suggestions and comments. The authors are grateful for the datasets shared by National Innovation Center for Digital Fishery.

Conflicts of Interest: The authors declare no conflicts of interest.

References

1. Asche, F.; Guttormsen, A.G.; Tveterås, R. Environmental problems, productivity and innovations in Norwegian salmon aquaculture. *Aquac. Econ. Manag.* **1999**, *3*, 19–29. [CrossRef]
2. Hishamunda, N.; Ridler, N.B.; Bueno, P.; Yap, W.G. Commercial aquaculture in Southeast Asia: Some policy lessons. *Food Policy* **2009**, *34*, 102–107. [CrossRef]
3. Chen, L.; Yang, X.T.; Sun, C.H.; Wang, Y.Z.; Xu, D.M.; Zhou, C. Feed intake prediction model for group fish using the MEA-BP neural network in intensive. *Inf. Process. Agric.* **2020**, *7*, 261–271. [CrossRef]
4. Llorens, S.; Pérez-Arjona, I.; Soliveres, E.; Espinosa, V. Detection and target strength measurements of uneaten feed pellets with a single beam echosounder. *Aquacult. Eng.* **2017**, *78*, 216–220. [CrossRef]
5. Simmonds, J.; MacLennan, D.N. *Fisheries Acoustics: Theory and Practice*; John Wiley & Sons: Hoboken, NJ, USA, 2008.
6. Terayama, K.; Shin, K.; Mizuno, K.; Tsuda, K. Integration of sonar and optical camera images using deep neural network for fish monitoring. *Aquacult. Eng.* **2019**, *86*, 102000. [CrossRef]
7. Foster, M.; Petrell, R.; Ito, M.R.; Ward, R. Detection and counting of uneaten feed pellets in a sea cage using image analysis. *Aquac. Eng.* **1995**, *14*, 251–269. [CrossRef]
8. Parsonage, K.D.; Petrell, R.J. Accuracy of a machine-vision pellet detection system. *Aquac. Eng.* **2003**, *29*, 109–123. [CrossRef]
9. Atoum, Y.; Srivastava, S.; Liu, X. Automatic Feeding Control for Dense Aquaculture Fish Tanks. *IEEE Signal Process. Lett.* **2015**, *22*, 1089–1093. [CrossRef]
10. Li, D.W.; Xu, L.H.; Liu, H.Y. Detection of uneaten fish food pellets in underwater images for aquaculture. *Aquacult. Eng.* **2017**, *78*, 85–94. [CrossRef]
11. LeCun, Y.; Bengio, Y.; Hinton, G. Deep learning. *Nature* **2015**, *521*, 436–444. [CrossRef]
12. Yang, X.; Zhang, S.; Liu, J.; Gao, Q.; Dong, S.; Zhou, C. Deep learning for smart fish farming: Applications, opportunities and challenges. *Rev. Aquac.* **2021**, *13*, 66–90. [CrossRef]
13. Rauf, H.T.; Lali, M.I.U.; Zahoor, S.; Shah, S.Z.H.; Rehman, A.U.; Bukhari, S.A.C. Visual features based automated identification of fish species using deep convolutional neural networks. *Comput. Electron. Agric.* **2019**, *167*, 105075. [CrossRef]
14. Fernandes, A.F.A.; Turra, E.M.; de Alvarenga, É.R.; Passafaro, T.L.; Lopes, F.B.; Alves, G.F.O.; Singh, V.; Rosa, G.J.M. Deep Learning image segmentation for extraction of fish body measurements and prediction of body weight and carcass traits in Nile tilapia. *Comput. Electron. Agric.* **2020**, *170*, 105274. [CrossRef]
15. Labao, A.B.; Naval, P.C. Cascaded deep network systems with linked ensemble components for underwater fish detection in the wild. *Ecol. Inform.* **2019**, *52*, 103–121. [CrossRef]
16. Zhao, J.; Bao, W.; Zhang, F.; Zhu, S.; Liu, Y.; Lu, H.; Shen, M.; Ye, Z. Modified motion influence map and recurrent neural network-based monitoring of the local unusual behaviors for fish school in intensive aquaculture. *Aquaculture* **2018**, *493*, 165–175. [CrossRef]
17. Xu, X.; Li, W.; Duan, Q. Transfer learning and SE-ResNet152 networks-based for small-scale unbalanced fish species identification. *Comput. Electron. Agric.* **2021**, *180*, 105878. [CrossRef]
18. Zou, Z.X.; Shi, Z.W.; Guo, Y.H.; Ye, J.P. Object Detection in 20 Years: A Survey. *arXiv* **2019**, arXiv:1905.05055. [CrossRef]
19. Cai, K.; Miao, X.; Wang, W.; Pang, H.; Liu, Y.; Song, J. A modified YOLOv3 model for fish detection based on MobileNetv1 as backbone. *Aquac. Eng.* **2020**, *91*, 102117. [CrossRef]
20. Hu, X.; Liu, Y.; Zhao, Z.; Liu, J.; Yang, X.; Sun, C.; Chen, S.; Li, B.; Zhou, C. Real-time detection of uneaten feed pellets in underwater images for aquaculture using an improved YOLO-V4 network. *Comput. Electron. Agric.* **2021**, *185*, 106135. [CrossRef]
21. Chen, J.; Er, M.J. Dynamic YOLO for small underwater object detection. *Artif. Intell. Rev.* **2024**, *57*, 165. [CrossRef]
22. Ge, L.; Singh, P.; Sadhu, A. Advanced deep learning framework for underwater object detection with multibeam forward-looking sonar. *Struct. Health Monit.* **2024**, 14759217241235637. [CrossRef]
23. Wang, Y.; Yu, X.; Liu, J.; Zhao, R.; Zhang, L.; An, D.; Wei, Y. Research on quantitative method of fish feeding activity with semi-supervised based on appearance-motion representation. *Biosyst. Eng.* **2023**, *230*, 409–423. [CrossRef]
24. Acker, T.; Burczynski, J.; Hedgepeth, J.; Ebrahim, A. Digital Scanning Sonar for Fish Feeding Monitoring in Aquaculture. In Proceedings of the Sixth European Conference on Underwater Acoustics ECUA 2002, Gdansk, Poland, 24–27 June 2002.

25. Xu, C.; Wang, Z.; Du, R.; Li, Y.; Li, D.; Chen, Y.; Li, W.; Liu, C. A method for detecting uneaten feed based on improved YOLOv5. *Comput. Electron. Agric.* **2023**, *212*, 108101. [CrossRef]
26. Huang, H.; Chen, Z.; Zou, Y.; Lu, M.; Chen, C.; Song, Y.; Zhang, H.; Yan, F. Channel prior convolutional attention for medical image segmentation. *Comput. Biol. Med.* **2024**, *178*, 108784. [CrossRef]
27. Girshick, R.; Donahue, J.; Darrell, T.; Malik, J. Rich feature hierarchies for accurate object detection and semantic segmentation. In Proceedings of the IEEE Conference on Computer Vision and Pattern Recognition, Columbus, OH, USA, 23–28 June 2014; pp. 580–587.
28. Ma, S.; Xu, Y. Mpdiou: A loss for efficient and accurate bounding box regression. *arXiv* **2023**, arXiv:2307.07662.
29. Kang, M.; Ting, C.M.; Ting, F.F.; Phan, R.C.W. ASF-YOLO: A novel YOLO model with attentional scale sequence fusion for cell instance segmentation. *Image Vis. Comput.* **2024**, *147*, 105057. [CrossRef]
30. Han, K.; Wang, Y.; Tian, Q.; Guo, J.; Xu, C.; Xu, C. Ghostnet: More features from cheap operations. In Proceedings of the IEEE/CVF Conference on Computer Vision and Pattern Recognition, Seattle, WA, USA, 13–19 June 2020; pp. 1580–1589.
31. Han, K.; Wang, Y.; Guo, J.; Wu, E. ParameterNet: Parameters Are All You Need for Large-scale Visual Pretraining of Mobile Networks. In Proceedings of the IEEE/CVF Conference on Computer Vision and Pattern Recognition, Seattle, WA, USA, 17–21 June 2024; pp. 15751–15761.
32. Chen, J.; Kao, S.H.; He, H.; Zhuo, W.; Wen, S.; Lee, C.H.; Chan, S.H.G. Run, don't walk: Chasing higher FLOPS for faster neural networks. In Proceedings of the IEEE/CVF Conference on Computer Vision and Pattern Recognition, Vancouver, BC, Canada, 18–22 June 2023; pp. 12021–12031.

biology

MDPI

Article

Dietary *Astragalus* Polysaccharides Can Improve the Immune Capacity and Reproductive Performance of the Lined Seahorse (*Hippocampus erectus*)

Siping Li [1,2], Xin Liu [1,3], Tingting Lin [1,2,*], Yuanhao Ren [1], Dong Zhang [1] and Keji Jiang [1]

1 Key Laboratory of Inland Saline-Alkaline Aquaculture, Ministry of Agriculture and Rural Affairs, East China Sea Fisheries Research Institute, Chinese Academy of Fishery Sciences, Shanghai 200090, China; lisiping@ecsf.ac.cn (S.L.); liux@ecsf.ac.cn (X.L.); renyh@ecsf.ac.cn (Y.R.); zhangdong@ecsf.ac.cn (D.Z.); jiangkj@ecsf.ac.cn (K.J.)
2 Wenchang Innovation Research Center, Fengjiawan Modern Fishery Industry Park, Wenchang 571300, China
3 Key Laboratory of Tropical Marine Ecosystem and Bioresource, Ministry of Natural Resources of China, Beihai 536000, China
* Correspondence: lintt@ecsf.ac.cn

Simple Summary: In aquaculture, due to the high stocking density, undiversified food, and monotonous environment, the decline in the immune function of cultured animals is a common issue. Therefore, some immunostimulants are usually used to improve the immunity and disease resistance of cultured animals. *Astragalus* polysaccharides (APSs) have been used as immunostimulants in aquaculture for more than 20 years, and their immune-promoting effects have been well-verified in a large number of cultured animals. Seahorses are widely cultivated in China, and recently also face the issue of decline in immune capacity. Up until now, there has been no report of APS application in this animal. In order to test whether APSs would also have immune-promoting effects on seahorses, we conducted several investigations in this study. The results showed that dietary APSs could significantly improve the growth, survivorship, plasma immune levels, and intestinal microbiota diversity of seahorse juveniles as well as the reproductive performance of seahorse broodstocks. These results are valuable for the healthy breeding of seahorses.

Abstract: Seahorse (*Hippocampus* spp.) is popular in the markets of traditional Chinese medicine, aquarium, and curio. In order to protect wild stocks and still meet the market demand, China attempted the large-scale cultivation of seahorses in the early 21st century and achieved it in the 2010s. However, in recent years, two new issues have gradually emerged in Chinese seahorse cultivation. One is that the juveniles are prone to disease during diet conversion, and the other is that the reproductive performance of broodstocks is significantly reduced. With the aim to provide some measures that can alleviate these issues, in the present study, we used lined seahorse (*Hippocampus erectus*, a species widely cultured in China) as the experimental subject and *Astragalus* polysaccharides (APSs) as the immunostimulant to test whether APSs could improve the immune-health status and reproductive performance of seahorses. The measured indices for reproductive performance included ovarian lipid content, assessment time required before mating for paired male and female seahorses, mating success rate, brood size, and newborn body height. The results showed that for juveniles during diet conversion, their body weight, survival rate, plasma immunocytokine contents (interleukin-2, interferon-α, and immunoglobulin M), and alpha diversity indices (Simpson and Pielou's-e) of the intestinal microbiota were significantly higher than those of the control group after dietary APSs. For broodstocks, compared with the control group, the expression of lipid substances in the ovary was significantly upregulated, the assessment time was significantly shortened, and the body height of their newborns was significantly increased in the APS group. These results demonstrate that APSs could indeed improve the immune-health status and reproductive performance of seahorses, providing guidance for addressing existing issues in seahorse cultivation.

Keywords: *Astragalus* polysaccharides; immune capacity; juveniles and broodstocks; reproductive performance; seahorse *Hippocampus erectus*

1. Introduction

Seahorse (*Hippocampus* spp.) is very popular in traditional Chinese medicine, aquarium, and curio markets. The growing global trade of seahorses has exerted a great pressure on wild stocks, and all species of this genus have been listed in Appendix A of CITES since 2004 [1]. In China, seahorses are mainly used as a traditional Chinese medicine and are in wide demand on the market [2,3]. In order to protect wild seahorses and still meet the strong market demand, China has increased its attempts at the artificial cultivation of seahorses since the early 21st century [4–6], and has successfully achieved the large-scale cultivation of some species such as the lined seahorse *H. erectus* [7,8] and the big-belly seahorse *H. abdominalis* [9,10]. The successful cultivation of seahorses not only brings considerable economic benefits to the enterprises engaged in aquaculture or traditional Chinese medicine, but also reduces the dependence on wild seahorses and protects them.

However, in recent years, there have been some new issues in seahorse cultivation, among which two are the most prominent. The first is that seahorse juveniles are prone to disease during diet conversion [8,11]. Seahorse juveniles after birth only feed on live food (e.g., larvae and adults of copepods and cladocerans) for the first two and a half months and start feeding on frozen food (e.g., *Artemia*, *Mysis* and *Acetes*) after two and a half months (their body height reaches at approximately 7.5 cm at this time) [12]. After 10–15 days of consuming frozen food, they begin to show symptoms of disease. The symptoms manifest as diseased seahorses floating on the water surface, weak swimming, inactive feeding, inability to hold on the holdfasts, abdominal swelling, and anal protrusion. After the onset of symptoms, the vast majority of seahorses die within 7 days, with a total mortality rate of up to 60%. The second is that the reproductive performance of seahorse broodstocks has significantly declined compared with the previous level. Specifically, the time required for male and female seahorses from meeting to first mating success is extended (this period is a process of mutual familiarity and assessment), and the continuous breeding frequency of pair-bonded male and female seahorses is also reduced. Even worse, after two or three pregnancies, the brood pouches of some male seahorses begin to contract and harden, even filling with gas that cannot be expelled, leading them to be unable to receive female eggs and temporarily stop breeding. The decline in reproductive performance greatly increases the management and maintenance cost of seahorse broodstocks for aquaculture enterprises.

Regarding the above two issues, some researchers have conducted relevant studies. Tian et al. [11] believe that the reason for the first issue may be the intestinal microbiota dysbiosis of seahorse juveniles caused by frozen food, thereby leading to diseases. Therefore, they suggest supplementing the diet with *Enterococcus faecium* (a probiotic), which can improve the survival rate and intestinal health of seahorse juveniles during diet conversion. Tang et al. [13] and Xue et al. [14] hold that the reason for the second issue may be related to the increase in pollutants in the marine environment. Through toxicological tests, they demonstrated that some marine pollutants (e.g., tributyltin, benzo[a]pyrene) do indeed have an inhibitory effect on the reproductive performance of seahorses. In addition to the possible reasons above-mentioned, we suppose that the decline in seahorse immunity may also be one. Therefore, an improvement in seahorse immunity by adding immunostimulants may help alleviate these two issues.

Astragalus polysaccharides (APSs) have been widely used as immunostimulants in aquaculture for more than 20 years. Their beneficial effects, such as the promotion of growth and digestion, enhancement of immune response, antioxidant capacity, and bacterial resistance, and improvement of intestinal homeostasis, have been well-verified in a large number of aquaculture animals such as large yellow croaker (*Larimichthys crocea*) [15], Asian seabass (*Lates calcarifer*) [16], giant freshwater prawn (*Macrobrachium rosenbergii*) [17], Nile tilapia (*Oreochromis niloticus*) [18], turbot (*Scophthalmus maximus* L.) [19], crucian carp (*Carassius auratus*) [20] and rainbow trout (*Oncorhynchus mykiss*) [21]. However, so far, there has been no verification in seahorses. The lined seahorse (*H. erectus*) is naturally distributed in Nova Scotia along the western Atlantic coast through the Gulf of Mexico and Caribbean to Venezuela [22]. Due to its large size, fast growth, diverse body color, and high temperature resistance, it was introduced into China for captive breeding in 2009 and successfully commercialized in about 2015 [6,23]. The lined seahorse is widely cultivated in southern China, such as in Hainan, Guangdong, Guangxi and Fujian, and the annual output (dried specimen) has exceeded 10 tons since 2020.

In the present study, we used the lined seahorse as the experimental subject and employed APSs to test whether it helped improve the immune-health status of juveniles during diet conversion and the reproductive performance of broodstocks during breeding. The measured indices for immune-health status included growth, survival rate, plasma immunocytokine contents (interleukin-2 (IL-2), interferon-α (IFN-α), and immunoglobulin M (Ig M)), and composition and diversity of intestinal microbiota. IL-2, IFN-α, and Ig M are widely reported to be three immunocytokines closely related to the immune status of the host, and their levels largely reflect the host's immune capacity [24,25]. The intestinal microbiota has been well-demonstrated to play an important role in maintaining the host's intestinal health and metabolic homeostasis [26]. The measured indices for reproductive performance included female ovarian lipid content, assessment time required before mating for paired male and female seahorses, mating success rate, brood size, and newborn body height. Ovarian lipids are the most important energy source for embryonic development and are an important indicator of female reproductive investment [27]. Paired male and female seahorses from their first encounter to successful mating need a period of time for mutual familiarity and assessment, and usually a longer assessment time is required between pairs with poor health status. The aim of the present study was to provide potential measures to the issues that exist in seahorse cultivation.

2. Materials and Methods

2.1. Experimental Seahorses and Astragalus Polysaccharides

The present study was carried out at the Seahorse Research Center of the East China Sea Fisheries Research Institute, Qionghai City, Hainan Province, China. The Center has hundreds of thousands of artificially cultivated lined seahorses of various sizes and several sets of standardized aquaculture facilities. These seahorses were fed live copepods for the first two and a half months after birth, and then switched to feeding frozen *Mysis*.

The gender-undifferentiated juvenile seahorses with a body height of approximately 7.5 cm who had not yet switched to feeding frozen *Mysis* were selected as the experimental seahorses for Experiment 1, and the male and female seahorses near sexual maturity (with a body height of approximately 11 cm) who had been feeding on frozen *Mysis* for a period of time were selected as the experimental seahorses for Experiment 2. These experimental seahorses were stored in their respective holding tanks.

Astragalus polysaccharides (APSs, Henghuibio^{TM}) were purchased from Beijing Rui Da Heng Hui Science & Technology Development Co., Ltd. (Beijing, China), and the purity was over 90% (Product No. HD-6198-25g).

2.2. Administration Route of APSs

In numerous aquaculture applications, the administration route of APSs is mostly through adding polysaccharides to the basal diet formula, then drying and pelleting them into formulated feed for feeding [15–17]. However, for seahorse cultivation, the current food used is mainly fresh or frozen natural feed, and there is no formulated feed as yet. Therefore, the route of the integration of APSs into the formulated feed is not workable in seahorses. As for the route of simply mixing APSs with frozen food, it may not be very effective. On the one hand, the taste of frozen food may be altered by APSs, thereby affecting the feeding of seahorses. On the other hand, the APSs adhered to the surface of frozen food is easily diluted by seawater, resulting in a limited intake of APSs by seahorses. Therefore, in the present study, we adopted the route of using *Artemia* as the polysaccharide carrier [28], that is, live *Artemia* were soaked in the APS suspension for a period of time, and when the APSs in their bodies was enriched to a certain level, they were then fed to the seahorses.

In order to determine the APS concentration used for soaking *Artemia* and the APS content inside the *Artemia* after soaking, we conducted a preparatory experiment. In this experiment, five different concentrations (i.e., 0, 2, 4, 8, 16% (w/v)) of APS suspensions were designed, and each concentration had three replicates, where each replicate contained one gram of live *Artemia* (body length of 4–5 mm). After soaking for 30 min, the APS content inside the live *Artemia* was measured using UV spectrophotometry (UV-2700, Shimadzu, Kyoto, Japan) [29]. The results showed that the APS contents inside the *Artemia* corresponding to 0, 2, 4, 8, and 16% APS suspensions were 0.114 ± 0.014, 0.176 ± 0.017, 0.241 ± 0.024, 0.229 ± 0.014, and 0.226 ± 0.012 g/g, respectively. Statistical analysis showed that there was no significant difference among 4, 8, and 16%, but all three were significantly higher than 0% and 2%. Therefore, in the present study, live *Artemia* (body length of 4–5 mm) soaking in the 4% APS suspension for 30 min was adopted as the administration protocol of APSs.

2.3. Experiment 1: Effects of APSs on the Survival Rate, Growth, and Immunity of Seahorse Juveniles During Diet Conversion

2.3.1. Experiment 1 Protocol

The experimental seahorses (with a body height of approximately 7.5 cm) were divided into two groups, namely the APS group and the control group. Each group consisted of three replicates, and each replicate contained 50 seahorses and placed in an independent rearing tank (0.7 m in diameter and 0.8 m in depth). The seahorses in two groups were both fed with live *Artemia* at 7:00 a.m., followed by frozen *Mysis* at 8:30 a.m. and 3:00 p.m. every day. In contrast, seahorses in the APS group were fed with *Artemia* that was soaked in 4% APS suspension for 30 min, and seahorses in the control group were fed with *Artemia* that was not soaked in APS suspension. The feces and uneaten frozen shrimp were siphoned out two hours after each feeding. The rearing tanks were continuously flowed with fresh seawater for 12 h a day at a flowrate of 1.2–1.5 L/min. The parameters of seawater were a temperature of 26–27 °C, salinity of 30–31‰, dissolved oxygen content of 6.0–6.5 mg/L, and pH of 8.0–8.2, respectively. The experiment lasted for 30 days.

2.3.2. Survival Rate and Growth Measurement

During the experiment, the dead seahorses (if any) from each replicate of each group were timely picked out every day and recorded for calculating the final survival rate. On the last day of the experiment, 20 surviving seahorses from each replicate of each group were taken for measuring their body height and wet body weight. Body height was measured with a ruler with a precision of 0.1 cm (i.e., 1 mm), and the wet body weight was measured using a balance (XPR303S/AC, Mettler Toledo, Zurich, Switzerland) with a precision of 0.001 g (i.e., 1 mg).

2.3.3. Plasma Immunocytokine Determination

On the last day of the experiment, six surviving seahorses were randomly sampled from each replicate of each group for blood collection. Each seahorse was first anesthetized with a solution of MS-222 (50 mg/L) (Sigma-Aldrich, St. Louis, MO, USA) in seawater, and then a quarter of the tail was cut off. The remaining tail of the amputated seahorse was immediately inserted into a 2-mL centrifuge tube containing 0.8 mL of anticoagulant (citric acid 0.48 g, sodium citrate 1.32 g, glucose 1.47 g, and distilled water 100 mL) and dipped into the anticoagulant. Blood spontaneously flowed out of the caudal artery and mixed with the anticoagulant. Due to the small volume of blood in each seahorse, the blood from each of the three seahorses was pooled in the same centrifuge tube. The mixture of blood and anticoagulant was left to stand for 10 min at 4 °C and then centrifuged at 840× *g* for 10 min at 4 °C (Sigma 3K18, Sigma-Aldrich, St. Louis, MO, USA) to collect the supernatant (i.e., plasma). The plasma was stored at −80 °C until use for the determinations of Ig M, IL-2, IFN-α, and plasma protein. Ig M, IL-2, IFN-α, and plasma protein were determined with commercial ELISA kits (Nanjing Jiancheng Bioengineering Institute, Nanjing, China) for fish plasma following the manufacturer's instructions. The units of Ig M, IL-2, and IFN-α were expressed as the amount contained in each milligram of plasma protein.

2.3.4. Intestinal Microbiota Analysis

After blood collection, the surface of the seahorse was disinfected with 75% (*v*/*v*) of alcohol, then placed on an ultra-clean workbench (BBS-H1500, BioBase, Shandong, China) and dissected aseptically for intestinal tract collection. The intestinal tracts of each of the three seahorses were pooled in the same frozen tube (Rnase/Dnase-free) and stored at −80 °C until use for the analysis of the intestinal microbiota composition. This analysis was conducted with 16S rDNA sequencing [30]. Six indices, including Chaol richness estimator (Chao1), observed species richness (Observed species), Simpson diversity index (Simpson), Shannon Wiener's diversity index (Shannon), Pielou's evenness index (Pielou's-e), and Good's nonparametric coverage estimator (Good's coverage), were used to assess the alpha diversity of the intestinal microbiota.

2.4. *Experiment 2: Effects of APSs on the Reproductive Performance of Seahorse Broodstocks*

2.4.1. Experiment 2 Protocol

The experimental seahorses (with a body height of approximately 11 cm) were divided into two groups, namely the APS group and the control group. The seahorses in each group were separated by gender and reared separately. The seahorses in each group were fed with live *Artemia* at 7:00 a.m., followed by frozen *Mysis* at 8:30 a.m. and 3:00 p.m. every day. In contrast, seahorses in the APS group were fed with APSs-soaked *Artemia*, while seahorses in the control group were fed with APSs-unsoaked *Artemia*. The daily management and rearing conditions were the same as in Experiment 1. After one month of feeding, six female seahorses were randomly taken from each group to collect their ovaries for metabolite component analysis, and another 20 males and 20 females were selected from each group to pair them in one male and one female, for a total of 20 pairs, for reproductive performance observation. Each pair was placed in an independent breeding tank (0.8 m in diameter and 1.0 m in depth), and the identities of these pairs were named after the tank numbers they were in, namely Tank #1 to Tank #20. During the pairing period, seahorses stopped feeding on live *Artemia* but continued to feed on frozen *Mysis* twice a day.

2.4.2. Ovarian Metabolite Component Analysis

The female seahorses were first anesthetized with a solution of MS-222 (50 mg/L) in seawater and then dissected to collect their ovaries for metabolite component analysis. Metabolite components were determined using ultra-high performance liquid chromatography-tandem mass spectrometry (UHPLC-MS/MS). Briefly, 60 mg of the ovary samples were homogenized in 200 μL of distilled water. Then, the metabolites were extracted with 800 μL of methanol:acetonitrile solution (1:1, *v*/*v*) and vacuum freeze-dried. The extract was dissolved in 100 μL of acetonitrile:water (1:1, *v*/*v*), and 2 μL of this solution was taken to add onto an ACQUITY UPLC BEH Amide column (1.7 μm, 2.1 mm × 100 mm) (Waters Technologies Ireland Ltd., Wexford, Ireland) for hydrophilic interaction liquid chromatography with an Agilent 1290 Infinity UHPLC system (Agilent Technologies, Santa Clara, CA, USA) [31].

2.4.3. Mating Success Rate, Assessment Time Required Before Mating, Brood Size, and Newborn Body Height

After pairing, the mating status of each pair was observed and checked daily for two weeks. The criterion for mating success was to simultaneously observe an obvious expansion of the male's pouch (due to the reception of eggs) and an obvious contraction of the female's abdomen (due to the expulsion of eggs). Once mating success was observed, the identity and mating date of the pair were recorded for calculating the mating success rate and the assessment time required before mating. After giving birth, its number of newborns was counted, and the body height of ten randomly-sampled newborns was measured. The mating success rate refers to the percentage of successfully-mated pairs among the 20 pairs, while the assessment time required before mating refers to the time from the day of pairing to the day of successful mating. Brood size refers to the total number of newborns in a brood. Body height, a key indicator to evaluate the quality of the newborn [32], refers to the straight-line distance from the tip of the head coronet to the tip of the uncurled tail.

2.5. Statistical Analysis

All data were expressed as the mean ± standard deviation (SD) and were analyzed using Origin PRO (v9.8.0.200) graphing and analysis software. The differences between the control and APS groups in both two experiments were analyzed using the independent sample *t*-test. The p value of 0.05 was taken as a significance level. Before running the *t*-test, the normality of all data was tested using the Shapiro–Wilk test.

3. Results

3.1. Effects of APSs on the Survival Rate, Growth, and Immunity of Seahorse Juveniles During Diet Conversion

3.1.1. Survival Rate and Growth

Throughout the entire experiment, both groups of seahorses died to varying degrees, and most of the deaths occurred between the 15th and 24th days of the experiment. However, the number of dead individuals in the APS group was obviously less than that in the control group. The final survival rates of the APS group and the control group were 84.7 ± 3.1% and 69.3 ± 3.1%, respectively, with a significant difference between the two (p = 0.004) (Figure 1). As for the growth, the body height of the APS group and the control group both reached about 9.8 cm at the end of the experiment, and there was no significant difference between them (p = 0.794). However, in terms of the wet body weight, the level of the APS group (4.217 ± 0.530 g) was significantly heavier than that of the control group (3.991 ± 0.565 g) (p = 0.026) (Figure 2).

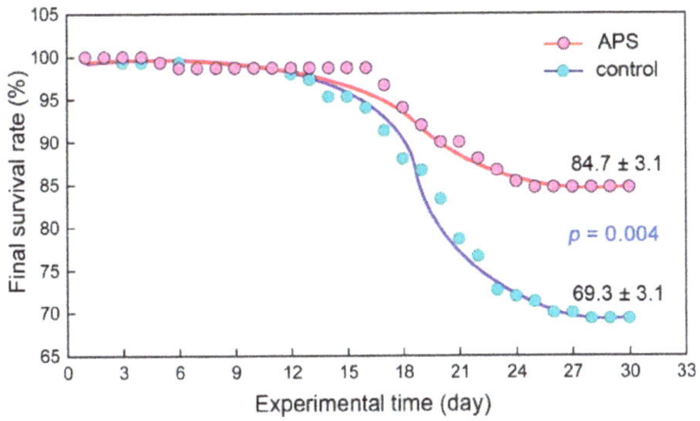

Figure 1. Final survival rates of seahorse juveniles in the control group and the APS group in Experiment 1.

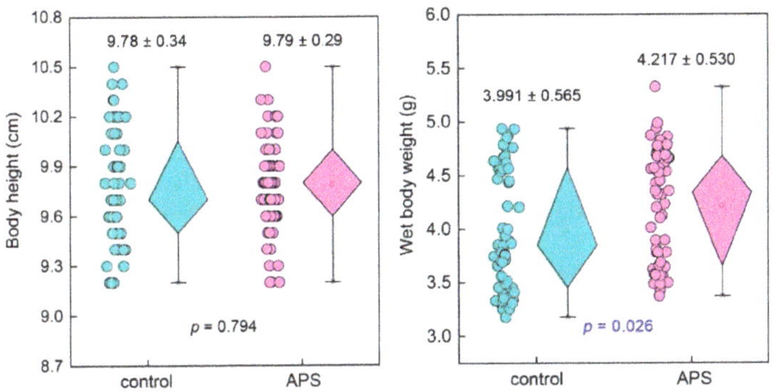

Figure 2. The body height and wet body weight of seahorse juveniles in the control group and the APS group in Experiment 1.

In addition, at the later stage of the experiment, the body color of the seahorses in the APS group was also different from that in the control group. Specifically, the seahorses in the APS group often appeared reddish brown in the morning and dusk (Figure 3B) while the seahorses in the control group were always black or gray black (Figure 3A).

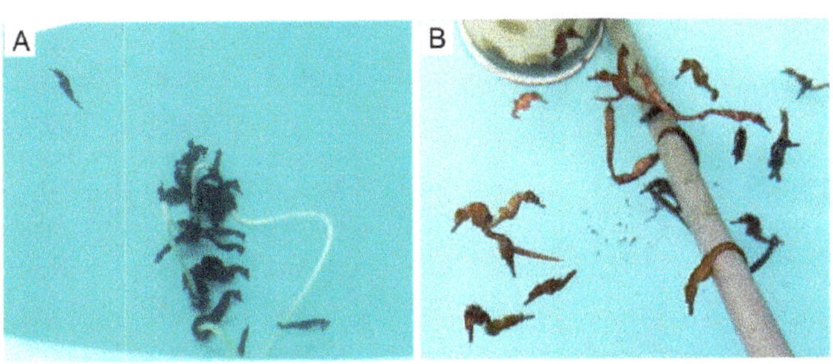

Figure 3. The body color of seahorse juveniles in the control group (**A**) and the APS group (**B**) in the later stage of Experiment 1.

3.1.2. Plasma Immunocytokines

At the end of the experiment, the plasma Ig M in the seahorses of the APS group was 0.64 ± 0.07 mg/mg plasma protein, which was significantly higher than that of the control group (0.55 ± 0.05 mg/mg plasma protein) ($p = 0.035$) (Figure 4A). Similarly, the plasma IFN-α (0.115 ± 0.012 vs. 0.094 ± 0.010 ng/mg plasma protein, $p = 0.007$) and plasma IL-2 (2.65 ± 0.29 vs. 2.13 ± 0.37 pg/mg plasma protein, $p = 0.022$) were also significantly higher in the APS group than in the control group (Figure 4B,C).

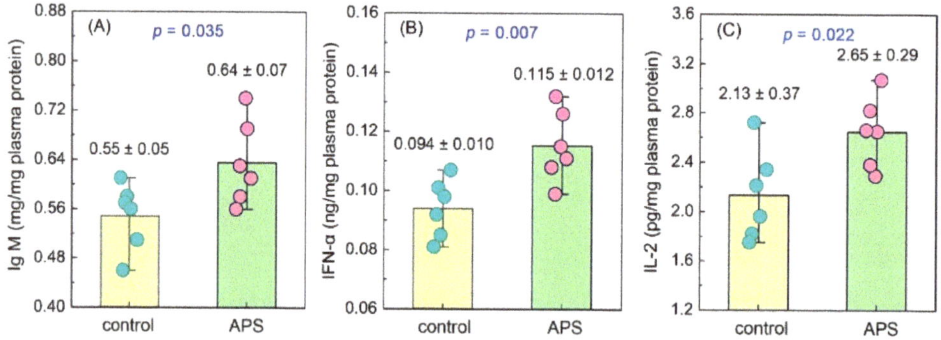

Figure 4. The contents of the plasma Ig M (**A**), IFN-α (**B**) and IL-2 (**C**) of seahorse juveniles in the control group and the APS group in Experiment 1.

3.1.3. Intestinal Microbiota

As for the alpha diversity of intestinal microbiota, two indices, the Simpson (indicating microbial community diversity) ($p = 0.026$) and Pielou's-e (indicating microbial community evenness) ($p = 0.036$) of the APS group were significantly higher than those of the control group, while the other four indices had no significant difference (Figure 5). Among the top 50 intestinal microbiota with average abundance ranking, the abundance of genera of *Zobellia* (mean value: 0.0139 vs. 0.0004), *Bacteriovorax* (0.0023 vs. 0), *Oceaniovalibus* (0.0368 vs. 0.0122), and *Thiothrix* (0.0017 vs. 0.0005) in the APS group was higher than that in the control group (Figure 6, red border), while the abundance of genera of *Roseobacter* (0.0054 vs. 0.0005), *Glaciecola* (0.0012 vs. 0), *Pseudoalteromonas* (0.0038 vs. 0), *Psychrobacter* (0.0063 vs. 0.0004), *Maribacter* (0.0168 vs. 0.0030), and *Tenacibaculum* (0.0065 vs. 0.0027) in the control group were higher than those in the APS group (Figure 6, blue border).

3.2. Effects of APSs on the Reproductive Performance of Seahorse Broodstocks

3.2.1. Ovarian Metabolite Components

The metabolomic data showed that there were 141 significantly differential metabolites (SDMs) in the ovaries between the APS group and the control group. Taking the control group as the baseline, the relative abundance of 128 SDMs in the APS group was upregulated, while the relative abundance of 113 SDMs was downregulated (Figure 7). Among the top 50 SDMs in VIP (variable importance in the projection) values, metabolites with upregulated relative abundance in the APS group included many lipids, such as glycerophospholipids (1-dodecanoyl-choline alfoscerate, LPC (13:0), PC (14:0/16:1) and PI (18:0/20:3)), steroids and steroid derivatives (CHEBI: 70313), prenol lipids (vitamin A2), sphingolipids (SM (d18:0/16:1)), and fatty acyls (tridec-8-enoylcarnitine and desthiobiotin) (Figure 8, red border), indicating that the lipids in the ovaries of the APS group were more abundant than those of the control group.

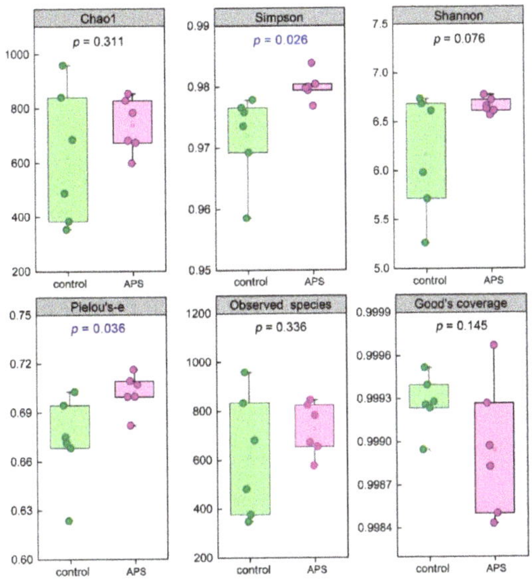

Figure 5. The alpha diversity of intestinal microbiota in seahorse juveniles of the control group and the APS group in Experiment 1.

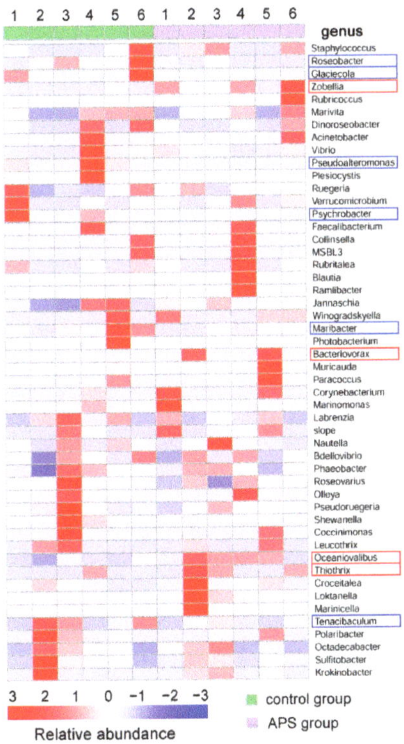

Figure 6. The composition of intestinal microbiota in seahorse juveniles of the control group and the APS group in Experiment 1. The intestinal microbiota listed in the heatmap are the top 50 genera in terms of average abundance. The bacterial genera with red borders are those with higher abundance in the APS group than in the control group, while the bacterial genera with blue borders are those with higher abundance in the control group than in the APS group.

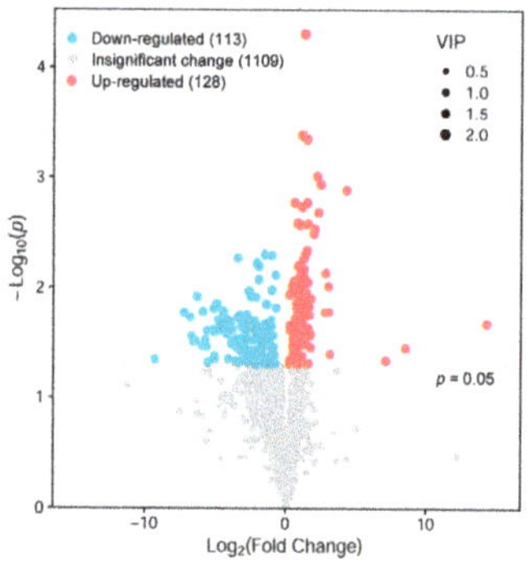

Figure 7. Volcano plot of significantly differential metabolites (SDMs) in the ovaries of seahorse broodstocks between the control group and the APS group in Experiment 2. Using fold change (FC) $\geq$1.5 as the baseline, the red plots indicate the abundance-upregulated SDMs in the APS group compared with the control group, and the blue plots indicate the abundance-downregulated SDMs in the APS group compared with the control group. The size of the plot indicates the VIP (variable importance in the projection) value of the OPLS-DA model. The dashed line is the critical line for $p = 0.05$.

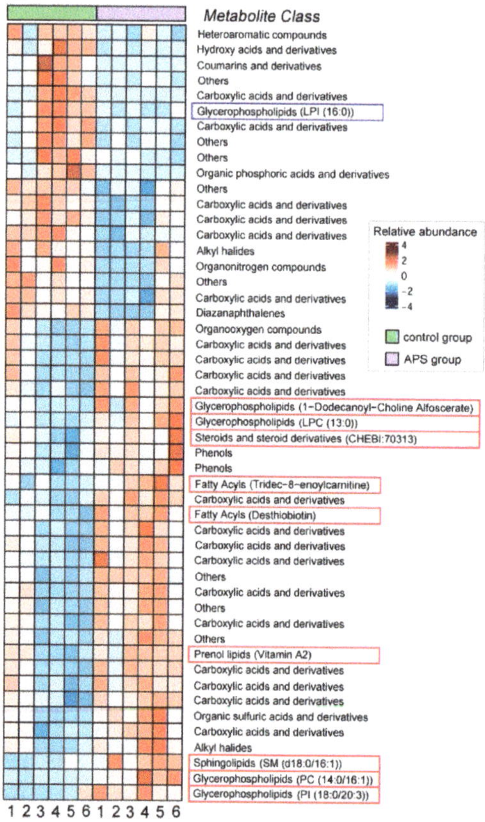

Figure 8. The top 50 significantly differential metabolites (SDMs) in VIP (variable importance in the projection) values in the ovaries of seahorse broodstocks between the control group and the APS group in Experiment 2. The SDMs with red borders are those lipids whose abundance was upregulated in the APS group, and the SDMs with blue borders were those lipids whose abundance was downregulated in the APS group.

3.2.2. Mating Success Rate and Assessment Time Required Before Mating

The mating success rate of the APS group was 65% (13/20), and that of the control group was 60% (12/20). The assessment time required before mating of the successfully-mated pairs in the APS group was 4.92 ± 1.55 day, which was significantly shorter than 7.92 ± 1.38 day in the control group ($p < 0.001$) (Table 1).

Table 1. The mating status and the assessment time required before the mating of seahorses in the control group and the APS group in Experiment 2.

Tank No.	Control		APS	
	Mating Status	[1] Time (Day)	Mating Status	Time (Day)
1#	√	6	–	
2#	√	7	√	6
3#	–		√	5
4#	–		√	3
5#	√	8	–	
6#	√	6	√	7
7#	–		–	
8#	√	10	√	6
9#	√	7	√	5
10#	√	9	√	3
11#	–		–	
12#	–		√	3
13#	√	7	√	5
14#	√	8	√	4
15#	√	8	–	
16#	–		√	8
17#	√	9	√	4
18#	–		–	
19#	√	10	–	
20#	–		√	5
Total	12		13	
Mean		7.92 ± 1.38	($p < 0.001$)	4.92 ± 1.55 *

[1] Time: the assessment time required before mating (i.e., the time from the day of pairing to the day of successful mating). "√": Successfully-mated pair; "–": Pair without successful mating within the two-week observation period. "*" indicates a significant difference between the control group and the APS group.

3.2.3. Brood Size and Newborn Body Height

The brood size of the APS group was 149.2 ± 31.0 inds, with no significant difference compared with the 144.0 ± 29.4 inds of the control group ($p = 0.670$). For the newborn body height, the value of the APS group was 0.99 ± 0.05 cm, which was significantly higher than 0.94 ± 0.05 cm of the control group ($p < 0.001$) (Figure 9).

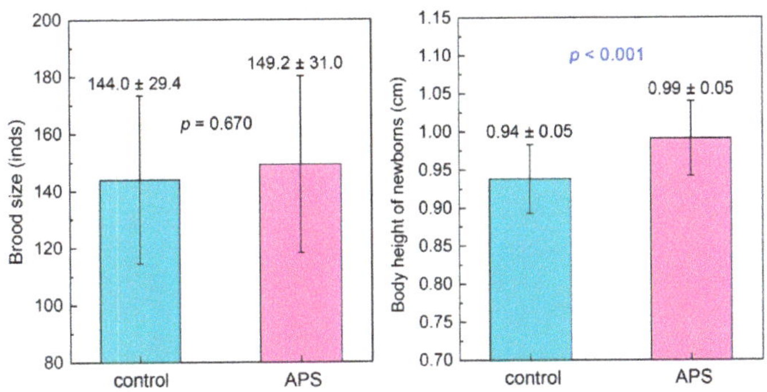

Figure 9. The brood size and the newborn body height of seahorse broodstocks in the control group and the APS group in Experiment 2.

4. Discussion

The results of the present study showed that dietary APS could significantly improve the growth, survivorship, plasma immune levels, and intestinal microbiota diversity of seahorse juveniles during diet conversion as well as the ovarian lipid content, mating assessment time, and newborn body height of seahorse broodstocks during breeding. These results indicate that the promoting effects of APSs on growth, immune function, antibacterial activity, and intestinal homeostasis are also applicable to the lined seahorse (*H. erectus*), just like other aquaculture animals. Conspicuously, these results also indicate that APSs have the effect of improving reproductive performance, which has rarely been reported in other aquaculture animals. Taken together, in the present study, we have demonstrated not only the applicability of APSs in the seahorse, but also the feasibility of using live *Artemia* as the carrier for APS administration.

4.1. APSs Improve the Immune Capacity of Seahorse Juveniles

One of the most prominent effects of APSs is immune improvement. Many applications have shown that it can improve the phagocytic ability and respiratory burst activity in the blood cells of tilapia (*Oreochromis niloticus*) [33], induce the mRNA expression of interleukin-1β in the head kidney of common carp (*Cyprinus carpio* L.) [34] and toll-like receptors in the liver of turbot (*Scophthalmus maximus* L.) [19], activate the pro-phenoloxidase and increase the immunocytokine content in the plasma of giant freshwater prawn (*Macrobrachium rosenbergii*) [17], and enhance the disease resistance of catla (*Catla catla*) [35], Asian seabass (*Lates calcarifer*) [16], and crucian carp (*Carassius auratus*) [20]. In recent years, APSs, as an immunostimulant, have been increasingly applied in aquaculture.

Ig M, IL-2, and IFN-α are three immunocytokines closely related to immune status. Their high levels usually mean that the fish has a relatively good immune status [24,25]. In the present study, after dietary APSs, the Ig M, IL-2, and IFN-α in the plasma of the seahorse increased, which intuitively indicated that the immune capacity of the seahorse was stimulated by APSs. In other words, the immune-promoting effect of APSs was also workable in the seahorse. In addition to the direct evidence of plasma immunocytokines, there were two indirect pieces of evidence that could also indicate APSs improving the immune status of the seahorse. The first was the survival rate. The immune enhancement stimulated by APSs largely synergistically enhanced the bacterial resistance, free-radical scavenging activity, and pathogen killing ability of the seahorse, thereby leading to a significant increase in survival rate. The second was body color. Seahorses are a kind of ornamental fish and can present a variety of beautiful body colors (e.g., bright yellow, orange, reddish brown, and light green). The presentation of beautiful body color usually means that the seahorse is in a good physical condition, because color presentation is a high-energy consumption behavior, and usually only healthy individuals can afford this behavior [36–38]. In our previous study, we observed that the seahorses fed live *Mysis* exhibited a brighter body color compared with the seahorses fed frozen *Mysis* [39], and in the present study, the body color of seahorses treated with APSs was brighter than that of the untreated seahorse, both suggesting that the immune-health status of the seahorses had been greatly improved to the extent that they could afford the behavior of color presentation.

As for the mechanism of APSs promoting the immunity of aquatic animals, many researchers have also carried out extensive studies [40,41]. After APSs are ingested and transferred to the intestinal tract, some intestinal bacteria are induced, and they break down the polysaccharides into fermentable monosaccharides by encoding numerous carbohydrate-active enzymes (CAZymes) [42]. These fermentable monosaccharides are further broken down by other microorganisms to form various bioactive metabolites. Among these metabolites, there is a type that can stimulate the proliferation of macrophages and lymphocytes, probably through the activation of the TLR4-MyD88-dependent pathway through TLR4, which subsequently activates key nodes in the pathway (e.g., TRAF-6 and NF-κB) [43]. Macrophages are essential immune cells that participate in stimulating innate immunity, which are also crucial for the adaptive immunological response. Depending on different external stimuli, macrophages are activated and differentiated into M1 or M2 macrophages. M1 macrophages can produce proinflammatory cytokines and synthesize inducible nitric oxide synthase, mainly participating in the elimination of pathogens and the initiation of inflammation [44,45]. Lymphocytes can produce various immune cytokines, including antibodies, interleukins, and interferons, through different cellular subsets, promoting the growth and activation of other immune cells, thereby killing and clearing intracellular pathogenic infections [46]. The immune mechanism of APSs provides a good explanation for why the content of plasma immunocytokines (i.e., Ig M, IL-2, and IFN-α) in the APS group increased in the present study.

4.2. APSs Improve the Growth of Seahorse Juveniles

In addition to the effect of immune-promoting, another prominent effect of APSs, namely growth-promoting, was also verified in the present study. We found that the body height of the APS group was not different from that of the control group, but the body weight was significantly heavier than that of the control group. The reason why there was no difference in body height but a difference in body weight may be that seahorse juveniles with a body height of approximately 9.5 cm begin to prepare for reproduction, and so they allocated more energy to their body weight for the development of gonads and brood pouches, resulting in a reduction in the energy allocated to body height [8]. As for the reason why APSs can promote growth, some studies believe that this is related to the fact that APSs have the property of improvement in digestive enzyme activities, digestive juice secretion, and absorption of nutrients [15,47].

4.3. APSs Improve the Intestinal Microbiota of Seahorse Juveniles

The homeostasis of intestinal microbiota has a profound impact on host health. The high richness, evenness, and diversity of the intestinal microbiota usually mean that the host has good intestinal homeostasis, which in turn represents that the host is in a healthy status [48,49]. In the present study, after dietary APSs, two alpha diversity indices of intestinal microbiota, the Simpson (indicating microbial community diversity) and Pielou's-e (indicating microbial community evenness), were significantly increased, which indicated that the health status of the seahorses had been significantly improved. As for the mechanism of APSs improving intestinal homeostasis, some researchers have also conducted relevant studies. Polysaccharides, acting as carbon sources, are involved in regulating the makeup, development, metabolism, biological activity, and metabolite synthesis of intestinal microbiota [42]. Moreover, polysaccharides can also be degraded by intestinal microbiota (especially *Bacteroidetes*, *Firmicutes*, and *Actinobacteria*) into prebiotics that can be absorbed by the host and are highly beneficial to the host's health [41,50].

4.4. APSs Improve the Reproductive Performance of Seahorse Broodstocks

Although the effects of APSs in promoting immunity and growth have been widely demonstrated in fish [15–18], its effect in promoting reproduction has rarely been reported. In the present study, we found that APSs also had the effect of promoting reproduction, which was specifically shown as follows: compared with the control group, the assessment time required before mating for male and female seahorses in the APS group was significantly shortened, and the body height of the newborns was significantly increased. This improvement in reproductive performance may mainly be attributed to the immune enhancement induced by APSs, that is, the improvement in reproductive performance is also one of the intuitive manifestations of seahorse immune enhancement. In our previous study, we found that immunity and health status are important criteria for seahorse mate choice. Individuals with strong immunity usually show a beautiful body color and active courtship behavior, and they spend significantly less time in assessing their mates and succeeding in mating than individuals with poor body color and inactive courtship behavior [39]. In addition, seahorses are monogamous species [51], and they rarely change their mates during their life, but once their mates have health problems, they will switch mates [52]. This fact also illustrates the importance of immune capacity in seahorse mate choice.

Moreover, it may be precisely due to the immune enhancement induced by APSs that the energy investment for seahorses to maintain their own immunity can be appropriately reduced, thereby allocating more energy to other aspects such as reproduction (e.g., lipid synthesis in ovaries). In the present study, we found that the expression abundance of lipids (e.g., glycerophospholipids, steroids, prenol lipids, sphingolipids, and fatty acyls) in the ovary of the APS group was significantly higher than those of the control group. Lipids in the ovaries or eggs are the most important substances for embryonic development. They not only provide the primary energy for developing embryos, but are also involved in many biological functions such as embryonic pigmentation, organogenesis (e.g., eyes and brain) and neurological development, embryonic immune regulation, and lipid metabolism regulation [53,54]. Eggs with high lipid content are widely regarded as high-quality eggs [27]. In the present study, it is precisely because the ovaries in the APS group contained more abundant lipids that their embryos developed better, and their newborns' body height was significantly higher than that in the control group.

5. Conclusions

In the present study, we used live *Artemia* as the carrier of APSs to feed the seahorses (*H. erectus*). The results showed that APSs could not only improve the survival rate, growth, and immune capacity of seahorse juveniles during diet conversion, but also improve the reproductive performance of seahorse broodstocks. These results provide guidance for the healthy cultivation of seahorses and the administration route of APSs in other cultured fish.

Author Contributions: Conceptualization, S.L. and T.L.; Methodology, S.L. and X.L.; Software, S.L. and Y.R.; Validation, S.L., X.L., and T.L.; Formal analysis, S.L. and T.L.; Investigation, K.J.; Resources, D.Z. and K.J.; Data curation, S.L.; Writing—original draft preparation, S.L.; Writing—review and editing, T.L.; Visualization, Y.R.; Supervision, D.Z.; Project administration, S.L. and X.L.; Funding acquisition, T.L. and D.Z. All authors have read and agreed to the published version of the manuscript.

Funding: This research was funded by the Key Research and Development Projects of Hainan (Grant No. ZDYF2022XDNY350), the Central Public-Interest Scientific Institution Basal Research Funds of CAFS (Grant No. 2022YJ09), the National Natural Science Foundation of China (Grant No. 32172950), and the Open Project of Key Laboratory of Tropical Marine Ecosystem and Bioresource, Ministry of Natural Resources of China (Grant No. 2023QN01).

Institutional Review Board Statement: All procedures were conducted in compliance with the recommendations in the Guide for the Management and Use of the Experimental Animals of China Science and Technology Commission. The animal study protocol was authorized by the Committee on the Ethics of Animal Experiments of Chinese Academy of Fishery Sciences, with approval reference number: CAFS-ECSF-2024-016.

Informed Consent Statement: Not applicable.

Data Availability Statement: The original data presented in the study are included in the article; further inquiries can be directed to the corresponding author.

Acknowledgments: We are very grateful to the Wenchang Innovation Research Center for the free provision of experimental seahorses as well as their staff's assistance in seahorse rearing and management.

Conflicts of Interest: The authors declare no conflicts of interest.

References

1. Koldewey, H.J.; Martin-Smith, K.M. A global review of seahorse aquaculture. *Aquaculture* **2010**, *302*, 131–152. [CrossRef]
2. Kuo, T.C.; Vincent, A. Assessing the changes in international trade of marine fishes under CITES regulations—A case study of seahorses. *Mar. Policy* **2018**, *88*, 48–57. [CrossRef]
3. Zeng, L.; Armani, A.; Wen, J.; Lin, H.; Xu, Y.; Fan, S.; Sun, Y.; Yang, C.; Chen, Z.; Chen, D.; et al. Molecular identification of seahorse and pipefish species sold as dried seafood in China: A market-based survey to highlight the actual needs for a proper trade. *Food Control* **2019**, *103*, 175–181. [CrossRef]
4. Sheng, J.Q.; Lin, Q.; Chen, Q.X.; Gao, Y.L.; Shen, L.; Lu, J.Y. Effects of food, temperature and light intensity on the feeding behavior of three-spot juveniles, *Hippocampus trimaculatus* Leach. *Aquaculture* **2006**, *256*, 596–607. [CrossRef]
5. Lin, Q.; Gao, Y.L.; Sheng, J.Q.; Chen, Q.X.; Zhang, B.; Lu, J.Y. The effects of food and the sum of effective temperature on the embryonic development of the seahorse, *Hippocampus kuda* Bleeker. *Aquaculture* **2007**, *262*, 481–492. [CrossRef]
6. Lin, Q.; Lin, J.; Zhang, D. Breeding and juvenile culture of the lined seahorse, *Hippocampus erectus*. *Aquaculture* **2008**, *277*, 287–292. [CrossRef]
7. Huang, J.; Qin, G.; Zhang, B.; Tan, S.; Sun, J.; Lin, Q. Effects of food, salinity, and ammonia-nitrogen on the physiology of juvenile seahorse (*Hippocampus erectus*) in two typical culture models in China. *Aquaculture* **2020**, *520*, 734965. [CrossRef]
8. Lin, T.; Li, S.; Zhang, D.; Liu, X.; Ren, Y. Effects of stocking density on the survival, growth, and stress levels of the juvenile lined seahorse (*Hippocampus erectus*) in recirculating aquaculture systems. *Biology* **2024**, *13*, 807. [CrossRef]
9. He, L.B.; Qi, J.F.; Lin, J.B.; Chen, X.X.; Luo, H.Y.; Wang, Q.; Zheng, L.Y. Artificial propagation and seeding technique of *Hippocampus abdominalis*. *J. Appl. Oceanogr.* **2022**, *41*, 701–707, (In Chinese with English Abstract).
10. Zhang, Y.; Dong, D.; Wang, H.; Wang, Y.; Gan, Q.; Teng, F. Research advances on the biological characteristics and breeding technology of seahorse (*Hippocampus abdominalis*). *Trans. Oceanol. Limnol.* **2024**, *46*, 190–198. (In Chinese with English Abstract)
11. Tian, Y.; Wang, F.; Su, X.; Zhang, L.; Ma, Z.; Gao, L.; Yan, H.; Xue, Y.; Lv, C.; Zhang, X.; et al. Supplementation with *Enterococcus faecium* enhances growth performance, intestinal health and immunity of big-belly seahorses (*Hippocampus abdominalis*) during diet conversion. *Aquacult. Rep.* **2023**, *28*, 101466. [CrossRef]
12. Lin, Q.; Lin, J.; Zhang, D.; Wang, Y. Weaning of juvenile seahorses *Hippocampus erectus* Perry, 1810 from live to frozen food. *Aquaculture* **2009**, *291*, 224–229. [CrossRef]
13. Tang, L.; Liu, Y.L.; Qin, G.; Lin, Q.; Zhang, Y.H. Effects of tributyltin on gonad and brood pouch development of male pregnant lined seahorse (*Hippocampus erectus*) at environmentally relevant concentrations. *J. Hazard. Mater.* **2021**, *408*, 124854. [CrossRef] [PubMed]
14. Xue, Y.; Wang, F.; Su, X.; Li, M.; Yan, H.; Zheng, S.; Ma, Y.; Dong, J.; Liu, Y.; Lin, Q.; et al. Reproductive toxicity and molecular mechanisms of benzo[a]pyrene exposure on ovary, testis, and brood pouch of sex-role-reversed seahorses (*Hippocampus erectus*). *Environ. Pollut.* **2025**, *367*, 125569. [CrossRef] [PubMed]
15. Liu, Y.; Miao, Y.; Xu, N.; Ding, T.; Cui, K.; Chen, Q.; Zhang, J.; Fang, W.; Mai, K.; Ai, Q. Effects of dietary *Astragalus* polysaccharides (APS) on survival, growth performance, activities of digestive enzyme, antioxidant responses and intestinal development of large yellow croaker (*Larimichthys crocea*) larvae. *Aquaculture* **2020**, *517*, 734752. [CrossRef]
16. Yu, W.; Yang, Y.; Zhou, Q.; Huang, X.; Huang, Z.; Li, T.; Wu, Q.; Zhou, C.; Ma, Z.; Lin, H. Effects of dietary *Astragalus* polysaccharides on growth, health and resistance to *Vibrio harveyi* of *Lates calcarifer*. *Int. J. Biol. Macromol.* **2022**, *207*, 850–858. [CrossRef]
17. Sun, S.; Bian, C.; Zhou, N.; Shen, Z.; Yu, M. Dietary *Astragalus* polysaccharides improve the growth and innate immune response of giant freshwater prawn *Macrobrachium rosenbergii*: Insights from the brain-gut axis. *Int. J. Biol. Macromol.* **2023**, *243*, 125158. [CrossRef]
18. Zahran, E.; Risha, E.; AbdelHamid, F.; Mahgoub, H.A.; Ibrahim, T. Effects of dietary *Astragalus* polysaccharides (APS) on growth performance, immunological parameters, digestive enzymes, and intestinal morphology of Nile tilapia (*Oreochromis niloticus*). *Fish Shellfish Immunol.* **2014**, *38*, 149–157. [CrossRef]
19. Sun, Y.; Wang, X.; Zhou, H.; Mai, K.; He, G. Dietary *Astragalus* polysaccharides ameliorates the growth performance, antioxidant capacity and immune responses in turbot (*Scophthalmus maximus* L.). *Fish Shellfish Immunol.* **2020**, *99*, 603–608. [CrossRef]
20. Liu, J.; Zhang, P.; Wang, B.; Lu, Y.; Li, L.; Li, Y.; Liu, S. Evaluation of the effects of *Astragalus* polysaccharides as immunostimulants on the immune response of crucian carp and against SVCV in vitro and in vivo. *Comp. Biochem. Phys. C* **2022**, *253*, 109249. [CrossRef]

21. Pan, Y.; Liu, Z.; Lu, J.; Quan, J.; Zhao, G.; Song, G.; Wang, J.; Ren, Z. *Astragalus* polysaccharide inhibits infectious hematopoietic necrosis virus damage to rainbow trout (*Oncorhynchus mykiss*) spleen by promoting the efffcacy of inactivated vaccine. *Fish Shellfish Immunol.* **2025**, *159*, 110180. [CrossRef]
22. Lourie, S.A.; Foster, S.J.; Cooper, E.W.T.; Vincent, A.C.J. A guide to the identification of seahorses. In *Project Seahorse and TRAFFIC North America*; University of British Columbia and World Wildlife Fund: Washington, DC, USA, 2004; pp. 46–47.
23. Zhang, D.; Lin, T.; Liu, X. A comparison of growth, survival, and fatty acid composition of the lined Seahorse, *Hippocampus erectus*, juveniles fed enriched *Artemia* and a calanoid copepod, *Schmackeria dubia*. *J. World Aquacult. Soc.* **2015**, *46*, 608–616. [CrossRef]
24. Tort, L.; Balasch, J.C.; Mackenzie, S. Fish immune system. A crossroads between innate and adaptive responses. *Inmunología* **2003**, *22*, 277–286.
25. Pestka, S.; Krause, C.D.; Walter, M.R. Interferons, interferon-like cytokines, and their receptors. *Immunol. Rev.* **2004**, *202*, 8–32. [CrossRef] [PubMed]
26. Ma, W.; Huang, Z.; Liu, K.; Li, D.; Mo, T.; Liu, Q. The role of intestinal microbiota and metabolites in intestinal inflammation. *Microbiol. Res.* **2024**, *288*, 127838. [CrossRef]
27. Brooks, S.; Tyler, C.R.; Sumpter, J.P. Egg quality in fish: What makes a good egg? *Rev. Fish Biol. Fish.* **1997**, *7*, 387–416. [CrossRef]
28. Aguila, A.A.; Mansir, A.T.; Manríquez, A.R. Using brine shrimp as a drug carrier for therapeutic applications in aquaculture. *Aquacult. Eng.* **1994**, *13*, 301–309. [CrossRef]
29. Yang, S.P. Detection of *Astragalus* polysaccharide content by UV spectrophotometry. *Mod. Livest. Technol.* **2019**, *8*, 7–9. (In Chinese with English Abstract)
30. Lin, T.; Liu, X.; Xiao, D.; Zhang, D.; Cai, Y.; Zhu, X. *Lactobacillus* spp. as probiotics for prevention and treatment of enteritis in the lined seahorse (*Hippocampus erectus*) juveniles. *Aquaculture* **2019**, *503*, 16–25. [CrossRef]
31. Sun, B.; Sun, B.; Zhang, B.; Sun, L. Temperature induces metabolic reprogramming in fish during bacterial infection. *Front. Immunol.* **2022**, *13*, 1010948. [CrossRef]
32. Zhang, D.; Yin, F.; Lin, J. Criteria for assessing juvenile quality of the lined seahorse, *Hippocampus erectus*. *Aquaculture* **2011**, *322*, 255–258. [CrossRef]
33. Ardó, L.; Yin, G.; Xu, P.; Váradi, L.; Szigeti, G.; Jeney, Z.; Jeney, G. Chinese herbs (*Astragalus membranaceus* and *Lonicera japonica*) and boron enhance the non-speciffc immune response of Nile tilapia (*Oreochromis niloticus*) and resistance against *Aeromonas hydrophila*. *Aquaculture* **2008**, *275*, 26–33. [CrossRef]
34. Yuan, C.; Pan, X.; Gong, Y.; Xia, A.; Wu, G.; Tang, J.; Han, X. Effects of *Astragalus polysaccharides* (APS) on the expression of immune response genes in head kidney, gill and spleen of the common carp, *Cyprinus carpio* L. *Int. Immunopharmacol.* **2008**, *8*, 51–58. [CrossRef]
35. Harikrishnan, R.; Devi, G.; van Doan, H.; Tapingkae, W.; Balasundaram, C.; Arockiaraj, J.; Ringø, E. Changes in immune genes expression, immune response, digestive enzymes -antioxidant status, and growth of catla (*Catla catla*) fed with *Astragalus polysaccharides* against edwardsiellosis disease. *Fish Shellfish Immunol.* **2022**, *121*, 418–436. [CrossRef]
36. Lim, K.C.; Yusoff, F.M.; Shariff, M.; Kamarudin, M.S. Astaxanthin as feed supplement in aquatic animals. *Rev. Aquacult.* **2018**, *10*, 738–773. [CrossRef]
37. Luo, M.; Lu, G.; Yin, H.; Wang, L.; Atuganile, M.; Dong, Z. Fish pigmentation and coloration: Molecular mechanisms and aquaculture perspectives. *Rev. Aquacult.* **2021**, *13*, 2395–2412. [CrossRef]
38. Vissio, P.G.; Darias, M.J.; Di Yorio, M.P.; Pérez Sirkin, D.I.; Delgadin, T.H. Fish skin pigmentation in aquaculture: The inffuence of rearing conditions and its neuroendocrine regulation. *Gen. Comp. Endocr.* **2021**, *301*, 113662. [CrossRef] [PubMed]
39. Lin, T.; Liu, X.; Li, S.; Zhang, D.; Shen, F.; Jiang, K. Females increase reproductive investment when mated to less sexually attractive males in a serially monogamous fish. *Sci. Rep.* **2024**, *14*, 19020. [CrossRef] [PubMed]
40. Shahrajabian, M.; Sun, W.; Cheng, Q. A review of *Astragalus* species as foodstuffs, dietary supplements, a traditional Chinese medicine and a part of modern pharmaceutical science. *Appl. Ecol. Environ. Res.* **2019**, *17*, 13371–13382. [CrossRef]
41. Chen, T.; Xie, L.; Shen, M.; Yu, Q.; Chen, Y.; Xie, J. Recent advances in *Astragalus polysaccharides*: Structural characterization, bioactivities and gut microbiota modulation effects. *Trends Food Sci. Technol.* **2024**, *153*, 104707. [CrossRef]
42. Gan, L.; Wang, J.; Guo, Y. Polysaccharides influence human health via microbiota-dependent and -independent pathways. *Front. Nutr.* **2022**, *9*, 1030063. [CrossRef] [PubMed]
43. Zhou, L.; Liu, Z.; Wang, Z.; Yu, S.; Long, T.; Zhou, X.; Bao, Y. *Astragalus* polysaccharides exerts immunomodulatory effects via TLR4-mediated MyD88-dependent signaling pathway in vitro and in vivo. *Sci. Rep.* **2017**, *17*, 44822. [CrossRef]
44. Wang, J.; Li, R.; Peng, Z.; Hu, B.; Rao, X.; Li, J. HMGB1 participates in LPS-induced acute lung injury by activating the AIM2 inflammasome in macrophages and inducing polarization of M1 macrophages via TLR2, TLR4, and RAGE/NF-κB signaling pathways. *Int. J. Mol. Med.* **2020**, *45*, 61–80. [CrossRef]
45. Li, C.; Liu, Y.; Zhang, Y.; Li, J.; Lai, J. *Astragalus* polysaccharide: A review of its immunomodulatory effect. *Arch. Pharm. Res.* **2022**, *45*, 367–389. [CrossRef]

46. Ren, L.; Zhang, J.; Zhang, T. Immunomodulatory activities of polysaccharides from Ganoderma on immune effector cells. *Food Chem.* **2021**, *340*, 127933. [CrossRef] [PubMed]
47. Chen, Y.; Zhou, H. Effect of three kinds of polysaccharide on protease activity, amylase activity in intestine and hepatopancreas of allogynogenetic silver crucian carp. *J. Shanghai Ocean. Univ.* **2005**, *14*, 468–471, (In Chinese with English Abstract).
48. Apper, E.; Weissman, D.; Respondek, F.; Guyonvarch, A.; Baron, F.; Boisot, P.; Rodiles, A.; Merrifield, D.L. Hydrolysed wheat gluten as part of a diet based on animal and plant proteins supports good growth performance of Asian seabass (*Lates calcarifer*), without impairing intestinal morphology or microbiota. *Aquaculture* **2016**, *453*, 40–48. [CrossRef]
49. Spragge, F.; Bakkeren, E.; Jahn, M.T.; Araujo, E.B.N.; Pearson, C.F.; Wang, X.; Pankhurst, L.; Cunrath, O.; Foster, K.R. Microbiome diversity protects against pathogens by nutrient blocking. *Science* **2023**, *382*, 6676. [CrossRef]
50. Niu, Y.; Liu, W.; Fan, X.; Wen, D.; Wu, D.; Wang, H.; Liu, Z.; Li, B. Pharmaceutical potential for bioactive plant polysaccharides in treating disease and gut dysbiosis. *Front. Microbiol.* **2023**, *14*, 1183130. [CrossRef]
51. Whiteman, E.A.; Côté, I.M. Monogamy in marine fishes. *Biol. Rev.* **2004**, *79*, 351–375. [CrossRef]
52. Lin, T.; Liu, X.; Zhang, D. Does the female seahorse still prefer her mating partner after a period of separation? *J. Fish Biol.* **2021**, *99*, 1613–1621. [CrossRef] [PubMed]
53. Bruce, M.P.; Shields, R.J.; Bell, M.V.; Bromage, N.R. Lipid class and fatty acid composition of eggs of Atlantic halibut, *Hippoglossus hippoglossus* (L.), in relation to egg quality in captive broodstock. *Aquac. Res.* **1993**, *24*, 417–422. [CrossRef]
54. Ohkubo, N.; Sawaguchi, S.; Nomura, K.; Tanaka, H.; Matsubara, T. Utilization of free amino acids, yolk protein and lipids in developing eggs and yolk-sac larvae of Japanese eel *Anguilla japonica*. *Aquaculture* **2008**, *282*, 130–137. [CrossRef]

animals

MDPI

Article

Effects of Habitat Differences and Invasive Species Competition on Age and Growth of *Triplophysa strauchii*

Ya-Han Meng [1,†], Wei-Zhen Gao [1,2,†], Yan Li [1] and Lei Shi [1,*]

1 Xinjiang Key Laboratory for Ecological Adaptation and Evolution of Extreme Environment Organism, College of Life Sciences, Xinjiang Agricultural University, Urumqi 830052, China; m15201008090@163.com (Y.-H.M.); gao7222@yeah.net (W.-Z.G.); a18119152762@163.com (Y.L.)
2 College of Animal Sciences, Xinjiang Agricultural University, Urumqi 830052, China
* Correspondence: leis@xjau.edu.cn
† These authors contributed equally to this work.

Simple Summary: Fish growth traits constitute key adaptive attributes to environmental conditions, with growth variation representing the most direct and common response to habitat challenges. Distinct aquatic environments shape divergent life-history strategies and population characteristics in *Triplophysa strauchii*, whose geographically isolated populations experience significant habitat variation in factors including resource availability and interspecific competition. Through lapillus microstructure analysis of stream versus oxbow lake populations, we demonstrate that oxbow lake habitats—characterized by abundant food resources and reduced competition—support greater longevity and larger body sizes, whereas stream populations exposed to biological invasions exhibit reduced lifespan and body size miniaturization. This differential susceptibility to environmental fluctuations highlights conservation vulnerabilities in this endemic species. Population-level analyses further reveal substantial growth potential in the Dacao Lake, contrasting with prevalent miniaturization in the Liutiao River. Effective conservation thus requires preserving native habitats, preventing and controlling invasive species, and implementing long-term monitoring of fishery resource dynamics.

Abstract: Accurate age determination is fundamental for investigating fish population dynamics and growth patterns. This study used the lapillus to determine age in *Triplophysa strauchii* populations from an oxbow lake and a stream. Growth patterns were evaluated using three models (the Von Bertalanffy, Gompertz, and Logistic models). The oxbow lake population showed faster growth and longer lifespan (6 years in Dacao Lake; 5 years in Liutiao Stream). Conversely, the stream population displayed a trend toward smaller size and younger age. Both populations exhibited higher Fulton's condition factor in juveniles than in adults. The species exhibited a fast-growth type, with similar fitting results across models. These findings reflect subtle differentiation in life history strategies across habitats, likely related to environmental conditions and competitive pressure from invasive species. These insights into *T. strauchii* life history underscore the importance of further research to support conservation and sustainable management of this endemic species.

Keywords: *Triplophysa strauchii*; lapillus; age identification; growth characteristics; invader competition

1. Introduction

Accurate determination of fish age is indispensable for elucidating population dynamics, assessing ecological adaptability, and forecasting population trends [1–3]. Calcified

https://doi.org/10.3390/ani15142128

tissue analysis is a well-established method for evaluating population age structure [2,4]. Environmental fluctuations influence the deposition of calcium in fish hard tissues, creating annual rings [5]. These rings are used to deduce fish age [6]. For precise age determination, it is crucial to rely on high-precision structural materials [7]. Otoliths are formed of $CaCO_3$ and protein [8,9] and grow continuously throughout a fish's lifespan [10]. They exhibit non-cellular and metabolic inertness [11] and lack the capacity for reabsorption [12]. These characteristics make otoliths highly reliable for fish age determination [13–15]. Otoliths are categorized into lapilli, asteriscus, or sagittae [16]. Lapilli are relatively stable, and the annual rings are easy to read [17], thus making age identification more accurate.

The growth patterns of fish typically vary among species and can diverge within species along environmental gradients, such as latitude [18], diet [19], temperature [20], or population density [21]. Currently, three main methods are used in fish growth research: the direct method, age-based statistical analysis, and the back-calculation method [22]. Among these, the age-based statistical analysis involves compiling data on body length and weight across different age groups to characterize growth patterns. Fitting growth models using mathematical approaches facilitates comparative analyses [23]. The Von Bertalanffy growth function (VBGF), currently the most widely applied model, estimates fish growth by utilizing either directly measured or inferred age data.

The genus *Triplophysa* represents a highly diverse group within the family Cobitidae (order Cypriniformes) [24] and is distributed in most water areas of Xinjiang [25]. The ongoing tectonic uplift of the Tibetan Plateau and its adjacent mountain ranges has served as a primary driver for both the evolutionary origin and subsequent diversification of this genus [26,27]. Owing to its exceptional morphological differentiation and ecological adaptability, *Triplophysa* has garnered considerable attention from scholars across various domains, including taxonomy [28,29], genomics [30], evolutionary biology [26,31], and basic biological characteristics [32]. Among them, there are relatively few basic biological studies, such as the research on the growth and reproduction, and genetic diversity of *T. yarkandensis* [33], the classification and morphology of *T. stoliczkae* [34], and the age and growth patterns of *T. orientalis* [35] and *T. markehenensis* [36]. In recent years, *Triplophysa* has experienced a precipitous decline due to escalating human-induced disturbances and the invasion of alien species [37]. Therefore, it is imperative to focus on the study and conservation of these species.

The distribution area of *T. strauchii* in China is mainly in the Xinjiang Uygur Autonomous Region, and the species inhabits rivers along the northern slopes of the Tianshan Mountains [38]. Current research on this species has been limited to preliminary studies of its biological traits [38,39] and a description of its mitochondrial genome [40]. *T. strauchii* typically occupies small, cold, and pristine stream habitats characterized by oligotrophic conditions [41]. These habitats have low nutrient availability, limited primary productivity, and scarce food resources. Notably, Guo et al. [38] reported that populations in Chaiwopu Lake exhibited miniaturization and high parasitic infection rates due to competition from invasive fish species.

In this paper, the age structure and growth pattern of two geographically distinct populations of *T. strauchii* (stream-dwelling vs. oxbow lake-dwelling) were analyzed. The effects of divergent aquatic environments on fish growth were discussed. Specifically, the Liutiao Stream population faces a serious invasion of alien species, including benthic *Abbottina rivularis* and *Misgurnus anguillicaudatus,* which compete for the living space of *T. strauchii* [42,43]. In addition, *M. anguillicaudatus* demonstrates three invasive traits: (1) rapid environmental adaptation, (2) high fecundity, and (3) generalist benthic feeding behavior [44]. Research indicates that under equivalent feeding conditions, dominant species exhibit faster growth rates compared to subordinate species [45]. This indirectly reflects that such competition can directly impact the normal growth characteristics of *T. strauchii*. [46]. Conversely, no exotic species were detected in Dacao Lake. *Phoxinus grumi*, a sympatric native species, is a mid-upper layer, diurnal fish. Dietary analysis showed that *P. grumi* mainly feeds on zooplankton, and *T. strauchii* is polyphagous (unpublished data). Consequently, significant niche differentiation exists between *P. grumi* and *T. strauchii* across spatial, temporal, and trophic dimensions. This niche partitioning suggests *T. strauchii* experiences relatively reduced ecological competition within this lake ecosystem (Table A1) [47].

Driven by different survival pressures (habitat utilization, interspecific competition, etc.). It is predicted that the growth patterns of *T. strauchii* from both the oxbow lake and stream might be significantly different. The lake population, which experiences sufficient resources and less competition, is likely to exhibit better growth. Research on their growth pattern would be conducive to understanding the current status of wild resources and the population growth potential of this native species. It may also serve as a theoretical foundation for the conservation of *T. strauchii*.

2. Materials and Methods

2.1. Sample Collection

A total of 218 and 101 specimens of *T. strauchii* were collected during May 2024 from two distinct localities: 1. Dacao Lake (DL), Dabancheng District, Urumqi, Xinjiang Uygur Autonomous Region (88°24′ E, 43°21′ N; altitude 1169 m); 2. Liutiao Stream (LS), Balikun County, Hami, Xinjiang Uygur Autonomous Region (92°59′ E, 43°39′ N; altitude 1650 m). The fish were captured using cage nets (nets with a length of 5 m, a width of 3 m, and a mesh size of 4 mm). For detailed information on the catch composition, please refer to Table A1. DL is an oxbow lake formed in the middle section of the Baiyang River. It has a relatively small water surface and a slow flow rate, and it connects with the main river course. LS has a muddy and sandy bottom. The captured individual fish were anesthetized, fixed with 10% formaldehyde, then stored and brought back to the lab for further analysis. All experiments and animal handling were conducted according to research protocols approved by the Animal Welfare and Ethics Committee of Xinjiang Agricultural University.

2.2. Lapillus Processing and Age Determination

Specimens were numbered after fixation, and sex identification was conducted based on the species' distinct secondary sexual characteristics. External morphological criteria included the presence of posterior or anterior processes on the lateral ethmoid bone (with cartilaginous buds) in males, which were covered by skin and projected below the anterior margin of the eyes (Figure 1A). In males, several outer pectoral fin rays were hardened and broadened, with dense, small spinules (villous nodules) on the dorsal surface (Figure 1B). Once developed, male secondary sexual characteristics persist throughout life. Females lacked these features (Figure 1C,D). Additionally, sexing was difficult in some small individuals, requiring dissection to observe gonad type and developmental stage. If both methods failed, the individual was classified as a juvenile [48].

Subsequently, with the left side as a reference, body length (BL, the horizontal distance from the tip of the snout to the base of the caudal fin) and total length (TL, the distance from the tip of the snout to the end of the caudal fin) were measured using an electronic digital caliper. The body weight was measured using an electronic balance [49]. The measurement data of the fish body length were recorded with an accuracy of 0.01 mm, and the weight measurements were recorded with an accuracy of 0.01 g.

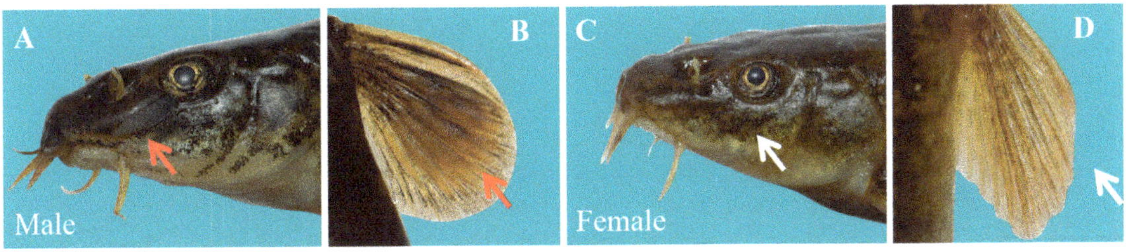

Figure 1. Comparison of female and male secondary sexual characteristics of *Triplophysa strauchii* in Dacao Lake. The arrow points to the prominent supra-opercular bone (**A**) and the thickened and enlarged fin rays (**B**) in males, while females do not have these features (**C**,**D**).

The lapilli were fixed on glass slides and polished with 1500-grit sandpaper. They were moistened with a small amount of water until the central primordium and growth rings were clearly visible under an optical microscope. Then they were sealed on slides using neutral gum [50], photographed under an inverted BA210 Digital LED-Motic Microscopes (Motic, Xiamen, China), and archived for subsequent age determination [51].

Yahan Meng and Yan Li conducted the initial reading of the otoliths (using double-blind reading). Weizhen Gao combined and evaluated the readings of the two experimental operators and re-read the inconsistent data, and his judgment was final (Figure 2).

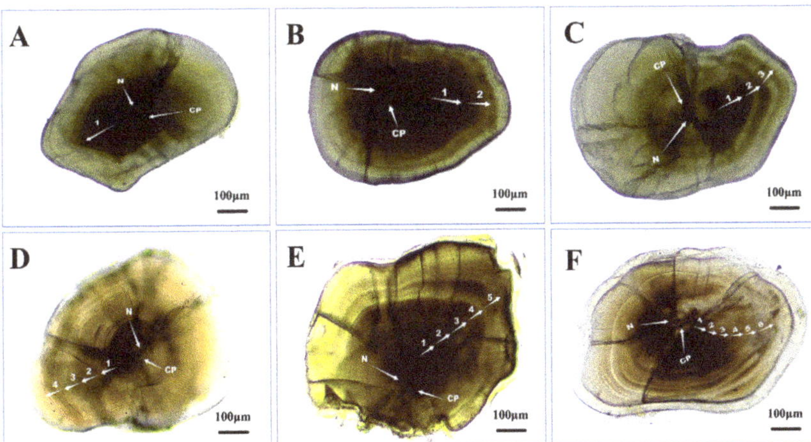

Figure 2. Lapillus of *Triplophysa strauchii* in Dacao Lake, showing different age groups (arrows show scale annuli). CP: central primordium; N: nucleus, 100×. (**A**): 1 year-old; (**B**): 2 year-old; (**C**): 3 year-old; (**D**): 4 year-old; (**E**): 5 year-old; (**F**): 6 year-old.

In this study, 218 lapilli were extracted from *T. strauchii* in DL, and clear annual rings suitable for age determination were found in 195 lapilli. Moreover, 101 lapilli were extracted from *T. strauchii* in LS, and clear annual rings suitable for age determination were found in 74 lapilli. The inability of some individuals to have their otoliths' annual rings read may be due to factors such as varying degrees of calcification in the otoliths. However, the age of the majority of individuals can be accurately determined from their otoliths. Based on a large amount of accurate body length and age data, we have drawn a body length–age frequency distribution chart. The results show that these data meet the assumption that "individuals of the same generation have similar body lengths, while those of different generations have different body lengths [15,52]." Therefore, we can match the body length range of some individuals whose age cannot be determined through otoliths with that of individuals of known ages to determine their age [53].

2.3. Data Analysis

2.3.1. Length–Weight Relationship

The Keys formula can be used to perform regression analysis on the length–weight relationship [22]. The formula is $W = aL^b$, where W is the body weight (g), L is the body length (mm), a is the condition factor, and b is the allometric growth factor. The Pauly t-test is used to determine if there is a significant difference between the b value and 3, which helps to identify the growth type of *T. strauchii*. The formula for the t-test is $t = \frac{SD(L)}{SD(W)} \times \frac{|b-3|}{\sqrt{1-r^2}} \times \sqrt{n-2}$, where $SD(L)$ and $SD(W)$ are the standard deviations of the body length and weight, respectively, n is the sample size, and r^2 is the correlation coefficient from the length–weight equation [54].

2.3.2. Fulton's Condition Factor

The Fulton's condition factor (K), also called Fulton's coefficient, serves as a crucial biometric index for assessing fish's nutritional status and body condition. It also helps evaluate their responses to environmental factors, providing valuable guidance for fisheries management. It was initially proposed by Fulton and expressed as a percentage, the formula is $K = (\frac{W}{L^3}) \times 10^5$, where W is the body weight (g), and L is the body length (mm) [55].

2.3.3. Growth Models

The growth pattern is one of the most crucial biological characteristics in the study of fish population dynamics [56,57]. In this study, we selected three commonly used growth models (Table 1). They were analyzed to compare the growth characteristics of *T. strauchii* in DL and LS, respectively. The equations were based on the premise that a linear correlation existed between the instantaneous growth rate and the logarithm of the body weight of the fish.

Table 1. The most commonly evaluated growth models.

Growth Model	Equation	Parameter Description	Reference
Von Bertalanffy	$L_t = L_\infty\left[1 - e^{-k(t-t_0)}\right]$	$L_{(t)}$ = standard length at age t L_∞ = theoretical asymptotic length t_0 = age at zero length k = growth parameter	[58]
Logistic	$t_{IP} = \frac{lnb}{k} + t_0$	t_{IP} = growth inflection point	[59]
Gompertz	$L_t = L_\infty\left[e^{-e^{-G(t-t_{IP})}}\right]$ $G = \ln(\frac{L_\infty}{L_0})$	G = instantaneous growth rate coefficient at age t_{IP} L_0 = the length at t_0	[60]

2.3.4. Statistical Analysis

Morphological data were expressed as mean ± standard deviation (mean ± SD). The data were tested for normal distribution and homogeneity of variance. If the data conformed to both normal distribution and homogeneity of variance, one-way ANOVA was used to analyze differences between age groups in each basin. The LSD method was applied for multiple comparisons, and the results were labeled using the alphabetical method. For non-normally distributed data, a non-parametric test (e.g., Mann–Whitney U test for two groups) was used to analyze differences between groups. Statistical significance was set at $p < 0.05$. Analyses were performed using IBM SPSS 21.0, and nonlinear fitting of the growth curve was conducted with Origin 2024 software (Version 10.1).

3. Results

3.1. Body Length Distribution

The body length of the specimens collected from DL ranged from 33.46 mm to 139.32 mm, with a mean (±SD) of 41.16 ± 12.72 mm. The dominant body length group ranged from 91 mm to 110 mm, comprising 90 individuals, accounting for 46.88% of the total. The second most frequent group was 71–90 mm, with 47 individuals, representing 24.48% of the population. In LS, the body length of the specimens ranged from 39.27 mm to 114.68 mm, with a mean (±SD) of 83.08 ± 12.92 mm. The dominant body length group ranged from 71 mm to 90 mm, comprising 54 individuals, and accounting for 53.47% of the total. The second most frequent group was 91–110 mm, with 31 individuals, representing 30.69% of the population (Figure A1).

3.2. Length–Weight Relationship

The body length of *T. strauchii* in DL was primarily distributed between 75 mm and 125 mm (Figure 3A). The results of the Pauly *t*-test, based on the corresponding *t* value and degrees of freedom ($t = 29.68$, $p < 0.05$), indicate that the population of *T. strauchii* in DL exhibited a negative allometric growth, implying the growth rate in length was faster than in weight. The *b* for males (Figure 3B) and sub-adult (Figure 3D) were 2.63 and 2.98, respectively, both significantly lower than 3 ($t_{♂} = 22.40$, $t = 2.20$, $p < 0.05$), indicating a negative allometric growth. In contrast, the *b* for females (Figure 3C) was 3.03, showed no significant difference from 3 ($t_{♀} = 1.40$, $p > 0.05$), suggesting an isometric growth.

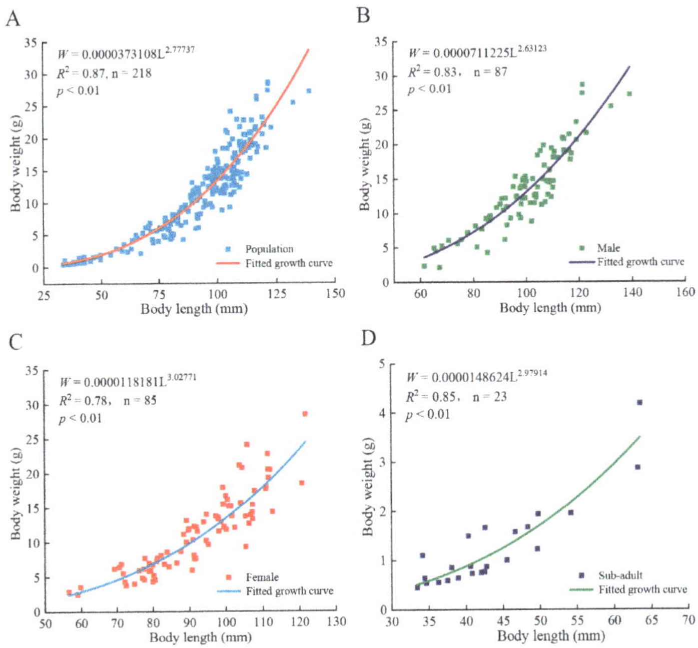

Figure 3. Length–weight relationships of *Triplophysa strauchii* in Dacao Lake. The relationship between body length and weight across the entire *Triplophysa strauchii* population (**A**), male (**B**), female (**C**), and juvenile (**D**) differentiation based on body length–weight relationships.

The body length of *T. strauchii* in LS was primarily distributed between 75 mm and 95 mm (Figure 4A). The *b* was $2.21 < 3$ ($t = 18.03$, $p < 0.05$), indicating that the population exhibited a negative allometric growth. The *b* for males (Figure 4B) and females (Figure 4C) were 2.02 and 2.33, respectively, both significantly lower than 3 ($t_{♂} = 5.14$, $t_{♀} = 3.79$, $p < 0.05$), indicating a negative allometric growth. In contrast, the *b* for sub-adult (Figure 4D) was 2.62 ($t = 0.04$, $p > 0.05$), suggesting an isometric growth.

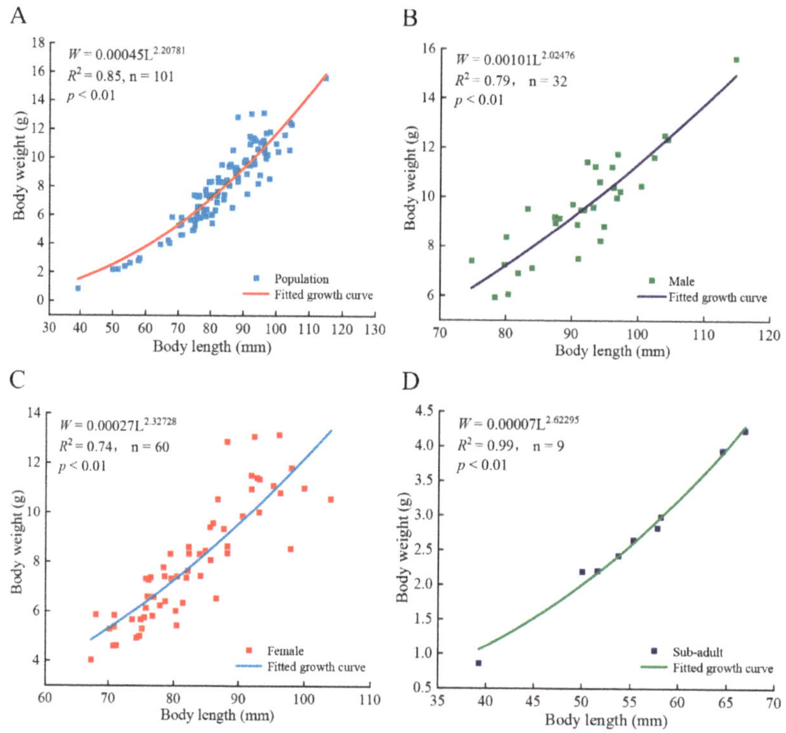

Figure 4. Length–weight relationships of *Triplophysa strauchii* in Liutiao Stream. The relationship between body length and weight across the entire *Triplophysa strauchii* population (**A**), male (**B**), female (**C**), and juvenile (**D**) differentiation based on body length–weight relationships.

3.3. Age Structure

The age of the specimens in DL ranged from 1 to 6 years. The dominant age was 3 years. In LS, the age of *T. strauchii* ranged from 1 to 5 years. The dominant age was 2 years (Figure 5). One-way ANOVA revealed significant differences in body length between populations in different basins. The results revealed that male *T. strauchii* aged 1 and 2 in LS were significantly larger than their counterparts in DL ($p = 0.001$, $0.038 < 0.05$). Conversely, males aged 3 or 4 from DL were significantly larger than those from LS ($p = 0.021$, $0.000 < 0.05$). Female *T. strauchii* aged 3, 4, or 5 from DL exhibited significantly greater body length compared to those from LS ($p = 0.000$, 0.000, $0.000 < 0.05$). Within the same basin, sexual dimorphism was observed in 2-year and 3-year individuals from LS, where males were significantly larger than females ($p = 0.008$, $0.036 < 0.05$) (Table A2).

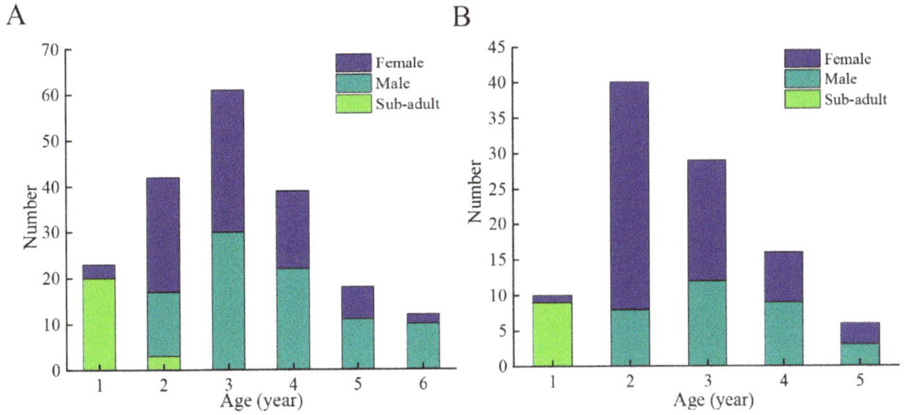

Figure 5. Age distribution of *Triplophysa strauchii* in Dacao Lake (**A**) or Liutiao Stream (**B**).

3.4. Fulton's Condition Factor

Both DL and LS populations of *T. strauchii* exhibited higher condition factors in 1-year juveniles compared to adults. Female condition factors of 2-year, *T. strauchii* in LS were significantly greater than those of 2-year males in DL ($p = 0.004 < 0.05$). The male condition factor of 5-year individuals in DL was significantly higher than that of 5-year males in LS ($p = 0.022 < 0.05$). Within the population, comparisons showed female condition factors of 4 years in LS displaying significantly higher than males ($p = 0.048 < 0.05$) (Figure 6).

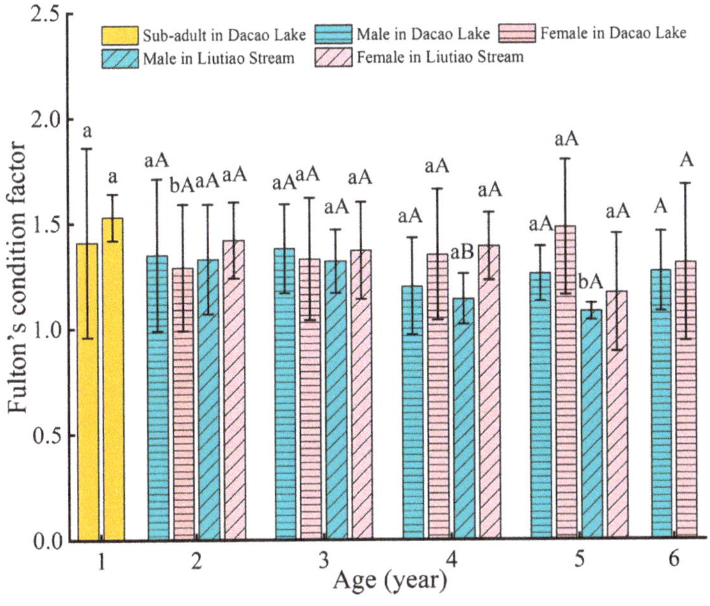

Figure 6. The condition factor of *Triplophysa strauchii* across different age groups. Significance difference was deemed at $p < 0.05$. Differences between the two geographic populations within the same sex were indicated by lowercase letters, while differences between sexes within each age group were denoted by uppercase letters. The same letter indicates no significant difference, whereas different letters represent a statistically significant difference.

3.5. Growth Models

Body length and age data were fitted using nonlinear regression with Von Bertalanffy, Logistic, and Gompertz growth models (Table A3 and Figure 7).

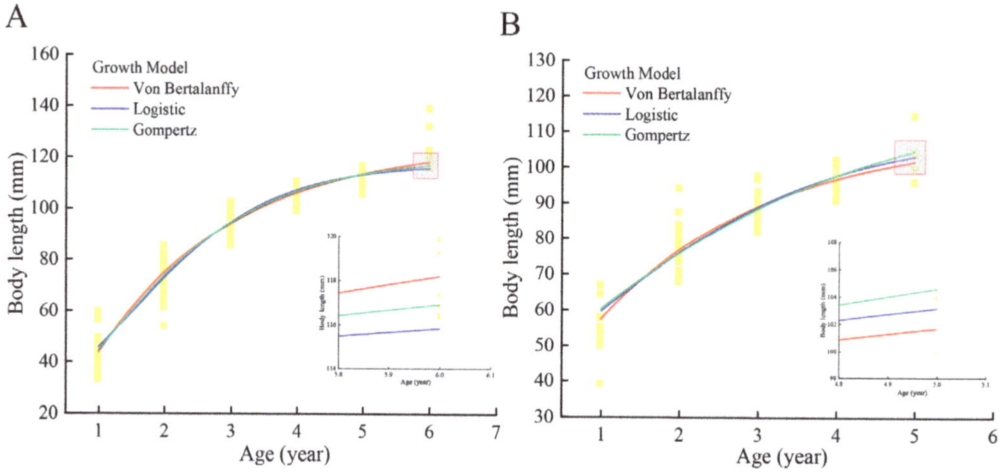

Figure 7. The different growth models fitted to *Triplophysa strauchii* in Dacao Lake (**A**) or Liutiao Stream (**B**).

The VBGM predicted the largest asymptotic body length L_∞ (125.667 mm), while the Logistic model yielded the smallest estimate (117.537 mm) in DL. Conversely, the Logistic model produced the highest L_∞ (115.647 mm), with the VBGM showing the lowest value (102.782 mm) in LS. The specific fitting results of each growth equation, along with the coefficient of determination (R^2) and residual sum of squares (RSS), were presented in Table A3. The fitting results of the three growth models for the species were similar and effectively described the species' growth pattern. The growth rate in body length was maximized at 1 year, and it slowed down with increasing age, approaching the asymptotic body length (Figure 7A,B).

4. Discussion

4.1. Growth Characteristics

The growth characteristics of fish are the results of the interaction between external (environmental) and internal (biological) factors, primarily reflected in body length and weight relationships. Even within the same species, population growth characteristics can diverge significantly under differing environmental conditions [2,61]. Nutritional conditions, habitat states, fishing pressure, and interspecific competition also affect the population characteristics [62–64]. Therefore, the growth parameter *b* always changes [65]. When *b* is close to 3, it typically suggests an isometric growth. When *b* is close to 3, this typically indicates isometric growth; when *b* is less than 3, it indicates negative allometric growth; when *b* is greater than 3, it indicates positive allometric growth [66]. In this study, the *b* value for *T. strauchii* in either DL ($b = 2.78$) or LS ($b = 2.21$) was significantly lower than 3 ($p < 0.05$), indicating negative allometric growth.

The DL was characterized by relatively sluggish water flow, a predominantly muddy substrate, and abundant aquatic weeds, thereby offering a plentiful food source. These conditions were likely conducive to the rapid growth of *T. strauchii*. In contrast, the LS predominantly had a gravel substrate, with fast-flowing water and barely any aquatic weeds. The food resources were scarce, and the species also faced direct competition from invasive species occupying the same ecological niche. Consequently, the growth rate of *T. strauchii* in the LS was lower, and their body size was smaller. It can be inferred that the same species can exhibit different growth characteristics in response to various habitat types and resource conditions [62,63].

4.2. *Age Structure*

The population of *T. strauchii* exhibited a relatively low average age and a straightforward age structure. Currently, *T. strauchii* in DL aged 1 to 2 years accounted for 29.36%, while individuals aged 3 to 6 years comprised 70.64%. This indicates that *T. strauchii* in this area faces lower environmental pressure, with a low proportion of young individuals. *T. strauchii* in LS aged 1 to 2 years accounted for 75.68%, while individuals aged 4 to 5 years comprised 24.32%, demonstrating a distinct younger age structure. The population age structure of *T. strauchii* in DL or LS was primarily composed of juveniles and young adults. These findings were similar to observations of *T. strauchii* in Sailimu Lake [39]. In contrast, individuals from Chaiwopu Lake showed considerable signs of parasitic infections [38]. These results are consistent with the growth patterns of *T. yarkandensis*, whose population exhibits faster growth in better habitats [13]. Field surveys revealed significant invasive species presence in the watershed of LS, where benthic competitors, including *M. anguillicaudatus* and *A. rivularis*, were observed to overlap ecologically with the native *T. strauchii.* This interspecific competition likely constrained habitat availability and food resources for *T. strauchii* [42,43]. As a consequence, the dominance of younger individuals and the smaller body size of *T. strauchii* in this water area indicated that the population is facing severe ecological pressure.

The growth inflection point was not only closely related to sexual maturity and aging, but also related to water temperature changes and nutritional conditions [67]. In this study, the growth inflection point for *T. strauchii* in DL and LS was 2.32 and 2.07 years, respectively, indicating that the majority of the individuals had surpassed the rapid growth and exhibited a relatively stable state [13]. In DL, the growth inflection points for males and females were 3.32 and 2.48, respectively. Their growth was obvious in the later stage, with delayed sexual maturity and minimal sexual dimorphism. Because females require continuous rapid growth to meet reproductive demands, whereas males show less observable growth in response to the development of sexual glands [68]. The LS population of *T. strauchii* exhibited more prominent male-bias sexual size dimorphism, with males significantly exceeding females in body length at 2 or 3 years ($p < 0.05$). The earlier growth inflection points ($t_{IP♀} = 1.57$; $t_{IP♂} = 1.81$) showed that the sexual maturity of fish in this population was early.

4.3. *Fulton's Condition Factor*

The condition factor (K) serves as a crucial biological indicator in fish studies [55]. This metric is influenced by both environmental factors and intrinsic physiological characteristics and is correlated with gonadal development [69]. In this study, both the DL and LS populations of *T. strauchii* displayed higher condition factors in 1-year-old individuals. This pattern likely reflected substantial energy reserves during early development stages, which facilitated survival and early growth [70]. Comparative analysis revealed that females generally exhibited higher condition factors than males in both watersheds. Therefore, we hypothesized that enhanced female condition might promote gonadal growth and development, thereby increasing reproductive investment and offspring output to improve population recruitment.

4.4. Growth Models

VBGM is widely regarded as a general model suitable for fitting growth patterns in most fish species [23], though it performs poorly in simulating the early growth stages of juveniles. The Gompertz model is often preferred for modeling juvenile [71] or female growth patterns [72]. The Logistic model is better fitted for males [73,74]. However, this study found consistent performance among these three models, suggesting comparable suitability for fitting the growth pattern of *T. strauchii*. Therefore, the conventional VBGM was selected to compare the growth pattern of *T. strauchii* across different watersheds. The growth coefficient *k* reflects the growth rate of fish. A *k* value between 0.05 and 0.10 indicates a slow growth; a *k* value between 0.10 and 0.20 suggests a moderate growth; and a *k* value between 0.20 and 0.50 signifies a rapid growth [75]. The *k* values for *T. strauchii* in DL and LS were 0.451 and 0.335, respectively, indicating a fast-growing species. In the case of limited resources or increased environmental pressure (fishing, biological invasion, etc.). *T. strauchii* tended to allocate energy preferentially to sexual maturity (reproduction) rather than growth, resulting in early sexual maturity and reduced maximum body length (smaller size). These results are consistent with predictions of the energy trade-off hypothesis [76].

Overall, by comparing the age and growth characteristics of different geographical populations of *T. strauchii*, this study reveals that environmental differences and ecological competitive pressure significantly affect the growth of this fish. However, this experiment only focused on the growth of *T. strauchii* in a single season (May) and did not consider the influence of seasonal changes on growth, which may result in an incomplete generalization of the population's overall growth patterns. In the future, we will continue to monitor the population dynamics of *T. strauchii*, adopt a longer time scale, expand to a wider range of geographical populations, and conduct controlled experiments to reveal how its growth characteristics change under different ecological competitive pressures, thereby offering more scientific support for the conservation of native fish species.

5. Conclusions

Fish normally develop unique life history strategies to adapt to specific aquatic environments, resulting in population-specific age structure and growth patterns. Our findings demonstrated that *T. strauchii* exhibited an allometric growth pattern across different habitats, indicating a fast-growing type. The population of *T. strauchii* in DL aged from 1 to 6 years. The dominant age group was 3-year-olds, with a growth inflection point at 2.32 years. *T. strauchii* in DL showed no signs of smaller size, younger age, and attained a larger asymptotic body length. In contrast, the population in LS exhibited an age range of 1–5 years, with a predominance of 2-year-old individuals and a growth inflection point at 2.07 years, demonstrating a distinct younger age structure and smaller size. The DL population experienced less ecological pressure and maintained a higher growth potential. In LS, the miniaturization and younger age of fish were more frequent, which might be caused by the deterioration of the water environment, the bait competition with alien species, or the limited nutritional resources. Therefore, the growth and population characteristics of *T. strauchii* are directly related to environmental conditions. Consequently, it is imperative to strengthen the conservation of the aquatic environment, safeguard against the invasion of alien species, maintain continuous surveillance of the fishery resources in the area, and implement scientifically informed conservation strategies.

Author Contributions: Conceptualization, W.-Z.G. and L.S.; methodology, W.-Z.G., Y.-H.M. and L.S.; software, Y.-H.M. and W.-Z.G.; validation, W.-Z.G. and Y.L.; formal analysis, Y.-H.M. and W.-Z.G.; investigation, W.-Z.G. and L.S.; resources, W.-Z.G. and L.S.; data curation, Y.-H.M. and Y.L.; writing—original draft preparation, W.-Z.G., Y.-H.M. and L.S.; writing—review and editing, W.-Z.G.,

Y.-H.M. and L.S.; visualization, Y.-H.M. and W.-Z.G.; supervision, L.S.; project administration, W.-Z.G. and L.S.; funding acquisition, L.S. All authors have read and agreed to the published version of the manuscript.

Funding: This study was supported by the Third Xinjiang Scientific Expedition Program (grant no. 2022xjkk1200), the College Student Innovation and Entrepreneurship Training Program of Xinjiang Uygur Autonomous Region (grant no. S202410758034).

Institutional Review Board Statement: All experimental procedures involving animals were approved by the Animal Welfare and Ethics Committee of Xinjiang Agricultural University, Urumqi, Xinjiang, China (animal protocol number: 2023014).

Informed Consent Statement: Not applicable.

Data Availability Statement: The dataset is available upon request from the authors.

Acknowledgments: We thank Lin Leng and Quan Zhang for their assistance in this study.

Conflicts of Interest: The authors declare no conflicts of interest.

Appendix A

Table A1. Fishing catch statistics table.

Sampling Site (Time)	Species	Number	Proportion of Catch	Type	Feeding Habits	Activity Rhythm	Habitat Water Layer	References
Dacao Lake (27 April 2024)	*Triplophysa strauchii*	218	57.80%	Native	Omnivory	Nocturnal	Demersal	Unpublished data; [48]
	Phoxinus grumi	92	42.20%	Native	Omnivory (zooplankton)	Diurnal	Pelagic	Unpublished data
Liutiao Stream (19 May 2024)	*Triplophysa strauchii*	101	56.11%	Native	Omnivory	Nocturnal	Demersal	Unpublished data; [48]
	Misgurnus anguillicaudatus	55	30.56%	Alien	Omnivory	Nocturnal	Demersal	[43]
	Abbottina rivularis	24	13.33%	Alien	Omnivory	Diurnal	Demersal	[42]

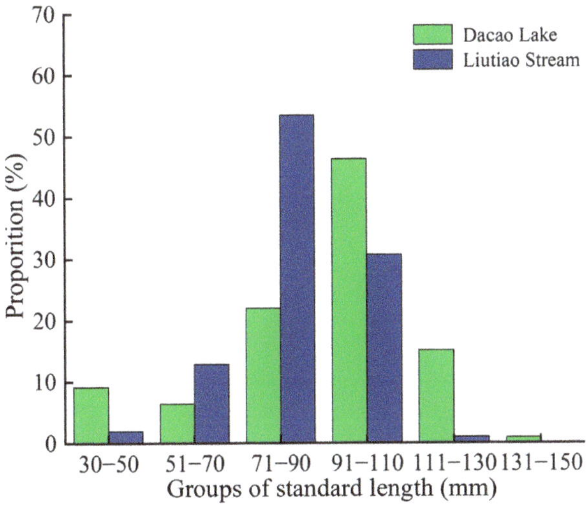

Figure A1. The distribution of body length of *Triplophysa strauchii*.

Table A2. Body length of *Triplophysa strauchii* in different age groups (mm).

Age	Sex	Dacao Lake		Liutiao Stream		p-Value
		Mean ± SD	Range	Mean ± SD	Range	
1	Sub-adult	43.29–7.75 ^{b} (n = 23)	33.46–59.99	55.24 ± 8.16 ^{a} (n = 9)	39.27–66.86	0.001
2	Male	74.48 ± 8.06 ^{bA} (n = 14)	61.54–85.45	81.69 ± 5.75 ^{aA} (n = 8)	74.80–94.32	0.038
	Female	77.88 ± 4.35 ^{aA} (n = 25)	69.46–84.59	76.62 ± 4.28 ^{aB} (n = 32)	68.03–87.64	0.279
3	Male	94.34 ± 4.97 ^{aA} (n = 30)	85.72–102.76	90.55 ± 3.54 ^{bA} (n = 12)	83.31–96.74	0.021
	Female	94.35 ± 4.50 ^{aA} (n = 31)	86.67–101.84	87.28 ± 4.18 ^{bB} (n = 17)	82.10–97.85	0.000
4	Male	105.90 ± 2.76 ^{aA} (n = 22)	100.25–109.83	96.58 ± 3.36 ^{bA} (n = 9)	91.01–102.45	0.000
	Female	104.81 ± 3.23 ^{aA} (n = 17)	99.39–110.97	94.16 ± 2.34 ^{bA} (n = 7)	91.77–97.95	0.000
5	Male	112.55 ± 2.50 ^{aA} (n = 11)	109.09–116.96	107.73 ± 6.02 ^{aA} (n = 3)	104.01–114.68	0.298
	Female	110.96 ± 2.28 ^{aA} (n = 7)	105.95–112.79	99.92 ± 3.91 ^{bA} (n = 3)	96.03–103.85	0.000
6	Male	122.68 ± 7.46 ^{A} (n = 10)	116.43–139.32			
	Female	121.40 ± 0.72 ^{A} (n = 2)	120.89–121.91			

Notes: Significance difference was deemed at $p < 0.05$. Differences between the two different geographic populations within the same sex (within the same row) were indicated by lowercase letters, while differences between sexes within each age group (within the same column) were denoted by uppercase letters. The same letter indicates no significant difference, whereas different letters represent a statistically significant difference.

Table A3. Growth function of *Triplophysa strauchii* after fitting.

Model	Dacao Lake	R^2	RSS	Liutiao Stream	R^2	RSS
Von Bertalanffy	$L_t = 125.667\left[1 - e^{-0.451(t-0.059)}\right]$	0.9393	6.346	$L_t = 105.307\left[1 - e^{-0.335(t+1.241)}\right]$	0.9982	2.9738
Logistic	$L_t = 117.537\left[1 + e^{-0.433(t-2.32}\right]^{-1}$	0.9309	7.234	$L_t = 115.647\left[1 + e^{-0.234(t-2.07}\right]^{-1}$	0.8352	2.7506
Gompertz	$L_t = 120.389\left[e^{-e^{-0.705(t-2.32)}}\right]$	0.9360	6.953	$L_t = 102.782\left[e^{-e^{-0.715(t-2.07)}}\right]$	0.8454	5.9596

References

1. Beamish, R.J.; McFarlane, G.A. The forgotten requirement for age validation in fisheries biology. *Trans. Am. Fish. Soc.* **1983**, *112*, 735–743. [CrossRef]
2. Morat, F.; Wicquart, J.; Schiettekatte, N.M.D.; de Sinéty, G.; Bienvenu, J.; Casey, J.M.; Brandl, S.J.; Vii, J.; Carlot, J.; Degregori, S.; et al. Individual back-calculated size-at-age based on otoliths from Pacific coral reef fish species. *Sci. Data* **2020**, *7*, 370. [CrossRef] [PubMed]
3. Legaki, A.; Leonhard, I.; Mytilineou, C.; Anastasopoulou, A. *Dentex maroccanus* Valenciennes, 1830 Otolith Morphology, Age, and Growth in the Aegean Sea (E. Mediterranean). *Animals* **2024**, *14*, 3151. [CrossRef] [PubMed]
4. Duan, Y.J.; Xie, C.X.; Zhou, X.J.; Ma, B.S.; Huo, B. Age and growth characteristics of *Schizomycosis younghusbandi* Regan, 1905 in the Yarlung Tsangpo River in Tibet, China. *J. Appl. Ichthyol.* **2014**, *30*, 948–954. [CrossRef]
5. Campana, S.E. Accuracy, precision and quality control in age determination, including a review of the use and abuse of age validation methods. *J. Fish Biol.* **2001**, *59*, 197–242. [CrossRef]
6. Weatherley, A.H.; Gill, H.S. *The Biology of Fish Growth*; Academic Press: Cambridge, MA, USA, 1987.
7. Polat, N.; Bostanci, D.; Yilmaz, S. Comparable age determination in different bony structures of *Pleuronectes flesus luscus* Pallas, 1811 inhabiting the Black Sea. *Turk. J. Zool.* **2001**, *25*, 441–446.
8. Willmes, M.; Sturrock, A.M.; Cordoleani, F.; Hugentobler, S.; Meek, M.H.; Whitman, G.; Evans, K.; Palkovacs, E.P.; Stauffer-Olsen, N.J.; Johnson, R.C. Integrating otolith and genetic tools to reveal intraspecific biodiversity in a highly impacted salmon population. *J. Fish Biol.* **2024**, *105*, 412–430. [CrossRef] [PubMed]
9. Meng, G.; Tang, J.B.; Wang, Q.L.; Sun, Z.H.; Yu, S.S.; Si, F. Effect of Temperature on the Early Development of *Paralichthys olivaceus* Otoliths. *Animals* **2025**, *15*, 814. [CrossRef] [PubMed]
10. Sapota, M.R.; Dąbrowska, V. Shapes of otoliths in some Baltic fish and their proportions. *Oceanol. Hydrobiol. Stud.* **2019**, *48*, 296–304. [CrossRef]
11. Phelps, Q.E.; Edwards, K.R.; Willis, D.W. Precision of five structures for estimating age of common carp. *N. Am. J. Fish. Manag.* **2007**, *27*, 103–105. [CrossRef]

12. Schulz-Mirbach, T.; Ladich, F.; Plath, M.; Heß, M. Enigmatic ear stones: What we know about the functional role and evolution of fish otoliths. *Biol. Rev.* **2019**, *94*, 457–482. [CrossRef] [PubMed]
13. Wang, X.Y.; Chen, S.G.; Zi, F.Z.; Ge, J.M.; Chang, D.S.; Song, Y.; Xie, C.X. Age and Growth of *Triplophysa* (*Hedinichthys*) *yarkandensis* (Day, 1877) in the Tarim River in Xinjiang, China. *Pak. J. Zool.* **2024**, *56*, 1771–1780. [CrossRef]
14. Zhang, S.Y.; Wang, J.L.; Li, P.L.; Liu, J.C.; Liu, Y.B.; Wang, T.; Liu, K.; Zhao, P.; Zhang, J. Age identification and growth model selection of *Triplophysa pseudoscleroptera* in the middle and upper reaches of the Yellow River. *J. N. Agric.* **2023**, *51*, 105–111.
15. Başusta, N.; Dürrani, Ö. Sexual dimorphism in the otolith shape of shi drum, *Umbrina cirrosa* (L.), in the eastern Mediterranean Sea: Fish size-otolith size relationships. *J. Fish Biol.* **2021**, *99*, 164–174. [CrossRef] [PubMed]
16. Xie, X.; Bao, Z.Y.; Wang, Q.Z. Advances on research and application of age determination by hard tissues in fish: A review. *J. Dalian Ocean. Univ.* **2021**, *36*, 1071–1080.
17. Shi, F.; Sun, J.; Lin, X.; Liu, H. Otolith ontogeny and increment formation in larval *Tanichthys albonubes*. *Chin. J. Zool.* **2006**, *41*, 10–16.
18. Midway, S.R.; Wagner, T.; Arnott, S.A.; Biondo, P.; Martinez-Andrade, F.; Wadsworth, T.F. Spatial and temporal variability in growth of Southern Flounder (*Paralichthys lethostigma*). *Fish. Res.* **2015**, *167*, 323–332. [CrossRef]
19. Noring, A.M.; Sass, G.G.; Midway, S.R.; VanDeHey, J.A.; Raabe, J.K.; Isermann, D.A.; Kampa, J.M.; Parks, T.P.; Lyons, J.; Jennings, M.J. Pelagic forage versus abiotic factors as drivers of Walleye growth in northern Wisconsin lakes. *Adv. Limnol.* **2021**, *66*, 207–223. [CrossRef]
20. Huntsman, B.M.; Martin, R.W.; Patten, K.A. Effects of temperature and spatial scale on Rio Grande cutthroat trout growth and abundance. *Trans. Am. Fish. Soc.* **2018**, *147*, 480–496. [CrossRef] [PubMed]
21. Caves, S.; Baumann, J.R.; Stich, D.S. Density-Dependent Changes in Grass Carp Growth and Mortality in Long-Term Aquatic Plant Management. *N. Am. J. Fish. Manag.* **2021**, *41*, 355–365. [CrossRef]
22. Yin, M.C. *Fish Ecology*; China Agriculture Press: Beijing, China, 1995.
23. Flinn, S.A.; Midway, S.R. Trends in growth modeling in fisheries science. *Fishes* **2021**, *6*, 1. [CrossRef]
24. Nelson, J.S.; Grande, T.C.; Wilson, M.V.H. *Fishes of the World*; John Wiley & Sons: Hoboken, NJ, USA, 2016.
25. Xiao, H.; Dai, Y.G. A review of study on diversity of *Triplophysa* in China. *Fish. Sci.* **2011**, *30*, 53–57.
26. Wang, Y.J.; Shen, Y.J.; Feng, C.G.; Zhao, K.; Song, Z.B.; Zhang, Y.P.; Yang, L.D.; He, S.P. Mitogenomic perspectives on the origin of Tibetan loaches and their adaptation to high altitude. *Sci. Rep.* **2016**, *6*, 29690. [CrossRef] [PubMed]
27. Wang, C.X.; Luo, S.T.; Yao, N.; Wang, X.Y.; Song, Y.; Chen, S.G. A comprehensive analysis of *Triplophysa labiata* (Kessler, 1874) mitogenome and its phylogenetic implications within the *Triplophysa* genus. *Genes* **2023**, *14*, 1356. [CrossRef] [PubMed]
28. Feng, C.G.; Zhang, Y.; Tong, C.; Zhou, B.Z.; Li, X.H.; Tang, Y.T.; Song, W.Z.; Zhao, K. A new species of *Triplophysa* (Cypriniformes, Nemacheilidae) from Weihe River in Gansu Province, China. *Zool. Res.* **2020**, *41*, 465–470. [CrossRef] [PubMed]
29. Sheraliev, B.; Kayumova, Y.; Peng, Z. *Triplophysa daryoae*, a new nemacheilid loach species (Teleostei, Nemacheilidae) from the Syr Darya River basin, Central Asia. *ZooKeys* **2022**, *1125*, 47–67. [CrossRef] [PubMed]
30. Wang, X.Y.; Song, Y.; Xie, H.Y.; Zi, F.Z.; Chen, S.G.; Luo, S.T. Complete mitogenome of the *Triplophysa bombifrons*: Comparative analysis and phylogenetic relationships among the members of *Triplophysa*. *Genes* **2023**, *14*, 128. [CrossRef] [PubMed]
31. Hu, J.; Liu, M.; He, D. Phylogeography of *Triplophysa stenura* (Nemacheilidae): Responded to the mid-pleistocene climate transition in the Qinghai-Tibetan plateau. *Zool. Stud.* **2020**, *59*, e67. [PubMed]
32. Xiong, H.L.; Yao, Y.H.; Wang, Z.J. Study on the structure of the digestive tract in *Triplophysa bleekeri*. *J. Southwest China Norm. Univ.* **2012**, *37*, 113–120.
33. Wang, X.Y. Age, Growth, Reproduction and Population Discrimination of *Triplophysa yarkandensis*. Master's Thesis, Tarim University, Alar, China, 2022.
34. Zhao, H.; Wang, X.B.; Zhao, N.H.; Wei, J.; Shen, J.Z. Morphology and age determination of loach *Triplophysa stoliczkae* in Kizil River, Xinjiang Uygur Autonomous Region. *Chin. J. Fish.* **2021**, *34*, 8–15.
35. Li, L.T. Study on the Age, Growth and Population Dynamics of *Triplophysa orientalis* in the Middle of the Yarlung Tsangpo River. Master's Thesis, Huazhong Agricultural University, Wuhan, China, 2016.
36. Zhang, X.F.; He, C.L.; Song, Z.B. Age and growth of *Triplophysa markehenensis* from the Markehe River in upper reaches of the Dadu River. *Chin. J. Zool.* **2010**, *45*, 11–20.
37. Feng, C.G.; Tong, C.; Zhang, R.Y.; Li, G.G.; Wanghe, K.Y.; Tang, Y.T.; Zhang, C.F.; Zhao, K. Biodiversity and distribution patterns of *Triplophysa* species in the northeastern margin of the Tibetan Plateau. *Biodivers. Sci.* **2017**, *25*, 53–61. [CrossRef]
38. Guo, Y.; Cai, L.G.; Aa, D.K.; Li, H. Study on biological characteristics of *Triplophysa strauchii* in Chaiwopu Lake. *J. Hydroecology* **2008**, *28*, 49–50+79.
39. Guo, Y.; Cai, L.G.; Tu, E.X.; Zhang, R.M.; Liu, K.L.; Zhang, B.P. The study on the biology of *Triplophysa strauchii* (Kessler) in the Sailimu Lake. *Chin. J. Fish.* **2002**, *15*, 6–11.
40. Han, M.M.; Lu, J.; Wang, L.; Mahboob, S.; Al-Ghanim, K.A.; Sun, X.W. Complete mitochondrial genome of the *Triplophysa bombifrons* and *Triplophysa strauchii*. *Mitochondrial DNA Part A* **2016**, *27*, 4710–4711. [CrossRef] [PubMed]

41. Friberg, N.; Dybkjaer, J.B.; Olafsson, J.S.; Gislason, G.M.; Larsen, S.E.; Lauridsen, T.L. Relationships between structure and function in streams contrasting in temperature. *Freshw. Biol.* **2009**, *54*, 2051–2068. [CrossRef]
42. Jang-Liaw, N.H.; Tominaga, K.; Zhang, C.; Zhao, Y.; Nakajima, J.; Onikura, N.; Watanabe, K. Phylogeography of the Chinese false gudgeon, *Abbottina rivularis*, in East Asia, with special reference to the origin and artificial disturbance of Japanese populations. *Ichthyol. Res.* **2019**, *66*, 460–478. [CrossRef]
43. Sun, B.; Huang, Y.; Castro, L.F.C.; Yang, S.; Huang, S.; Jin, W.; Zhou, H.; Ijiri, S.; Luo, Y.; Gao, J.; et al. The chromosome-level genome and key genes associated with mud-dwelling behavior and adaptations of hypoxia and noxious environments in loach (*Misgurnus anguillicaudatus*). *BMC Biol.* **2023**, *21*, 18. [CrossRef] [PubMed]
44. Cano-Barbacil, C.; Haubrock, P.J.; Radinger, J. Asian Loaches: An Emerging Threat as Global Invaders. *Freshw. Biol.* **2025**, *70*, e70026. [CrossRef]
45. Gardin, A.; Otero, O.; Réveillac, E.; Lafitte, A.; Valentin, X.; Lapalus, F.; Bouchon, D.; Garcia, G. Seasonality and growth in tropical freshwater ectotherm vertebrates: Results from 1-year experimentation in the African gray bichir, giraffe catfish, and the West African mud turtle. *Ecol. Evol.* **2023**, *13*, e9936. [CrossRef] [PubMed]
46. Wang, Y.J.; Hu, M.H.; Wang, W.M.; Cao, L.; Yang, Y.; Lü, B.P.; Yao, R.R. Transpositional feeding rhythm of loach *Misgurnus anguillicaudatus* from larvae to juveniles and its ontogenesis under artificial rearing conditions. *Aquac. Int.* **2008**, *16*, 539–549. [CrossRef]
47. Xie, P. Morphological Comparison and Temperature Adaptation of *Phoxinus* Species in China. Ph.D. Thesis, Huazhong Agricultural University, Zhejiang, China, 2019.
48. Zhu, S.Q. *The Loaches of the Subfamily Nemacheililinae in China*; Jiangsu Science and Technology Publishing House: Nanjing, China, 1989.
49. Wu, Y.F.; Wu, C.Z. *The Fishes of the Qinghai-Xizang Plateau*; Sichuan Publishing House of Science & Technology: Chengdu, China, 1992.
50. Zeng, L.; Tang, W.Q. Discussion on age determination methods for two esquamate *Triplophysa* fishes. *Chin. J. Zool.* **2010**, *45*, 94–103.
51. Tian, N.N.; Yang, R.B.; Tan, B.Z.; Zeng, X.L.; He, L.Q.; Xu, Z.L.; Zhu, Z.; Liu, H.P.; Yang, X.F. Age, growth, and reproductive characteristics of *Triplophysa stewarti* in Lake Chugutso, Tibet. *J. Fish. Sci. China* **2022**, *29*, 1013–1021.
52. Dinh, Q.M.; Qin, J.G.; Tran, D.D. Population and age structure of the goby *Parapocryptes serperaster* (Richardson, 1864; Gobiidae: Oxudercinae) in the Mekong Delta. *Turk. J. Fish. Aquat. Sci.* **2015**, *15*, 341–352. [CrossRef] [PubMed]
53. Zhu, X.; Zhao, Y.; Mathers, A.; Corkum, L.D. Length frequency age estimations of American Eel recruiting to the upper St. Lawrence River and Lake Ontario. *Trans. Am. Fish. Soc.* **2013**, *142*, 333–344. [CrossRef]
54. Pauly, D. *Fish Population Dynamics in Tropical Waters: A Manual for Use with Programmable Calculators*; WorldFish: Penang, Malaysia, 1984; Volume 8.
55. Liang, W.T. The Biology Research About Several Species of *Triplophysa*. Master's Thesis, Wuhan Polytechnic University, Wuhan, China, 2014.
56. Maciel, T.R.; Vaz-dos-Santos, A.M.; Barradas, J.R.d.S.; Vianna, M. Sexual dimorphism in the catfish *Genidens genidens* (Siluriformes: Ariidae) based on otolith morphometry and relative growth. *Neotrop. Ichthyol.* **2019**, *17*, e180101. [CrossRef]
57. Wang, Z.W.; Hao, H.M.; Wei, J.; Wu, H.; Hamid, S.M.; Lv, R.X.; Lu, H.L.; Nie, Z.L. Morphology, Age, and Growth of *Triplophysa strauchii* in Sayram Lake, Xinjiang, China. *Animals* **2025**, *15*, 1039. [CrossRef] [PubMed]
58. Von Bertalanffy, L. Quantitative laws in metabolism and growth. *Q. Rev. Biol.* **1957**, *32*, 217–231. [CrossRef] [PubMed]
59. Ricker, W.E. *Computation and Interpretation of Biological Statistics of Fish Populations (Fisheries Research Board of Canada Bulletin)*; The Blackburn Press: Caldwell, NJ, USA, 1975; Volume 191, pp. 1–382.
60. Gompertz, B. On the nature of the function expressive of the law of human mortality and on a new mode of determining the value of life contingencies. *Philos. Trans. R. Soc.* **1825**, *115*, 513–583.
61. Araya, P.R.; Agostinho, A.A.; Bechara, J.A. The influence of dam construction on a population of *Leporinus obtusidens* (Valenciennes, 1847) (Pisces, Anostomidae) in the Yacyreta Reservoir (Argentina). *Fish. Res.* **2005**, *74*, 198–209. [CrossRef]
62. Quinn, T.P.; Wetzel, L.; Bishop, S.; Overberg, K.; Rogers, D.E. Influence of breeding habitat on bear predation and age at maturity and sexual dimorphism of sockeye salmon populations. *Can. J. Zool.* **2001**, *79*, 1782–1793. [CrossRef]
63. Michaletz, P. Variation in characteristics among gizzard shad populations: The role of impoundment size and productivity. *Fish. Manag. Ecol.* **2017**, *24*, 361–371. [CrossRef]
64. He, J.Y.; Wu, Z.Q.; Huang, L.L.; Sun, Y.Y.; Wang, D.J.; Lin, Y.; He, A.Y.; Feng, J.; Liu, H. Age, growth, reproduction and status of resource development of *Ptychidio jordani*, a critically endangered freshwater fish in the Hongshui River, China. *J. Fish Biol.* **2024**, *104*, 399–409. [CrossRef] [PubMed]
65. Xie, C.X. *Ichthyology*; China Agriculture Press: Beijing, China, 2010.
66. Verreycken, H.; Van Thuyne, G.; Belpaire, C. Length–weight relationships of 40 freshwater fish species from two decades of monitoring in Flanders (Belgium). *J. Appl. Ichthyol.* **2011**, *27*, 1416–1421. [CrossRef]

67. Magloo, A.H.; Bhat, F.A.; Shah, T.H.; Wanjari, R.N.; Bazaz, A.I.; Mir, S.A.; Gul, S.; Sidiq, J. Length-weight relationship and condition factor of an indigenous fish, *Triplophysa marmorata* (Heckel, 1838), from Kashmir Valley. *Int. J. Vet. Sci. Anim. Husb.* **2024**, *9*, 766–770.
68. Rennie, M.D.; Purchase, C.F.; Lester, N.; Collins, N.C.; Shuter, B.J.; Abrams, P.A. Lazy males? Bioenergetic differences in energy acquisition and metabolism help to explain sexual size dimorphism in percids. *J. Anim. Ecol.* **2008**, *77*, 916–926. [CrossRef] [PubMed]
69. Cao, X.Q.; Li, Y.H.; Wei, J.; Nie, Z.L.; Yang, Z.J. Relationship of body mass, body length, and fatness of *Schizothorax eurystomus*. *Acta Agric. Boreali-Occident. Sin.* **2019**, *28*, 1380–1386.
70. Șerban, D.A.; Barbacariu, C.A.; Burducea, M.; Ivancia, M.; Creangă, Ș. Comparative Analysis of Growth Performance, Morphological Development, and Physiological Condition in Three Romanian *Cyprinus carpio* Varieties and Koi: Implications for Aquaculture. *Life* **2024**, *14*, 1471. [CrossRef] [PubMed]
71. Gamito, S. Growth models and their use in ecological modelling: An application to a fish population. *Ecol. Model.* **1998**, *113*, 83–94. [CrossRef]
72. Murphy, M.D.; Taylor, R.G. Age, growth, and mortality of spotted seatrout in Florida waters. *Trans. Am. Fish. Soc.* **1994**, *123*, 482–497. [CrossRef]
73. Lin, Y.J.; Tzeng, W.N. Modelling the growth of Japanese eel Anguilla japonica in the lower reach of the Kao-Ping River, southern Taiwan: An information theory approach. *J. Fish Biol.* **2009**, *75*, 100–112. [CrossRef] [PubMed]
74. Dippold, D.A.; Leaf, R.T.; Hendon, J.R.; Franks, J.S. Estimation of the length-at-age relationship of Mississippi's Spotted Seatrout. *Trans. Am. Fish. Soc.* **2016**, *145*, 295–304. [CrossRef]
75. Branstetter, S. Biological Parameters of the Sharks of the Northwestern Gylf of Mexico in Relation to Their Potential as a Commercial Fishery Resource. Ph.D. Thesis, Texas A&M University, College Station, TX, USA, 1986.
76. Cox, R.M.; Parker, E.U.; Cheney, D.M.; Liebl, A.L.; Martin, L.B.; Calsbeek, R. Experimental evidence for physiological costs underlying the trade-off between reproduction and survival. *Funct. Ecol.* **2010**, *24*, 1262–1269. [CrossRef]

biology

MDPI

Review

Threats of Climate Change to Freshwater Ecosystems in Pakistan: eDNA Monitoring Will Be the Next-Generation Tool Used in Biodiversity, Conservation, and Management

Ghazanfer Ali [1], Sidra Abbas [2], Satoshi Nagai [3], Norhafiza Mohd Arshad [4] and Subha Bhassu [1,4,*]

[1] Animal Genetics and Genome Evolutionary Laboratory, Institute of Biological Sciences, Universiti Malaya, Kuala Lumpur 50300, Malaysia; s2125068@siswa.um.edu.my
[2] Department of Zoology, University of Jhang, Jhang 35200, Pakistan; dr.sidraabbas@uog.edu.pk
[3] Fisheries Technology Institute, Japan Fisheries Research and Education Agency, Yokohama 220-6115, Japan; snagai@affrc.go.jp
[4] Center for Research in Biotechnology for Agriculture, Universiti Malaya, Kuala Lumpur 50300, Malaysia; norhafiza@um.edu.my
* Correspondence: subhabhassu@um.edu.my or dr.sidraabbas@oug.edu.pk

Simple Summary: The current rates of climate change are unprecedented, and biological responses to these changes are occurring rapidly at the ecosystem, community, and species levels. Climate change is a significant threat to freshwater biodiversity, resulting in higher extinction and local extinction rates among freshwater species than terrestrial taxa. In the current era of rapid environmental change and species loss, environmental DNA (eDNA) offers a more efficient and rapid method to monitor and conserve freshwater biodiversity. The use of eDNA metabarcoding has been recognized as a powerful technique for obtaining extensive data on biodiversity.

Abstract: Freshwater ecosystems are a significant entity that govern the livelihood of people and are an important source of food, employment, and recreation. However, climate change is impacting freshwater ecosystems by altering their natural habitats. The purpose of this review is to highlight the vulnerability of freshwater fish to climate change. Climate change is invariably affecting natural ecosystems everywhere and in every part of the world, but these threats are more severe in Pakistan. Freshwater fish are important biotic drivers of freshwater ecosystems. Unfortunately, uncertain climate changes and anthropogenic activities have led to a decline in the diversity of these fishes. Rising temperatures, melting glaciers, changes in seasonal patterns, disturbances in the natural flow of rivers, pollution, and invasive species are major threats to native freshwater fish fauna, leading to a decline in fish diversity and population. *Tor putitora*, *Glyptothorax kashmirensis*, and *Triplophysa kashmirensis* are some of the species that are critically endangered in Pakistan due to these factors. In recent decades, insufficient attention has been paid to the freshwater ecosystem. This review of threats to the endemic fish species in this region is presented so that the government and policymakers can use this information as part of their management and conservation policy, thus safeguarding Pakistan's fish industry. Environmental DNA (eDNA) biomonitoring is a new technique for assessing biodiversity and species distribution and can be useful for conserving biodiversity in this region. Another purpose of this review is to introduce this new conservation strategy to Pakistan.

Keywords: climate change; freshwater ecosystem; invasive species; eDNA as biomonitoring tool; biodiversity conservation

1. Climate Change

Climate is the result of the interaction of weather components under different circumstances of time and space, including air temperature, pressure, solar radiation, precipitation, humidity, wind speed, etc., and alterations in these components are termed climate change [1,2]. Climate change is directly or indirectly associated with human activities that modify the composition of the overall atmosphere, in addition to natural climatic changes observed over comparable periods. Climate change has turned into a global environmental problem that has taken over the international agenda and is one of the biggest challenges for humanity [3]. The global average temperature has increased by 0.7 °C over the past century and is expected to continue to rise [4]. According to the predictions of the Intergovernmental Panel on Climate Change (IPCC), temperatures are anticipated to rise by 1.1 to 6.4 °C by the end of the 21st century compared to the 1980–1999 baselines [5]. It is predicted that average global precipitation has increased by 2% in the last 100 years and it is expected to increase in the future [5]. The impact of global climate change on hydro-ecological, freshwater, and agricultural systems are expected to vary across regions due to the influence of multiple environmental processes and human activities. [6,7]. In general, there is increasing evidence of climate change affecting ecosystems, including groundwater compatibility, surface water, and wetland ecosystem performance. However, its impact on water resources, including important hydrological metrics such as the magnitude, intensity, and frequency of rainfall, varies significantly at local and regional levels [8]. Globally, climate change is a significant threat that has already had profound impacts on natural ecosystems and biodiversity [4]. Evidence shows that increasing climatic change is responsible for geographic range shifts in species distributions, modifying phenology and migration patterns, diminishing the accessibility of suitable habitats for species, and disturbing important ecological interactions in communities [9,10]. Even though a small number of recent species extinctions have, as yet, been officially or tentatively attributed to climate change [11,12], data from the fossil record provide evidence that rapid climate change can be a fundamental driver of mass extinction, capable of eliminating up to 90% of all species [13–16]. This raises concerns about the adaptive potential of existing species to cope with future climatic changes [17,18]. The duration, frequency, and magnitude of extreme events have already increased in the last few years and are expected to worsen in the future with the intensification of anthropogenic activities [5]. There is sound scientific agreement that climatic change is not only factual but also has [19] and will continue to have continuous major impacts globally, including species extinction [20,21], seasonal changes in key ecological events, disturbances to the provision of natural goods and services, and even conflict in human societies [20]. These effects will expand to living things across all biological groups, including freshwater ecosystems. The effects of warming, and the consequent impacts on freshwater fish as a food source and environments providing employment, recreation, and tourism worldwide, [22] will be a tragedy.

2. Climate Change in Pakistan

It is very evident that climatic changes altering natural systems are happening everywhere in the world, but the alterations in climate extremes and their effect on Pakistan will be significant and have more drastic impacts [23]. These climate changes are anticipated to have adverse impacts on Pakistan [24] and are expected to have extensive and long-lasting effects on Pakistan's environment and society [25]. Harsh weather conditions, drought, the melting of glaciers in the Himalayas, and increased heat in some parts of the country pose severe threats to many of the important rivers of Pakistan (Figures 1 and 2). In terms of how extreme changes in climatic conditions will impact nations, Pakistan was placed fifth among the most affected countries between 1999 and 2018 [25]. Even though the contribution of Pakistan to global greenhouse gas (GhG) emissions is minimal, e.g., less than 1% of the world total [26], it is very susceptible to the adverse impacts of climate change happening in other parts of the world. Because Pakistan is poor in terms of technology and has poorer financial resources to acclimate to the harmful effects of climate change, its susceptibility has worsened [27]. Major threats faced by countries due to climate change include threats to food and water safety and the large migration of populations [28]. Pakistan's economy is largely dependent on agriculture, which is the reason for its vulnerability to the circumstances caused by increasing climate change. Like other South Asian countries, Pakistan is facing a great threat because of climate change [29]. Extreme monsoons and cyclones are expected to increase in Pakistan as temperatures in the sea and the atmosphere rise [30]. Government forecasts highlight extensive developments in the intensity and frequency of life-threatening events, with irregular monsoon rains causing alternating floods and extreme droughts [27,29]. For instance, between 1998 and 2018, more than 150 life-threatening weather events were reported in Pakistan [31]. In 2022, catastrophic floods hit the country. The main causes of these were increased precipitation and glaciers melting, fueled by climate change. One-third of the country was underwater. According to Climate Minister Sherry Rehman, this "has exceeded every boundary, every norm we've seen in the past [32]" and 33 million people were affected [33] (Figure 2).

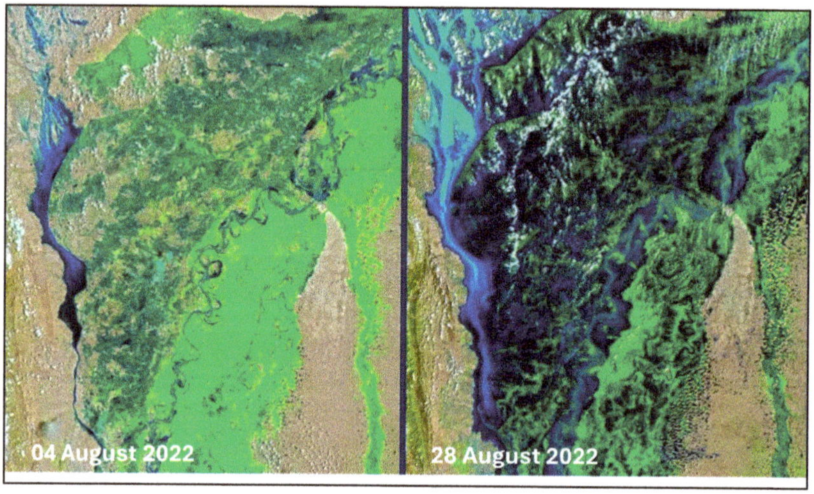

Figure 1. Satellite imagery showing a side-by-side comparison of southern Pakistan on 4 August 2022 and 28 August 2022. False-color satellite images from August 28th, showing floodwaters (blue) covering large portions of Pakistan's normally arid, brown landscape (right). Infrared and visible light are combined to enhance the contrast between water and land. In the image of August 28th, how floodwaters dramatically altered the country's landscape between early and late August is revealed. For instance, on August 4th, there was a stretch of approximately 30 miles (48 kilometers) of land separating the Indus River from Hamal Lake, situated west of the river in Pakistan's Qambar Shahdadkot District (left). However, by August 28th, the floodwaters had caused the two bodies of water to merge (right). (Image credit: NASA Earth Observatory images by Joshua Stevens, using Landsat data from the U.S. Geological Survey and VIIRS data from NASA EOSDIS LANCE, GIBS/Worldview, and the Joint Polar Satellite System (JPSS). (https://earthobservatory.nasa.gov/images/150279/devastating-floods-in-pakistan, accessed on 31 August 2022).

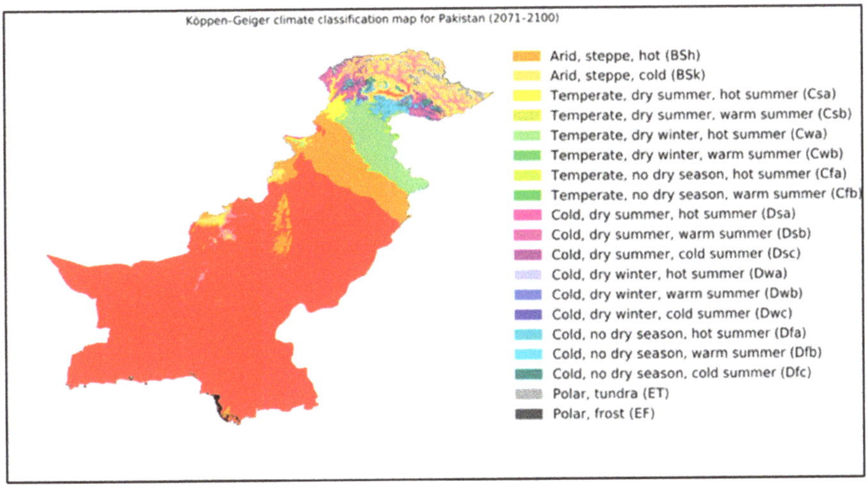

Figure 2. Future Koppen–Geiger climate classification map for Pakistan (2071–2100) [34]. A map for 2071–2100 under the most extreme climate change scenario, though mid-range scenarios are currently considered more likely [35].

(**a**) Balochistan, Pakistan (Fida Hussain/AFP via Getty Images)

(**b**) Karachi, Pakistan (AP Photo/Fareed Khan)

(**c**) Sukkur, Pakistan (ZUMA Press Wire 2022)

(**d**) Devastation caused by floods in Swat, Pakistan 2022

(**e**) Sehwan, Pakistan (Adeel Ahmed via Getty Images)

(**f**) Lasbella, Pakistan (Hamdan Khan/AP)

Figure 2. (**a**) Balochistan province on 29 August 2022. Image source: https://jordantimes.com/news/world/third-pakistan-under-water-flood-aid-efforts-gather-pace (accessed on 5 December 2023). (**b**) Pakistan's largest city, Karachi, inundated with an entire summer's worth of rain in one day, 25 July 2022. Image source: https://www.accuweather.com/en/severe-weather/pakistans-largest-city-inundated-with-an-entire-summers-worth-of-rain-in-one-day/1221899 (accessed on 5 December 2023). (**c**) A family navigates floodwater on Monday after a monsoon downpour in Sukkur, Pakistan. Image source: https://www.motherjones.com/environment/2022/08/pakistan-floods-flooding-monsoon-climate-change/ (accessed on 5 December 2023). (**d**) Screengrab of a video showing the devastation caused by floods in Swat, Pakistan 2022. Image source: https://www.geo.tv/latest/436128-emergency-declared-in-flood-ravaged-swat (accessed on 5 December 2023). (**e**) Flash flooding has devasted the lives of millions across Pakistan. Image source: https://www.axios.com/2022/09/15/pakistan-rains-increased-33-percent-climate-change (accessed on 5 December 2023). (**f**) Rescue workers help villagers evacuate from a flooded area after heavy rains fell in Lasbella, a district in Pakistan's southwest Balochistan province, on 26 July 2022. Image source: https://www.theatlantic.com/photo/2022/08/photos-monsoon-flooding-in-pakistan/671278/ (accessed on 5 December 2023).

3. Drivers of Global Climate Change

Emissions of greenhouse gases have boosted the greenhouse effect, causing the surface temperature of the earth to increase. The burning of fossil fuels is causing the climate to change more rapidly than any other human activity [36,37]. In 2022, the global levels of major greenhouse gasses—carbon dioxide (CO_2), methane (CH_4), and nitrous oxide (N_2O)—continued to increase, underscoring the persistent impact on climate change (Figure 3 and Table 1). CO_2, the predominant greenhouse gas, was responsible for approximately 64% of the climate's warming effect, with concentrations in 2022 surpassing pre-industrial levels by 50%, a pattern that extended into 2023. Almost all this increase is due to human activities [38]. The rise in CO_2 from 2021 to 2022 was measured at 2.2 parts per million (ppm), a slight decrease from the previous year's increase of 2.46 ppm and the average growth rate observed over the last decade. This decline may be attributed to the enhanced absorption of CO_2 by natural carbon sinks, including terrestrial ecosystems and oceans, particularly following several La Niña years. Nonetheless, the anticipated onset of an El Niño event in 2023 could potentially disrupt this trend [39]. Methane, accounting for approximately 19% of the greenhouse warming effect, experienced a notable increase. While the growth from 2021 to 2022 was somewhat less than the unprecedented rate recorded between 2020 and 2021, it still exceeded the average growth rate observed over the past decade [39]. Human activities, particularly agriculture and fossil fuel extraction, are responsible for about 60% of methane emissions, with the remaining 40% stemming from natural sources [39,40]. Nitrous oxide is a significant greenhouse gas and a contributor to ozone depletion, representing approximately 7% of the radiative forcing attributed to long-lived greenhouse gasses [39]. The rise in N_2O levels from 2021 to 2022 marked the largest annual increase ever documented, exceeding all prior growth rates. Emissions of N_2O originate from natural sources (60%) and human activities (40%), including the application of fertilizers, biomass combustion, and various industrial operations [39,41]. Pakistan was identified by the United Nations Environment Program (UNEP) as one of the countries responsible for global greenhouse gas (GHG) emissions, contributing 1.02% to the world's total [42,43]. In 2018, Pakistan emitted approximately 504.59 million tons of GHGs. The country is among the most affected by both the adverse impacts of climate change and deteriorating air quality due to pollution. GHG emissions are a primary driver of global warming, making them a critical focus of research in both the scientific and political arenas [42]. In 2019, the burning of fossil fuels, including coal, oil, and gas, released around 80 million tons of GHG, with carbon dioxide being the predominant component [26]. These emissions significantly contribute to air pollution and the decline in air quality. In response to these challenges, Prime Minister Imran Khan announced in 2020 that no further coal-fired power plants would be permitted in Pakistan [44].

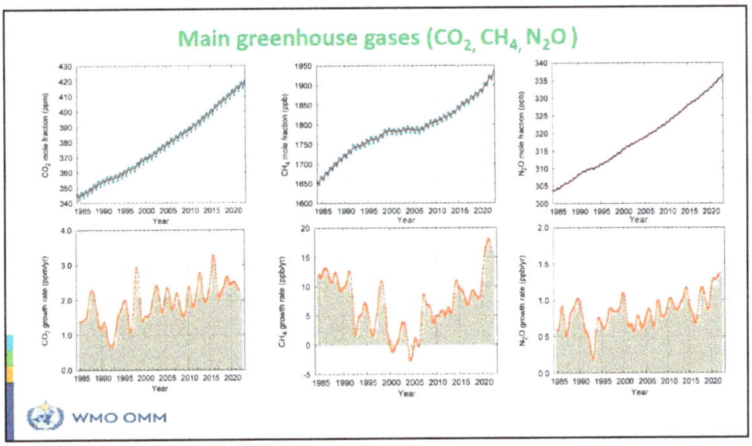

Figure 3. Graph showing the main greenhouse gas emissions for November 2023. Source: WMO 2024. (https://wmo.int/news/media-centre/greenhouse-gas-concentrations-hit-record-high-again. Accessed on 3 September 2024).

Table 1. Greenhouse gasses reach new levels.

	CO_2	CH_4	N_2O
2022 global mean abundance	417.9 ± 0.2 ppm	1923 ± 2 ppb	335.8 ± 0.1 ppb
2022 abundance relative to 1750 [a]	150%	264%	124%
2021–2022 absolute increase	2.2 ppm	16 ppb	1.4 ppb
2021–2022 relative increase	0.53%	0.84%	0.42%
Mean annual absolute increase over the past 10 years	2.46 ppm yr^{-1}	10.2 ppb yr^{-1}	1.05 ppb yr^{-1}

[a] Assuming a pre-industrial mole fraction of 278.3 ppm for CO_2, 729.2 ppb for CH_4, and 270.1 ppb for N_2O. Source: WMO 2024. (https://wmo.int/news/media-centre/greenhouse-gas-concentrations-hit-record-high-again accessed 3 September 2024).

4. Effects of Climate Change on Freshwater Ecosystems and Fishes

Perhaps, climate change has been affecting ecosystems for the last few years, and like many other unnoticed sources of environmental system degradation [45], may have altered our understanding of the 'baseline' conditions in freshwater ecosystems and important life history characteristics of freshwater fishes [46,47]. Over the last century, the global mean surface temperature has risen by 0.75 °C (possible range: 0.56–0.92), with the rate of increase accelerating since the 1970s [48]. This warming has also been documented in freshwater ecosystems worldwide [49–51]. Warming cannot be completely controlled, even if emissions are stopped or reduced, due to inertia in the climate system. Projections of the scale and rate of change differ between models, but by 2100, global atmospheric temperatures might rise 6.4 °C beyond existing levels [52] due to the abnormal weather conditions of the current biological times [53]. The scale of change is such that air temperatures once considered extremely high will rapidly become normal by mid-century [54]. In addition, unprecedented weather conditions will occur first in tropical areas and developing countries, where freshwater fishing is an important source of protein and income [55].

Freshwater ecosystems are considered sensitive indicators of climate change [56], with extinction representing the greatest risk to freshwater species resulting from its effects. [21]. There is scientific agreement that the increasing levels of greenhouse gas emissions have enhanced climatic change [20,57]; one effect of the past and present emissions of greenhouse gasses could be an increase in air temperature of around 6.4 °C by the end of the 21st century, even if firm mitigation policies are implemented. Many harmful effects have directly or indirectly impacted freshwater fish populations and their habitats [58,59]. The impacts of climate change are predicted to occur in temperate regions through increases in water temperatures, changes in rainfall systems, the beginning and extent of snow cover, changes in flow rates, and disturbance events, such as floods, wildfires, and pest infestations [59]. Since freshwater fishes are ectotherms, rising water temperatures might cause increased physiological stress, which in turn directly raises metabolic demand, affecting metabolism, productivity, reproduction, growth, and survival [60,61]. It was also found that changes in global rainfall patterns can affect the seasonal flow of rivers and their phenology [62]. Important life phases of some freshwater fishes, as well as their population dynamics, are dependent on the projected seasonal flow patterns [63]. Therefore, projected precipitation effects could lead to decreased reproductive success, alterations in species composition, and local extinction [64]. In freshwater ecosystems, certain indirect effects might occur through modifications in biotic processes, such as an increase in exotic species [65,66], a higher risk of parasites and diseases [67], and increased predation and competition [68]. Overall, this could alter food webs, fish communities, and, ultimately, ecosystems.

Various non-climatic anthropogenic stressors, such as overexploitation, water pollution, habitat changes and degradation, the generation of hydropower, flow modifications, and deforestation [69–71], are existing drivers of the freshwater fish decline worldwide [72]. The effects of these stressors on freshwater fish are expected to become more complex due to climate change [73]. In the current era of warming, terrestrial areas have warmed more rapidly than oceans, leading to an increase in water temperature in rivers, lakes, and wetlands globally [74]. This reflects a stronger relationship between water and air temperature in freshwater systems [75], which are already under pressure from a range of non-climatic stressors such as the over abstraction of water, eutrophication, habitat destruction, and the introduction of alien species [76]. Climate change has increasingly and continuously threatened the global environment [77], sustainable human development, and biodiversity, primarily by altering the water cycle and global thermal regimes and causing acidification [12,77].

Pakistan, a subtropical country, faces significant fluctuations in climate-related impacts due to its geographic and climatic positioning [78,79]. The country has experienced considerable climate change impacts in recent years [80], particularly its freshwater ecosystems, which support diverse fish populations. The heavy rainfall events and subsequent flooding in 2022 and 2023, exacerbated by climate change, resulted in the widespread disruption of freshwater habitats. Flooding significantly impacts the physical characteristics of rivers and streams, resulting in the erosion of riverbanks, alterations in river pathways, and the destruction of habitats essential for fish nesting and breeding [81–83]. These alterations can facilitate the introduction of new or invasive species into freshwater ecosystems, intensifying competition for resources and displacing indigenous species [82,84,85]. Additionally, the arrival of invasive species, along with the decline in water quality caused by increased sedimentation and pollution from floodwaters, can exacerbate the stress on native fish populations, adversely affecting their health and reproductive viability [86–88].

Pakistan's freshwater ecosystems face not only these immediate challenges but also significant long-term threats from climate change. The disruption of hydrological cycles, particularly due to the accelerated glacial melting in the Himalayas and Hindu Kush, altered the flow dynamics of the Indus River, which depends heavily on glacial meltwater during dry periods [89]. As the temperature rises, glacial melting accelerates, resulting in temporarily increased river flows; however, this trend poses the long-term danger of a diminished water supply as glaciers continue to shrink [90]. Such alterations disrupt essential seasonal flow patterns that are vital for agriculture and ecosystem stability, while also increasing sedimentation in rivers, which negatively impacts water quality and the habitats of aquatic species [91].

Alterations in monsoon patterns have caused erratic rainfall, resulting in extended periods of drought as well as severe flooding [92,93]. This variability significantly impacts freshwater ecosystems, causing water-level changes that can devastate breeding sites for fish and other aquatic life [94]. While flooding may restore certain wetlands, it frequently introduces pollutants and sediments that compromise water quality [95,96], thereby affecting the health and diversity of aquatic habitats [97]. On the other hand, drought conditions diminish the water supply, leading to higher pollutant concentrations [98] and intensified competition among species for scarce resources [99]. For example, rivers like the Ravi and Sutlej have experienced reduced water flow and sections that have dried up, conditions worsened by rising temperatures and shifting precipitation patterns resulting from climate change.These factors, compounded by climate change, contribute to the decline in freshwater fish diversity, with some species becoming endangered or extinct in certain regions [100,101]. The complex interplay of these threats underlines the urgent need for integrated water management and climate adaptation strategies to protect Pakistan's freshwater ecosystems and biodiversity.

The freshwater biodiversity of Pakistan, especially its fish species, is under significant threat due to climate change. The country is home to a wide variety of fish, including numerous endemic species that thrive in specific environmental conditions. As temperature increases and water flow patterns shift, these species encounter unprecedented challenges. An elevated water temperature can lead to lower levels of dissolved oxygen [102,103], putting stress on aquatic organisms [104,105] and potentially resulting in a shift in species dynamics, where more resilient and often invasive species may outcompete native ones [106]. Additionally, changes in river flow can interfere with migration patterns that are essential for the reproductive cycles of many fish species [107]. The overall effects of climate change on Pakistan's freshwater ecosystems are significant, endangering biodiversity, water security, and the livelihoods of countless individuals. To tackle these issues, it is essential to implement thorough monitoring, adopt sustainable water management practices, and create adaptive strategies to lessen the adverse effects of climate change on these crucial ecosystems [108,109].

Fish biologists have been discussing climatic changes and their effects on freshwater fish for nearly four decades [110–120]. Three significant phenomena have been identified as widespread responses by fishes to the current warming of their habitats: (a) shifts in latitudinal distributions, with temperate and subtropical fishes moving poleward [121];

(b) alterations in the phenology of important life events, such as the timing of spawning, the length of reproductive periods, and migration [122,123]; and (c) a reduction in average body size, linked to a decline in the size of adult individuals, leading to an increased proportion of smaller and younger individuals in populations [124]. The underlying mechanisms of these phenomena are not fully understood; however, they are believed to be driven, at least in part, by the biological functions and responses of individual animals, reflecting evolutionary processes or phenotypic plasticity [125,126].

As freshwater ecosystems provide vital services to humanity and are highly threatened by climate change (Figure 4), further studies are necessary to enhance our understanding of how the climate impacts species within these systems. Once the significance of this issue is formally recognized, legislative organizations worldwide should prioritize climate change as a primary area of concern; additionally, international and national funding institutions are increasingly supporting research into climatic change. The Food and Agriculture Organization (FAO) of the United Nations has recognized the importance of climate change's impacts on fisheries [127], and the IPCC assessment reports include references to freshwater systems [72,128]. Interested readers should refer to these resources for additional data and information.

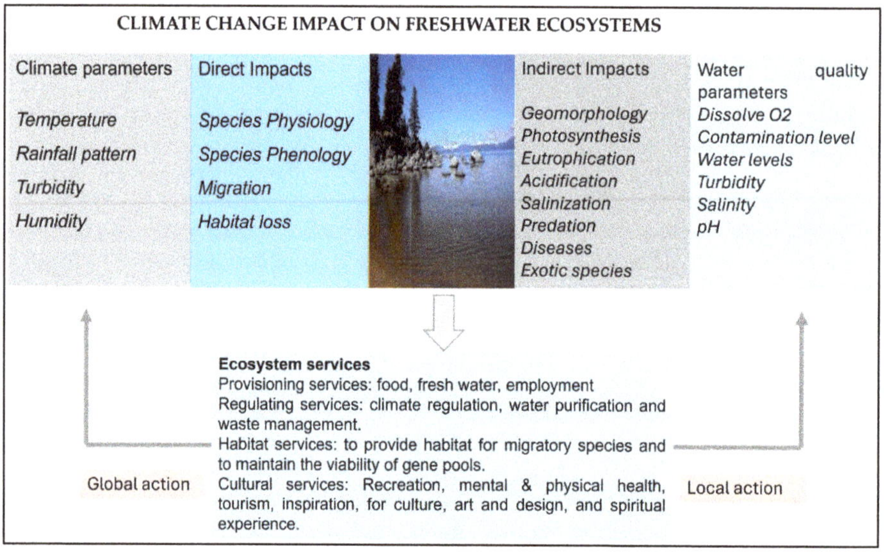

Figure 4. This Figure provides an overview of ecosystem services directly and indirectly impacted by climate change and local human activities. It illustrates the complex, cyclical relationship where the use of ecosystem services can, through direct and indirect mechanisms, influence the sustainability of those same services. Implementing global measures to mitigate harmful climate change impacts, along with local efforts to reduce disturbances in water quality, can help preserve these vital ecosystem services. Figure adapted from [129].

5. Significance of Freshwater Ecosystems

Freshwater ecosystems play an important ecological role by providing economically essential products and services, and they provide vital habitats for a large number of reptiles, mammals, birds, fishes, and aquatic plants. They serve as habitats for various migratory and threatened species of fishes, reptiles, and birds. Additionally, freshwater ecosystems attract tourists by providing recreational areas for bird-watching, sports, and games. Freshwater ecosystems, particularly wetlands, play a key role in mitigating climate change through various ecosystem functions, including CO_2 sequestration, shoreline stabilization, water purification, and flood control [130]. They support a wide range of ecosystem services that are threatened by climate change and invasive species, among other factors. Fortunately, there are many options for protecting and conserving freshwater ecosystems, which need to be addressed immediately [131] (Figure 4).

6. Key Threats to Freshwater Ecosystems

Globally, freshwater ecosystems and their associated biodiversity are at risk. Pollution, the introduction and spread of invasive species, habitat loss, overexploitation, the construction of instream barriers, hydrological modifications, excessive water abstraction, and the intensification of agricultural activities are the main threats to freshwater ecosystems and biodiversity [69]. Due to these compound risks and effects, the freshwater ecosystem is attracting continuous global attention in the quest to determine practical and adequate ways to combat the expected large-scale extinction of freshwater fish [132]. It is estimated that about 30% of freshwater fish species are at risk of extinction [133].

7. Freshwater Fishes

Fish are vital and diverse components of ecosystems, offering services that directly or indirectly improve human well-being. They are a rich source of proteins, vitamins, and minerals for human consumption [134]. Additionally, fish are also vital as they provide a means of employment and income to millions of people worldwide [134]. Freshwater fish are found on all continents except Antarctica. They are harvested from major rivers, lakes, wetlands, small ponds, and streams, making significant contributions to global fish production relative to their size [135]. Millions of people in developing countries rely on freshwater fisheries [55]. Freshwater fish differ from marine fish in terms of function, including differences in their trophic level, scale, and responses to natural and anthropogenic drivers [53,136]. To date, approximately 32,500 fish species have been described worldwide [137], with nearly 41% being freshwater species, 58% being marine species, and 1% being diadromous species [138]. Although freshwater makes up only about 0.3% of the Earth's total water, it surprisingly hosts about 15,000 species of freshwater fish [139]. In Pakistan, 193 freshwater species have been reported [140]. The Indus River alone is home to more than 180 fish species [141].

Freshwater fisheries provide important ecosystem services to humanity [22] by offering food, income, and recreation through recreational, touristic, and commercial fisheries [137,142]. Each sector varies in its susceptibility to and ability to adapt to climate change. Freshwater fish make significant economic and social contributions worldwide [143]. According to an FAO report, 11.5×10^6 tons of fish were captured from inland waters in 2011, marking an increase of more than 30 per cent since the early 2000s, with these catches increasing every year [144]. These statistics are potentially low and highlight the pressure placed on freshwater fisheries due to the world's growing population. Globally, the supply of animal protein to humans is dominated by freshwater fishes, especially in Africa [142], where larger fisheries are found (estimated at 2,567,427 tons in 2010), mainly in the Great Lakes of Africa, as well as in productive rivers in Egypt and Nigeria, and the lakes, reservoirs, and wetland fisheries in the Sahel [145]. The FAO [146] observed that the recent developments in inland catches were linked with India, Bangladesh, China, and Myanmar and that more than 70 per cent of the world's production comes from Asian countries. In 2022, global fisheries and aquaculture production reached a record 223.2 million tons, with 185.4 million tons coming from aquatic animals, and for the first time, aquaculture production surpassed capture fisheries. Asia dominated this production, contributing 167.1 million tons—75 per cent of the global total—and producing 70 per cent of the world's aquatic animals [146]. Some commercially important freshwater fishes of Pakistan, along with their IUCN status, are listed in Table 2.

Table 2. Some commercial freshwater fishes of Pakistan and their IUCN status.

Name of Fish	IUCN Status	Distributional Status	Commercial Value
Butter catfish (*Ompok bimaculatus*)	Near Threatened	Indigenous	NA
Pabdah catfish (*Ompok panda*)	Near Threatened	Indigenous	NA
Freshwater shark (*Wallago attu*)	Near Threatened	Indigenous	Very high
Gangetic ailia (*Ailia coila*)	Near Threatened	Indigenous	NA
Humped featherback (*Chitala chitala*)	Near Threatened	Indigenous	High
Gangetic goonch (*Bagarius bagarius*)	Near Threatened	Indigenous	High
Himalayan snowtrout (*Schizothorax plagiostomus*)	Vulnerable	Indigenous	High
Hilsa Shad (*Tenualosa ilisha*)	Not Evaluated	Indigenous	Very high
Mirgal (*Cirrhinus mrigala*)	Least Concern	Indigenous	Very high
Catla (*Gibelion catla*)	Least Concern	Indigenous	Very high
Thicklip labeo (*Labeo dyocheilus pakistanicus*)	Least Concern	Indigenous	High
Orangefin labeo (*Labeo calbasu*)	Least Concern	Indigenous	High
Kuria labeo (*Labeo gonius*)	Least Concern	Indigenous	High
Rahu (*Labeo rohita*)	Least Concern	Indigenous	Very high
Scaly Osman (*Diptychus maculatus*)	Not evaluated	Indigenous	High
Indus snowtrout (*Ptychobarbus conirostris*)	Not evaluated	Indigenous	High
Kunar snowtrout (*Racoma labiate*)	Vulnerable	Indigenous	High
Zig-zag eel (*Mastacembelus armatus*)	Least Concern	Indigenous	High
Sperata seenghala	Least Concern	Indigenous	Very high
Naziri bachcha (*Clupisoma naziri*)	Not evaluated	Indigenous	Very high
Rita Catfish (*Rita rita*)	Least Concern	Indigenous	Very high
Garua bachcha (*Clupisoma garua*)	Least Concern	Indigenous	Very high
Walking Catfish (*Clarias batrachus*)	Least Concern	Indigenous	High
Great snakehead (*Channa marulius*)	Least Concern	Indigenous	Very high
Chirruh snowtrout (*Schizopyge esocinus*)	Not evaluated	Indigenous	High
Ladakh snowtrout (*Schizopygopsis stoliczkai*)	Not evaluated/very rare	Indigenous	NA
Stoliczka triplophysaloach (*Triplophysa stoliczkai*)	Not evaluated/very rare	Indigenous	NA
Zebra Fish (*Danio rerio*)	Least Concern/very rare	Indigenous	NA
Whiptail Catfish (*Sisor rabdophorus*)	Not evaluated/very rare	Indigenous	NA
Bengala barb (*Megarasbora elonga*)	Least Concern/very rare	Indigenous	NA
Gangetic leaf fish (*Nandus nandus*)	Not evaluated/very rare	Indigenous	NA
Chameleon fish (*Badis badis*)	Not evaluated/very rare	Indigenous	NA
Gangetic mud eel (*Monopterus cuchia*)	Not evaluated/very rare	Indigenous	NA
One-stripe spiny eel (*Macrognathus aral*)	Vulnerable	Indigenous	NA
Punjab razorbellyminnow (*Salmophasia punjabensis*)	Not evaluated	Endemic	NA
Naseeri baril (*Barilius naseeri*)	Not evaluated	Endemic	NA
Pakistani baril (*Barilius Pakistanicus*)	Least Concern	Endemic	NA
Blue rahu (*Labeo caeruleus*)	Least Concern	Endemic	NA
Days' labeo (*Labeo nigripinnis*)	Not evaluated	Endemic	NA
Balochistan labeo (*Labeo gedrosicus*)	Not evaluated	Endemic	NA
Macmahons' labeo (*Labeo macmahoni*)	Not evaluated	Endemic	NA
Zhob mahasheer (*Naziritor zhobensis*)	Not evaluated	Endemic	NA
Punjab barb (*Puntius punjabensis*)	Least Concern	Endemic	NA
Salt Range barb (*Puntius waageni*)	Vulnerable	Endemic	NA
Javeds' loach (*Botia javedi*)	Not evaluated	Endemic	NA
Swat loach (*Schistura alepidota*)	Not evaluated	Endemic	NA
Anambar loach (*Schistura anambarensis*)	Least Concern	Endemic	NA
Arifs' loach (*Schistura arifi*)	Least Concern	Endemic	NA
Panjgur loach (*Schistura baluchiorum*)	Not evaluated	Endemic	NA
Kurram loach (*Schistura curtistigma*)	Vulnerable	Endemic	NA
Hangu loach (*Schistura fascimaculata*)	Not evaluated	Endemic	NA
Pishin loach (*Schistura kessleri*)	Not evaluated	Endemic	NA
Parachinar loach (*Schistura lepidocaulis*)	Not evaluated	Endemic	NA

Table 2. *Cont.*

Name of Fish	IUCN Status	Distributional Status	Commercial Value
Kohat loach (*Schistura kohatensis*)	Not evaluated	Endemic	NA
Mach loach (*Schistura machensis*)	Vulnerable	Endemic	NA
Khyber loach (*Schistura microlabra*)	Vulnerable	Endemic	NA
Rawlakot loach (*Schistura nalbanti*)	Not evaluated	Endemic	NA
Zhob loach (*Schistura pakistanica*)	Vulnerable	Endemic	NA
Pakhtunkhwa loach (*Schistura parashari*)	Near Threatened	Endemic	NA
Chenab loach (*Schistura shadiwalensis*)	Least Concern	Endemic	NA
Hazara Loach (*Triplophysa hazaraensis*)	Vulnerable	Endemic	NA
Nazir triplophysaloach (*Triplophysa naziri*)	Least Concern	Endemic	NA
Yasin triplophysaloach (*Triplophysa yasinensis*)	Least Concern	Endemic	NA
Horas' mystus (*Mystus horai*)	Vulnerable	Endemic	NA
Pakistani gagata (*Gagata pakistanica*)	Not evaluated	Endemic	NA
Naziri catfish (*Glyptothorax naziri*)	Near Threatened	Endemic	NA
Kalabagh nangra (*Nangra robusta*)	Not evaluated	Endemic	NA
Sindh catfish (*Ompok Sindhensis*)	Not evaluated	Endemic	NA

NA: Commercial status not valuated; source: www.fishbase.se/search.php. Accessed on 22 November 2023; www.iucnredlist.org. Accessed on 5 December 2023.

8. Freshwater Fishes Response to Climate Change

The most direct, fast, and common way in which fish respond to climate change is alterations in growth [147]. For example, the rapid somatic growth rates of fish are directly affected by temperature variations, including both air and water temperature [148,149]. Furthermore, climate change is predicted to alter the life history of plankton and benthos [150,151], which ultimately impacts the energy consumption and development of fish [152]. Alterations in fish growth are likely to have predictable long-term effects on population characteristics and features (e.g., egg size, age size structure, and reproductive phenology), recruitment, and dynamics (e.g., overwintering mortality) [153,154], and subsequently affect other biological processes at higher organization levels [151,155]. Therefore, understanding the fish growth responses to changing environmental conditions is a crucial step in predicting the impacts climate change will have on ecosystems and communities [156,157]. In response to the current climatic changes, freshwater fishes experience significant shifts and changes in their distribution, which seriously affect fish assemblage and composition [158]. About half of the species of freshwater fish are projected to become extinct in the coming decades, with a marked decrease in tropical areas and a higher likelihood of extinction for species with small body sizes and/or limited geographical ranges [159]. The response of species to climate change has been demonstrated globally in both terrestrial and aquatic environments. Changes in species composition, a reduction in geographic distribution, and alterations in ranges along gradients of altitude are important responses of species to climate change [160].

9. Biotic Components of Freshwater Ecosystems and Climate Change

The biotic component of freshwater ecosystems can also be altered by climate change, which has indirect effects on freshwater fish species. These effects can be produced by increasing the size of populations, the distribution of alien invasive species [66], changes in competition and predation rates [68], and an increase in pest occurrence and disease risk [67]. Exotic species or alien species, as part of the anthropogenic activity, also affect local aquatic systems. Exotic species can adversely impact local species and profoundly change aquatic ecosystems through alterations in trophic levels (competition, predation, changes in the food web), the introduction of diseases and parasites, modifications to fishes' habitat, and spatial shifts (overcrowding and aggressive effects) [161]. The impact of alien gam fish, largely unpredictable in terms of time and space, and the establishment of relatively few species will lead to the extinction of many endemic species worldwide [162]. Along with other climatic factors, the blind introduction of alien species is one of the major problems decreasing the number of native fish species [161,163]. Worldwide, invasive species have been recognized as agents of local biodiversity loss. IUCN Pakistan [139] has also examined the risk factors for endemic fishes, with about 493 that are critically endangered, and provides recommendations to restore the habitats of endangered local fish species. The common carp is known as a pest due to its abundance and tendency to eliminate aquatic plants and reduce water clarity in the habitats of various aquatic species. In the United States, habitat destruction was a major issue affecting many fish species. However, the invasion of alien species was the main factor [164].

Climate change and the interaction between alien invasive and diseased species are concerns, and both are intensified by global trade [12]. Climate change is thought to be beneficial for invasive species. These changes speed up the rates of colonization via adaptive migration and undermine the integrity of in situ biotic assemblages. This allows species to reside in novel locations and climates, thus increasing the chances of colonization. If the invasive species is pathogenic, the likelihood of other species developing new diseases may increase [165]. Changes in climate and environmental conditions lead to changes in disease vectors (e.g., parasites and malaria mosquitos), which are likely released from natural control. Anthropogenic factors are responsible for the decline in commercially important fish populations in Pakistan. *Glyptothorax kashmirensis,* found in the Jhelum River drainage, has been declared Critically Endangered by the IUCN. Currently, the river Jhelum is blocked at many locations. This will alter this species' habitat and consequently affect this fast-moving river species. Due to these severe, irreversible threats, a more than 80% decline is expected in the next five to ten years. *Tor putitora* has also been declared Critically Endangered in Pakistan by the IUCN due to severe population decline. This species is under pressure from habitat loss, overfishing, and alterations in habitat quality, which result in the demolishing of breeding grounds. Six species—*Ompok bimaculatus, Ompok pabda, Ailia coila, Wallago attu, Chitala chitala,* and *Bagarius bagarius*—are Near Threatened and one species, *Schizothorax plagiostomus,* is Vulnerable [166] (Tables 3 and 4).

Table 3. Exotic fishes found in the river system of Pakistan.

Species	Occurrence	IUCN Status	Commercial Value
Goldfish (*Carassius auratus*)	Introduced	Least Concern	High
Grass carp (*Ctenopharyngodon idella*)	Introduced	Not evaluated	Very high
Common carp (*Cyprinus carpio*)	Introduced	Least Concern	Very high
Mosquito fish (*Gambusia affinis*)	Introduced	Least Concern	Minor commercial
Silver carp (*Hypophthalmichthys molitrix*)	Introduced	Not evaluated	Very high
Bighead carp (*Hypophthalmichthys nobilis*)	Introduced	Not evaluated	Very high
Rainbow trout (*Oncorhynchus mykiss*)	Introduced	Least Concern	Very high
Blue tilapia (*Oreochromis aureus*)	Introduced	Not evaluated	Very high
Mozambique tilapia (*Oreochromis mossambicus*)	Introduced	Vulnerable	Highly commercial
Nile tilapia (*Oreochromis niloticus*)	Introduced	Least Concern	Highly commercial
Sea trout (*Salmo trutta*)	Introduced	Least Concern	Commercial

Table 4. Endangered and Critically Endangered species.

Common Name	Scientific Name	IUCN Status	Occurrence	Commercial Value
Kashmir catfish	*Glyptothorax kashmirensis*	Critically Endangered	Endemic	NA
Verinag triplophysaloach	*Triplophysa kashmirensis*	Critically Endangered	Endemic	NA
Golden mahasheer	*Tor putitora*	Endangered	Indigenous	Commercial
Wana garra	*Garra wanae*	Endangered	Endemic	NA
Havelian loach	*Schistura afasciata*	Endangered	Endemic	NA
Harnai loach	*Schistura harnaiensis*	Endangered	Endemic	NA
Dera loach	*Schistura macrolepis*	Endangered	Endemic	NA
Pakistans' batasio	*Batasio pakistanicus*	Endangered	Endemic	NA
Punjab catfish	*Glyptothorax punjabensis*	Endangered	Endemic	NA
Bhed catfish	*Glyptothorax stocki*	Endangered	Endemic	NA
	Glyptothorax punjabensis	Endangered	Endemic	NA
Harnai käärtrull	*Schistura harnaiensis*	Endangered	Endemic	NA

10. Climate Change and Abiotic Components of Ecosystem

Freshwater fish, as ectotherms, face greater physiological stress and metabolic requirements when exposed to high temperatures, which directly influence their survival, development, productivity, and reproduction [60,61]. Additionally, spatial and temporal changes to precipitation will alter climatic patterns, which will negatively impact life cycles, population dynamics, phenology, and reproductive successes [62,63]. In some cases, freshwater fishes have responded to the effects of climate change through modifications in their demographic processes, spatial distribution alterations, shifts in the timing of seasonal migration and breeding, and evolutionary changes [167]. The combined effects of climate change and other human stressors have affected the fishes' ability to deal with or react to these alterations at a rate matching that of climate change [58,168]. Species are in danger when they have a restricted ability to adjust or react to rapid climatic changes [169]. These species have a higher chance of extinction and are typically the focus of adaptation efforts [168,169]. The unparalleled rate and severity of climate change poses a threat to the world's freshwater fish population [20,170].

11. Climate Change and Hydrological Cycle

River networks are under the influence of climate change; many are also affected by existing stressors or have been exposed to a legacy of stressors. The general predictions regarding climate change's effects on the river networks include changes in the discharge timing and intensity, an increase in extreme events of both peak and low flows, and an increase in river temperature [171]. The hydrological cycle is significantly affected by climate change in Pakistan's rivers, and there have clearly been significant alterations in the flow systems of various rivers. These alterations in river and stream flow distress the productive capacity of freshwater ecosystems and aquatic biodiversity [8]. Among the freshwater reservoirs in Pakistan, several small and large tributaries join to form the Indus rivers system. The area occupied by Indus River system is about 1.12 million Km2, of which 47% is located in Pakistan, 39% in India, 8% in China, and only 6% in Afghanistan [136]. In the upper Indus basin, 13% of the mountains are covered by glaciers [172], and water flow in the Indus River is affected by the temperature, solar radiation, and precipitation [161]. In Pakistan, there is a monsoon season from July to September and 60 to 70 percent of the total precipitation takes place in this season, affecting the river flow [173]. The Indus River originates in Tibet from the Kailash range and is the backbone of the river systems in Pakistan and India. It flows westward and eventually falls into the Arabian Sea, draining an area of about 945,345 km^2 and reaching a length of about 3180 km. The Indus River system comprises 27 tributaries and seven chief rivers: Zhob River, Sawat River, Kurraam

River and Gomal River. The Ravi, Chenab, Jhelum, Sutlej, and Beas rivers flow from India to Pakistan and the Kabul River flows from Afghanistan toward Pakistan. Pakistan's well known and extensive canal system covers an area of about 780,000 hectares, of which 9.7% is wetland: 27% is coastal wetland and 73% contains fresh water [174]. Good water quality and purity are vital for human survival, and their availability is shaped by both anthropogenic and human-induced factors. Unfortunately, water quality is declining rapidly day by day due to anthropogenic influences, industrialization, agricultural growth, and rapid urbanization [175]. Heavy metal pollutants can contaminate the environment for a long period and have irreparable impacts on the productive capacity of aquatic ecosystem [176]. Heavy metals, even at very low concentrations, are dangerous for freshwater and aquatic biodiversity [177]. The increasing intensity of pollution affects water quality and threatens the aquatic ecosystem's integrity. The habitat of aquatic organisms has been destroyed in Pakistan due to transboundary management in the Satluj and Ravi. Over the past two decades in India, rapid economic growth and increasing agricultural, industrial, and urban development in the upper catchment areas have increased the level of toxic compounds in rivers [177].

12. eDNA Biomonitoring a Conservation Tool

Altering ecological communities in response to climate change and other anthropogenic activities has become a global issue [178]. A sharp decline in biodiversity has been one of the most critical challenges all over the world since the 20th century [179,180]. Due to these anthropogenic activities, increased rates of biodiversity loss, species extinction, and loss of ecosystem functioning have been observed worldwide [181], and the worldwide species extinction rate during this period exceeds that of prehuman periods [182]. This severely threatens the sustainability of ecosystems and human health [179,183]. Specifically, in aquatic ecosystems, a rapid decrease in fish diversity is one of the major problems in the management of fishery resources. This decrease is caused by overfishing, water pollution, habitat degradation, genetic pollution, climate change, and invasive species [184,185]. Therefore, it is necessary to take effective measures for fish protection. These measures are based on detailed information about fish distribution and population characteristics, as well as the physiological characteristics and ecological niche of different fish. However, due to the complexity of aquatic ecosystems and the diversity of fish migration routes [186], it is extremely difficult to accurately assess fish diversity. Traditional methods of investigating fish diversity, such as trawling, seining, electric fishing [180], underwater acoustics [187], and visual methods, may underestimate fish diversity, since some rare fish, e.g., endangered species and invasive taxa, are extremely difficult to detect [180].

In terms of the above-mentioned issues, environmental DNA (eDNA) provides an alternative means of detecting various types of fish. The use of eDNA, particularly metabarcoding, has gained attention as a powerful tool for assessing and monitoring aquatic biodiversity. By avoiding direct sampling, capture, and visual species observation, eDNA metabarcoding can reduce time and cost requirements, while supporting the conservation and management of ecosystems through improved species detection [188,189]. eDNA analysis using high-throughput sequencing has proven to be a very sensitive assay for the identification and assessment of single or multiple species in different samples [190]. The application of the eDNA method in freshwater ecosystems started in 2008, through the process of detecting the existence of American bullfrogs in a pond [191]. Since then, the eDNA method has increasingly been used for monitoring aquatic macro-organisms in both marine and freshwater ecosystems [192,193]. A big advantage of biodiversity assessment using eDNA metabarcoding is that it allows for effective and non-invasive species detection with little effort and minimum cost, even when conducting extensive surveys [194,195].

The eDNA method is defined as the sum of DNA fragments extracted directly from environmental components like soil, sediments, or water. The DNA fragments contain the intracellular genetic material released into water by biological cells, and the extracellular DNA released into water after cell structure lysis or death [196]. The eDNA method is based on amplicon and/or shotgun sequencing technologies. DNA fragments extracted from environmental components are identified using molecular biological means and the specific gene detection of target fish species. This can be used to analyze the distributional characteristics of target fish species. The eDNA method has progressed in terms of fish resource management, gathering information on endangered fish [197,198], investigating invasive species, and evaluating fish diversity [180,199] and fish prey diversity [200]. eDNA metabarcoding detected more fish species than those captured by traditional surveys like electrofishing [201] and underwater visual censuses [202]. In addition, eDNA metabarcoding has proven to be an efficient biomonitoring tool for surveying fish diversity, especially in regions that are often overlooked or difficult to access [203]. Studies have proven that eDNA metabarcoding can detect alien and endangered species that are not easily detected [188] (Figure 5). Traditional biomonitoring programs require vast efforts with repeated sampling over time and space. These methods are invasive because they use electrofishing and nets, which can destroy habitats. Moreover, they often fail to detect all species in an environment, and to identify species in collected samples, taxonomic experts are required. Due to these limitations, eDNA-based programs are an attractive approach for the conservation, management, and restoration of biodiversity [204,205].

The exploration of eDNA's role in monitoring freshwater biodiversity can be greatly enhanced by examining its implications for Pakistan's freshwater ecosystems, especially in light of the challenges presented by climate change. The country's freshwater environments are increasingly vulnerable to severe weather phenomena, including intense rainfall and flooding, which disrupt these ecosystems and promote the spread of invasive species. These non-native species frequently outcompete indigenous fish populations, resulting in reduced biodiversity and a shift in the ecological equilibrium of these habitats. In this context, eDNA serves as a groundbreaking and effective method for assessing and preserving fish biodiversity, providing numerous benefits compared to conventional survey techniques [206,207]. A significant advantage of eDNA is its high sensitivity in detecting the presence of species, even when they are rare or difficult to find [207,208]. This holds particular significance in Pakistan, where extensive freshwater ecosystems are prevalent and challenging to monitor thoroughly through traditional approaches. For example, rivers such as the Ravi and Indus face considerable human-induced pressures, further intensified by habitat alterations driven by climate change [209,210]. eDNA addresses these challenges by enabling species detection at various spatial scales, eliminating the necessity for the physical capture or direct observation of organisms [211,212].

In Pakistan, eDNA metabarcoding could be utilized to evaluate fish biodiversity in key river systems like the Indus, Ravi, and Jhelum. These rivers are not only essential for maintaining biodiversity but also support the livelihoods of local communities [213]. By integrating eDNA techniques, researchers can generate detailed inventories of fish species, including those that are endangered or invasive, thus providing essential data for conservation and management efforts [214–216]. Moreover, eDNA serves as a valuable tool for tracking the proliferation of invasive species that pose risks to native fish populations, providing an early detection mechanism to inform prompt management strategies [217,218]. The effects of climate change in Pakistan, characterized by heightened flooding and alterations in habitats, highlight the necessity for advanced monitoring technologies such as environmental DNA (eDNA). Flood events can facilitate the introduction of new species, including invasive ones, into freshwater ecosystems, significantly impacting species diversity and ecological interactions [218,219]. eDNA serves as a vital tool for monitoring these swift ecological shifts by delivering immediate insights into species distribution and population levels [220–223]. This functionality is essential for effective biodiversity management amid environmental upheavals and for maintaining the resilience of freshwater ecosystems [224].

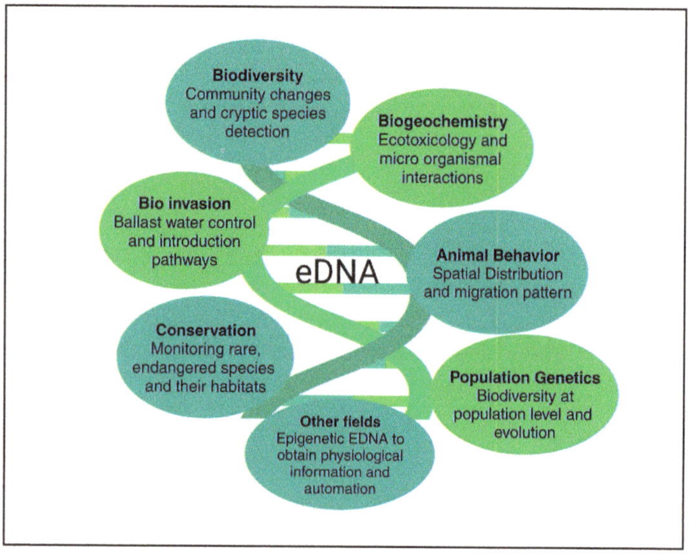

Figure 5. Environmental DNA and its services in biodiversity management and conservation.

eDNA has the potential to transform ecological risk assessments, particularly regarding pollution and habitat degradation [225]. In vital aquatic ecosystems like the wetlands of Head Islam, eDNA can be utilized to evaluate the impact of pollutants on fish populations. While eDNA does not directly indicate pollution, its analysis provides a highly sensitive method for tracking biodiversity changes, serving as a valuable indicator of environmental health. This allows for the indirect estimation of the scope of pollution and its impact on ecosystems [226]. Such an approach would deepen our understanding of pollution's influence on biodiversity and aid in formulating strategies to alleviate these effects. Overall, the application of eDNA in the monitoring of freshwater fish in Pakistan is not only feasible but also highly advantageous. By leveraging the strengths of eDNA, researchers can gain deeper insights into the status of fish populations, track changes over time, and inform conservation strategies. The adoption of eDNA in Pakistani freshwater ecosystems thus represents a forward-thinking approach that could significantly enhance efforts to conserve the country's aquatic biodiversity.

13. Conclusions

Like other developing countries, Pakistan is extremely susceptible to the effects of climate change. Pakistan's technological and financial limitations, along with its low GHS emission status, provide no protection from the negative effects of global climate change. As an agricultural country, consisting of arid and semi-arid regions and highly dependent on irrigated agriculture, Pakistan has been extremely threatened by climatic changes. Prolonged drought, higher levels of glacial melting, early and long summers, and short winters are some of the negative impacts of climate change, which also affect overall agricultural productivity and water sources in the country. The carbon sinks are declining rapidly because the forest cover is low in the country, at about 4.5%, with a high deforestation rate of about 0.2–0.4% per year. The long-lasting impacts of climate change are predicted to threaten our biodiversity, food security, water availability, human health, and overall well-being. Climate change in Pakistan seriously affects ecological fragility, freshwater ecosystems and fisheries, the agricultural sector, and human health. The increased warming will have consequences, leading to more rain, and could accelerate the hydrological cycles in most parts of Pakistan; this could lead to floods and drought in the region. In recent years, human activities and environmental changes have posed serious threats to freshwater habitats, resulting in an extensive decline in freshwater biodiversity. In the future, anthropogenic climate change and human pressure on freshwater ecosystems, including water diversion, abstraction, dam construction, and pollution, are predicted to have progressively important implications. Increases in the air temperature and alterations in precipitation patterns change water temperature and flow regimes globally, consequently affecting two important aspects of habitat for the survival of freshwater species. Fish are ectotherms, directly affected by water temperature, and the hydrological regime regulates the dynamics and structure of the freshwater habitat. Only a few studies have assessed the potential impacts of climate change on freshwater fishes, in contrast to many studies that examine the potential impacts of climate change on species in terrestrial systems. Climate change and other anthropogenic activities, such as pollution, dam construction, disturbances to the natural flow of rivers, and the introduction of invasive species, are responsible for the decline in local biodiversity. Many freshwater fishes are highly vulnerable, including *Tor putitora* and *Glyptothorax kashmirensis*, which are critically endangered in Pakistan. eDNA biomonitoring is a promising tool for the conservation and assessment of these species in the future. Environmental DNA (eDNA) refers to genetic material obtained directly from environmental samples, such as water, soil, or air, rather than from an individual organism. In the context of Pakistan's freshwater ecosystems, eDNA can be a valuable tool for biodiversity conservation, especially as the country faces climate change, pollution, erratic monsoons, and floods. By enabling the non-invasive and comprehensive monitoring of aquatic species, eDNA helps to detect shifts in biodiversity, including the presence of invasive species and declines in native species. These real-time data can inform conservation strategies, guide habitat restoration efforts, and support adaptive management practices to mitigate the impacts of these environmental stressors.

Author Contributions: G.A., S.A. and S.B. constructed the initial framework of the manuscript. G.A. wrote the manuscript. S.B., S.A., S.N. and N.M.A. edited and improved the manuscript. All authors have read and agreed to the published version of the manuscript.

Funding: There was no funding available, but a grant for the initiative of eDNA as a tool in a national agenda to implement eDNA as a tool for conservation was awarded to Professor Subha Bhassu by the Ministry of Natural Resources and Environment, Malaysia (GA01-2021). One part of the grant is education through a citizen science approach to increase awareness of invasive alien species as a threat to biodiversity and conservation.

Institutional Review Board Statement: Not applicable.

Informed Consent Statement: Not applicable.

Data Availability Statement: All data are available upon request, and we will be collaborating with the FishBase Consortium on depositing the data via the World Fish Centre.

Conflicts of Interest: The authors declare no conflicts of interest.

References

1. Ding, Y.J.; Li, C.Y.; Xiaoming, W.A.N.G.; Wang, Y.; Wang, S.X.; Chang, Y.P.; Qin, J.; Wang, S.P.; Zhao, Q.D.; Wang, Z.R. An overview of climate change impacts on the society in China. *Adv. Clim. Chang. Res.* **2021**, *12*, 210–223. [CrossRef]

2. Yalew, S.G.; van Vliet, M.T.; Gernaat, D.E.; Ludwig, F.; Miara, A.; Park, C.; Byers, E.; De Cian, E.; Piontek, F.; Iyer, G.; et al. Impacts of climate change on energy systems in global and regional scenarios. *Nat. Energy* **2020**, *5*, 794–802. [CrossRef]
3. Rahman, H.A. Climate change scenarios in Malaysia: Engaging the public. *Int. J. Malay-Nusant. Stud.* **2018**, *1*, 55–77.
4. Sintayehu, D.W. Impact of climate change on biodiversity and associated key ecosystem services in Africa: A systematic review. *Ecosyst. Health Sustain.* **2018**, *4*, 225–239. [CrossRef]
5. IPCC. *Climate Change 2013: The Physical Science Basis. Contribution of Working Group I to the Fifth Assessment Report of the Intergovernmental Panel on Climate Change*; Stocker, T.F., Qin, D., Plattner, G.-K., Tignor, M.M.B., Allen, S.K., Boschung, J., Nauels, A., Xia, Y., Bex, V., Midgley, P.M., Eds.; Cambridge University Press: Cambridge, UK; New York, NY, USA, 2013.
6. Ndehedehe, C.E.; Ferreira, V.G.; Agutu, N.O.; Onojeghuo, A.O.; Okwuashi, O.; Kassahun, H.T.; Dewan, A. What if the rains do not come? *J. Hydrol.* **2021**, *595*, 126040. [CrossRef]
7. Haile, G.G.; Tang, Q.; Leng, G.; Jia, G.; Wang, J.; Cai, D.; Sun, S.; Baniya, B.; Zhang, Q. Long-term spatiotemporal variation of drought patterns over the Greater Horn of Africa. *Sci. Total Environ.* **2020**, *704*, 135299. [CrossRef]
8. Usman, M.; Ndehedehe, C.E.; Farah, H.; Manzanas, R. Impacts of climate change on the streamflow of a large river basin in the Australian tropics using optimally selected climate model outputs. *J. Clean. Prod.* **2021**, *315*, 128091. [CrossRef]
9. Lenoir, J.; Bertrand, R.; Comte, L.; Bourgeaud, L.; Hattab, T.; Murienne, J.; Grenouillet, G. Species better track climate warming in the oceans than on land. *Nat. Ecol. Evol.* **2020**, *4*, 1044–1059. [CrossRef]
10. Pecl, G.T.; Araújo, M.B.; Bell, J.D.; Blanchard, J.; Bonebrake, T.C.; Chen, I.C.; Clark, T.D.; Colwell, R.K.; Danielsen, F.; Evengård, B.; et al. Biodiversity redistribution under climate change: Impacts on ecosystems and human well-being. *Science* **2017**, *355*, eaai9214. [CrossRef]
11. Cahill, A.E.; Aiello-Lammens, M.E.; Fisher-Reid, M.C.; Hua, X.; Karanewsky, C.J.; Yeong Ryu, H.; Sbeglia, G.C.; Spagnolo, F.; Waldron, J.B.; Warsi, O.; et al. How does climate change cause extinction? *Proc. R. Soc. B Biol. Sci.* **2013**, *280*, 20121890. [CrossRef]
12. IPCC. *Climate Change 2014: Synthesis Report. Contribution of Working Groups I, II and III to the Fifth Assessment Report of the Intergovernmental Panel on Climate Change*; IPCC: Geneva, Switzerland, 2014. [CrossRef]
13. Benton, M.J. Hyperthermal-driven mass extinctions: Killing models during the Permian–Triassic mass extinction. *Philos. Trans. R. Soc. A Math. Phys. Eng. Sci.* **2018**, *376*, 20170076. [CrossRef] [PubMed]
14. Dunhill, A.M.; Foster, W.J.; Azaele, S.; Sciberras, J.; Twitchett, R.J. Modelling determinants of extinction across two Mesozoic hyperthermal events. *Proc. R. Soc. B* **2018**, *285*, 20180404. [CrossRef] [PubMed]
15. Foster, G.L.; Hull, P.; Lunt, D.J.; Zachos, J.C. Placing our current 'hyperthermal' in the context of rapid climate change in our geological past. *Philos. Trans. R. Soc. A Math. Phys. Eng. Sci.* **2018**, *376*, 20170086. [CrossRef] [PubMed]
16. Bond, D.P.; Grasby, S.E. On the causes of mass extinctions. *Palaeogeogr. Palaeoclimatol. Palaeoecol.* **2017**, *478*, 3–29. [CrossRef]
17. Radchuk, V.; Reed, T.; Teplitsky, C.; van de Pol, M.; Charmantier, A.; Hassall, C.; Adamík, P.; Adriaensen, F.; Ahola, M.P.; Arcese, P.; et al. Adaptive responses of animals to climate change are most likely insufficient. *Nat. Commun.* **2019**, *10*, 3109. [CrossRef]
18. Roy, H.E.; Peyton, J.; Aldridge, D.C.; Bantock, T.; Blackburn, T.M.; Britton, R.; Clark, P.; Cook, E.; Dehnen-Schmutz, K.; Dines, T.; et al. Horizon scanning for invasive alien species with the potential to threaten biodiversity in Great Britain. *Glob. Chang. Biol.* **2014**, *20*, 3859–3871. [CrossRef]
19. Cook, J.; Nuccitelli, D.; Green, S.A.; Richardson, M.; Winkler, B.; Painting, R.; Way, R.; Jacobs, J.; Skuce, A. Quantifying the consensus on anthropogenic global warming in the scientific literature. *Environ. Res. Lett.* **2013**, *8*, 024024. [CrossRef]
20. Harrod, C. Climate change and freshwater fisheries. In *Freshwater Fisheries Ecology*; John Wiley & Sons: Hoboken, NJ, USA, 2015; pp. 641–694.
21. Collen, B.; Whitton, F.; Dyer, E.E.; Baillie, J.E.; Cumberlidge, N.; Darwall, W.R.; Pollock, C.; Richman, N.I.; Soulsby, A.M.; Böhm, M. Global patterns of freshwater species diversity, threat and endemism. *Glob. Ecol. Biogeogr.* **2014**, *23*, 40–51. [CrossRef]
22. Holmlund, C.M.; Hammer, M. Ecosystem services generated by fish populations. *Ecol. Econ.* **1999**, *29*, 253–268. [CrossRef]
23. Ali, S.; Eum, H.I.; Cho, J.; Dan, L.; Khan, F.; Dairaku, K.; Shrestha, M.L.; Hwang, S.; Nasim, W.; Khan, I.A.; et al. Assessment of climate extremes in future projections downscaled by multiple statistical downscaling methods over Pakistan. *Atmos. Res.* **2019**, *222*, 114–133. [CrossRef]
24. Aziz, N. Hydrological Modeling and Impact of Climate Change:(Historical Overview on Northern Areas of Pakistan). *Int. J. Res. Appl. Sci. Eng. Technol.* **2020**, *8*, 326–328. [CrossRef]
25. Eckstein, D.; Künzel, V.; Schäfer, L.; Winges, M. *Global Climate Risk Index 2020*; Germanwatch: Bonn, Germany, 2019.
26. Ritchie, H.; Roser, M.; Rosado, P. CO_2 and Greenhouse Gas Emissions. Our World in Data. 2020. Available online: https://ourworldindata.org/co2-and-greenhouse-gas-emissions (accessed on 5 January 2024).
27. Pakistan National Policy on Climate Change. Archived from the original on 5 March 2016. Retrieved 10 May 2015. Available online: https://www.gcisc.org.pk/National_Climate_Change_Policy_2012.pdf (accessed on 5 January 2024).
28. Zaheer, K.; Colom, A. How the People of Pakistan Live with Climate Change and What Communication Can Do. 2012. Available online: https://downloads.bbc.co.uk/mediaaction/pdf/climateasia/reports/ClimateAsia_PakistanReport.pdf (accessed on 17 December 2023).

29. Chaudhry, Q.U.Z. *Climate Change Profile of Pakistan*; Asian Development Bank: Metro Manila, Philippines, 2017.
30. Pakistan | UNDP Climate Change Adaptation. Available online: www.adaptation-undp.org (accessed on 25 November 2020).
31. Kouser, S.; Kausar, S.; Ghani, M. COVID-19 lockdown: Current situation and challenges facing in Pakistan during lockdown. *Biomedica* **2020**, *36*, 138–144. [CrossRef]
32. Rannard, G. How Pakistan floods are linked to climate change. *BBC*, 2 September 2022.
33. Sands, L. Pakistan floods: One third of country is under water—Minister. *BBC News*, 29 August 2022.
34. Beck, H.E.; Zimmermann, N.E.; McVicar, T.R.; Vergopolan, N.; Berg, A.; Wood, E.F. Present and future Köppen-Geiger climate classification maps at 1-km resolution. *Sci. Dat.* **2018**, *5*, 1–12. [CrossRef] [PubMed]
35. Hausfather, Z.; Peters, G.P. Emissions–the 'business as usual' story is misleading. *Nature* **2020**, *577*, 618–620. [CrossRef] [PubMed]
36. Nadeau, K.C.; Agache, I.; Jutel, M.; Annesi Maesano, I.; Akdis, M.; Sampath, V.; d'Amato, G.; Cecchi, L.; Traidl-Hoffmann, C.; Akdis, C.A. Climate change: A call to action for the United Nations. *Allergy* **2022**, *77*, 1087–1090. [CrossRef]
37. Audi, M.; Ali, A.; Kassem, M. Greenhouse gases: A review of losses and benefits. *Int. J. Energy Econ. Policy* **2020**, *10*, 403–418. [CrossRef]
38. Marvel, K.; Su, W.; Delgado, R.; Aarons, S.; Chatterjee, A.; Garcia, M.E.; Hausfather, Z.; Hayhoe, K.; Hence, D.A.; Jewett, E.B.; et al. Chapter 2: Climate trends. In *Fifth National Climate Assessment*; USGCRP (U.S. Global Change Research Program): Washington, DC, USA, 2023.
39. World Metrological Organization (WMO). Greenhouse Gas Concentrations Hit Record High Again. 2024. Available online: https://wmo.int/news/media-centre/greenhouse-gas-concentrations-hit-record-high-again (accessed on 28 July 2023).
40. IPCC (Intergovernmental Panel on Climate Change). *Climate Change 2022—Mitigation of Climate Change. Working Group III Contribution to the Sixth Assessment Report of the Intergovernmental Panel on Climate Change*; Shukla, P.R., Skea, J., Slade, R., Fradera, R., Pathak, M., Khourdajie, A.A., Belkacemi, M., van Diemen, R., Hasija, A., Lisboa, G., et al., Eds.; Cambridge University Press: Cambridge, UK, 2023. [CrossRef]
41. Bolster, C.H.; Mitchell, R.; Kitts, A.; Campbell, A.; Cosh, M.; Farrigan, T.L.; Franzluebbers, A.J.; Hoover, D.L.; Jin, V.L.; Peck, D.E.; et al. Chapter 11: Agriculture, food systems, and rural communities. In *Fifth National Climate Assessment*; USGCRP (U.S. Global Change Research Program): Washington, DC, USA, 2023. Available online: https://nca2023.globalchange.gov/chapter/11 (accessed on 16 January 2024).
42. United Nations. Climate Actions Note. 2021. Available online: https://www.unep.org/explore-topics/climate-action/what-we-do/climate-action-note/state-of-climate.html (accessed on 16 January 2024).
43. Tawiah, K.; Daniyal, M.; Qureshi, M. Pakistan CO_2 emission modelling and forecasting: A linear and nonlinear time series approach. *J. Environ. Public Health* **2023**, *2023*, 5903362. [CrossRef]
44. Opinion: Is Pakistan Really Phasing out Coal? The Third Pole. Retrieved 4 March 2021. Available online: https://dialogue.earth/en/energy/opinion-is-pakistan-really-phasing-out-coal/ (accessed on 25 November 2023).
45. Hermoso, V.; Clavero, M. Revisiting ecological integrity 30 years later: Non-native species and the misdiagnosis of freshwater ecosystem health. *Fish Fish.* **2013**, *14*, 416–423. [CrossRef]
46. Crozier, L.G.; Hutchings, J.A. Plastic and evolutionary responses to climate change in fish. *Evol. Appl.* **2014**, *7*, 68–87. [CrossRef]
47. Logez, M.; Pont, D. Global warming and potential shift in reference conditions: The case of functional fish-based metrics. *Hydrobiologia* **2013**, *704*, 417–436. [CrossRef]
48. Trenberth, K.E.; Jones, P.D.; Ambenje, P.; Bojariu, R.; Easterling, D.; Tank, A.K.; Parker, D.; Rahimzadeh, F.; Renwick, J.A.; Rusticucci, M.; et al. Observations: Surface and atmospheric climate change. Chapter 3. In *Climate Change 2007: The Physical Science Basis. Contribution of Working Group 1 to the 4th Assessment Report of the Intergovernmental Panel on Climate Change*; Cambridge University Press: Cambridge, UK, 2007.
49. Lough, J.M.; Hobday, A.J. Observed climate change in Australian marine and freshwater environments. *Mar. Freshw. Res.* **2011**, *62*, 984–999. [CrossRef]
50. Bates, B.C.; Kundzewicz, Z.W.; Wu, S.; Palutikof, J.P. (Eds.) *Climate Change and Water. Technical Paper of the Intergovernmental Panel on Climate Change*; IPCC Secretariat: Geneva, Switzerland, 2008; 210p.
51. Rosenzweig, C.; Casassa, G.; Karoly, D.J.; Imeson, A.; Liu, C.; Menzel, A.; Rawlins, S.; Root, T.L.; Seguin, B.; Tryjanowski, P.; et al. Assessment of observed changes and responses in natural and managed systems. In *Climate Change 2007: Impacts, Adaptation and Vulnerability. Contribution of Working Group II to the Fourth Assessment Report of the Intergovernmental Panel on Climate Change*; Cambridge University Press: Cambridge, UK, 2007.
52. Meehl, G.A.; Stocker, T.F.; Collins, W.D.; Friedlingstein, P.; Gaye, A.T.; Gregory, J.M.; Kitoh, A.; Knutti, R.; Murphy, J.M.; Noda, N.; et al. Global climate projections. Chapter 10. In *Climate Change 2007. Contribution of Working Group I to the Fourth Assessment Report of the Intergovernmental Panel on Climate Change*; Cambridge University Press: Cambridge, UK, 2007.
53. Solomon, S.; Qin, D.; Manning, M.; Chen, Z.; Marquis, M.; Averyt, K.B.; Tignor, M.; Miller, H.L. (Eds.) *Climate Change 2007: The Physical Science Basis. Contribution of Working Group I to the Fourth Assessment Report of the Intergovernmental Panel on Climate Change*; Cambridge University Press: Cambridge, UK, 2007.

54. Mora, C.; Frazier, A.G.; Longman, R.J.; Dacks, R.S.; Walton, M.M.; Tong, E.J.; Sanchez, J.J.; Kaiser, L.R.; Stender, Y.O.; Anderson, J.M.; et al. The projected timing of climate departure from recent variability. *Nature* **2013**, *502*, 183–187. [CrossRef] [PubMed]
55. Welcomme, R.L.; Cowx, I.G.; Coates, D.; Béné, C.; Funge-Smith, S.; Halls, A.; Lorenzen, K. Inland capture fisheries. *Philos. Trans. R. Soc. B Biol. Sci.* **2010**, *365*, 2881–2896. [CrossRef] [PubMed]
56. Jeppesen, E.; Meerhoff, M.; Holmgren, K.; Gonza'lez-Bergonzoni, I.; Teixeira-de Mello, F.; Declerck, S.J.A.; De Meester, L.; Sondergaard, M.; Lauridsen, T.L.; Bjerring, R.; et al. Impacts of climate warming on lake fish community structure and potential effects on ecosystem function. *Hydrobiologia* **2010**, *646*, 73–90. [CrossRef]
57. IPCC. *Climate Change 2007: Synthesis Report. Contribution of Working Groups I, II and III to the Fourth Assessment Report of the Intergovernmental Panel on Climate Change*; IPCC: Geneva, Switzerland, 2007.
58. Lynch, A.J.; Myers, B.J.; Chu, C.; Eby, L.A.; Falke, J.A.; Kovach, R.P.; Krabbenhoft, T.J.; Kwak, T.J.; Lyons, J.; Paukert, C.P.; et al. Climate change effects on North American inland fish populations and assemblages. *Fisheries* **2016**, *41*, 346–361. [CrossRef]
59. Fawzy, S.; Osman, A.I.; Doran, J.; Rooney, D.W. Strategies for mitigation of climate change: A review. *Environ. Chem. Lett.* **2020**, *18*, 2069–2094. [CrossRef]
60. Trumbo, B.A.; Nislow, K.H.; Stallings, J.; Hudy, M.; Smith, E.P.; Kim, D.Y.; Wiggins, B.; Dolloff, C.A. Ranking site vulnerability to increasing temperatures in southern Appalachian brook trout streams in Virginia: An exposure-sensitivity approach. *Trans. Am. Fish. Soc.* **2014**, *143*, 173–187. [CrossRef]
61. Warren, D.R.; Robinson, J.M.; Josephson, D.C.; Sheldon, D.R.; Kraft, C.E. Elevated summer temperatures delay spawning and reduce redd construction for resident brook trout (S alvelinus fontinalis). *Glob. Chang. Biol.* **2012**, *18*, 1804–1811. [CrossRef]
62. Ward, E.J.; Anderson, J.H.; Beechie, T.J.; Pess, G.R.; Ford, M.J. Increasing hydrologic variability threatens depleted anadromous fish populations. *Glob. Chang. Biol.* **2015**, *21*, 2500–2509. [CrossRef]
63. Poff, N.L.; Brinson, M.M.; Day, J.W. *Aquatic Ecosystems and Global Climate Change*; Pew Center on Global Climate Change: Arlington, VA, USA, 2002; Volume 44, pp. 1–36.
64. Brooks, R.T. Potential impacts of global climate change on the hydrology and ecology of ephemeral freshwater systems of the forests of the northeastern United States. *Clim. Chang.* **2009**, *95*, 469–483. [CrossRef]
65. Lawrence, D.J.; Stewart-Koster, B.; Olden, J.D.; Ruesch, A.S.; Torgersen, C.E.; Lawler, J.J.; Butcher, D.P.; Crown, J.K. The interactive effects of climate change, riparian management, and a nonnative predator on stream-rearing salmon. *Ecol. Appl.* **2014**, *24*, 895–912. [CrossRef] [PubMed]
66. Muhlfeld, C.C.; Kovach, R.P.; Jones, L.A.; Al-Chokhachy, R.; Boyer, M.C.; Leary, R.F.; Lowe, W.H.; Luikart, G.; Allendorf, F.W. Invasive hybridization in a threatened species is accelerated by climate change. *Nat. Clim. Chang.* **2014**, *4*, 620–624. [CrossRef]
67. Karvonen, A.; Rintamäki, P.; Jokela, J.; Valtonen, E.T. Increasing water temperature and disease risks in aquatic systems: Climate change increases the risk of some, but not all, diseases. *Int. J. Parasitol.* **2010**, *40*, 1483–1488. [CrossRef] [PubMed]
68. Abdel-Fattah, S. (Ed.) *Impacts of Climate Change on Fish Species and Aquatic Ecosystems in the Great Lakes and Prairie Regions of Canada: A Compilation of Reports*; Canaian Manuscript Report of Fisheries and Aquatic Sciences 3108; Fisheries and Oceans Canada: Québec City, QC, Canada, 2016.
69. Shen, M.; Xiao, N.; Zhao, Z.; Guo, N.; Luo, Z.; Sun, G.; Li, J. eDNA metabarcoding as a promising conservation tool to monitor fish diversity in Beijing water systems compared with ground cages. *Sci. Rep.* **2022**, *12*, 11113. [CrossRef] [PubMed]
70. Abbas, S.; Javed, M. Relationships among increase in biomass, planktonic productivity and physico-chemistry of ponds stocked with Zn+ Pb+ Mn mixture stressed fish. *JAPS J. Anim. Plant Sci.* **2019**, *29*, 125–131.
71. Strayer, D.L.; Dudgeon, D. Freshwater biodiversity conservation: Recent progress and future challenges. *J. N. Am. Benthol. Soc.* **2010**, *29*, 344–358. [CrossRef]
72. Almond, R.E.; Grooten, M.; Peterson, T. *Living Planet Report 2020-Bending the Curve of Biodiversity Loss*; World Wildlife Fund: Glans, Switzerland, 2020.
73. Williams, J.E.; Neville, H.M.; Haak, A.L.; Colyer, W.T.; Wenger, S.J.; Bradshaw, S. Climate change adaptation and restoration of western trout streams: Opportunities and strategies. *Fisheries* **2015**, *40*, 304–317. [CrossRef]
74. Kundzewicz, Z.W.; Mata, L.J.; Arnell, N.W.; Doll, P.; Kabat, P.; Jimenez, B.; Miller, K.A.; Oki, T.; Sen, Z.; Shiklomanov, I. Freshwater resources and their management. In *Climate Change 2007: Impacts, Adaptation and Vulnerability. Contribution of Working Group II to the Fourth Assessment Report of the Intergovernmental Panel on Climate Change*; Cambridge University Press: Cambridge, UK, 2007.
75. Caissie, D. The thermal regime of rivers: A review. *Freshw. Biol.* **2006**, *51*, 1389–1406. [CrossRef]
76. Carpenter, S.R.; Stanley, E.H.; Vander Zanden, M.J. State of the World's freshwater ecosystems: Physical, chemical, and biological changes. *Annu. Rev. Environ. Resour.* **2011**, *36*, 75–99. [CrossRef]
77. Huang, M.; Ding, L.; Wang, J.; Ding, C.; Tao, J. The impacts of climate change on fish growth: A summary of conducted studies and current knowledge. *Ecol. Indic.* **2021**, *121*, 106976. [CrossRef]
78. Ahmad, M.; Hashmi, R.S. Global Climatic Transformation: Implications For Pakistan. *Pak. J. Soc. Res.* **2023**, *5*, 1113–1123. [CrossRef]

79. Fahad, S.; Wang, J. Climate change, vulnerability, and its impacts in rural Pakistan: A review. *Environ. Sci. Pollut. Res.* **2020**, *27*, 1334–1338. [CrossRef] [PubMed]
80. Otto, F.E.; Zachariah, M.; Saeed, F.; Siddiqi, A.; Kamil, S.; Mushtaq, H.; Arulalan, T.; AchutaRao, K.; Chaithra, S.T.; Barnes, C.; et al. Climate change increased extreme monsoon rainfall, flooding highly vulnerable communities in Pakistan. *Environ. Res. Clim.* **2023**, *2*, 025001. [CrossRef]
81. Talbot, C.J.; Bennett, E.M.; Cassell, K.; Hanes, D.M.; Minor, E.C.; Paerl, H.; Raymond, P.A.; Vargas, R.; Vidon, P.G.; Wollheim, W.; et al. The impact of flooding on aquatic ecosystem services. *Biogeochemistry* **2018**, *141*, 439–461. [CrossRef] [PubMed]
82. Death, R.G.; Fuller, I.C.; Macklin, M.G. Resetting the river template: The potential for climate-related extreme floods to transform river geomorphology and ecology. *Freshw. Biol.* **2015**, *60*, 2477–2496. [CrossRef]
83. Aldous, A.; Fitzsimons, J.; Richter, B.; Bach, L. Droughts, floods and freshwater ecosystems: Evaluating climate change impacts and developing adaptation strategies. *Mar. Freshw. Res.* **2011**, *62*, 223–231. [CrossRef]
84. Siddha, S.; Sahu, P. Impact of climate change on the river ecosystem. In *Ecological Significance of River Ecosystems*; Elsevier: Amsterdam, The Netherlands, 2022; pp. 79–104.
85. Fausch, K.D.; Taniguchi, Y.; Nakano, S.; Grossman, G.D.; Townsend, C.R. Flood disturbance regimes influence rainbow trout invasion success among five holarctic regions. *Ecol. Appl.* **2001**, *11*, 1438–1455. [CrossRef]
86. Bănăduc, D.; Curtean-Bănăduc, A.; Barinova, S.; Lozano, V.L.; Afanasyev, S.; Leite, T.; Branco, P.; Gomez Isaza, D.F.; Geist, J.; Tegos, A.; et al. Multi-Interacting Natural and Anthropogenic Stressors on Freshwater Ecosystems: Their Current Status and Future Prospects for 21st Century. *Water* **2024**, *16*, 1483. [CrossRef]
87. Carosi, A.; Lorenzoni, F.; Lorenzoni, M. Synergistic Effects of Climate Change and Alien Fish Invasions in Freshwater Ecosystems: A Review. *Fishes* **2023**, *8*, 486. [CrossRef]
88. Hulme, P.E. Climate change and biological invasions: Evidence, expectations, and response options. *Biol. Rev.* **2017**, *92*, 1297–1313. [CrossRef]
89. Lutz, A.F.; Immerzeel, W.W.; Kraaijenbrink, P.D.; Shrestha, A.B.; Bierkens, M.F. Climate change impacts on the upper Indus hydrology: Sources, shifts and extremes. *PLoS ONE* **2016**, *11*, e0165630. [CrossRef] [PubMed]
90. Radić, V.; Hock, R. Glaciers in the Earth's hydrological cycle: Assessments of glacier mass and runoff changes on global and regional scales. *Surv. Geophys.* **2014**, *35*, 813–837. [CrossRef]
91. Nie, Y.; Pritchard, H.D.; Liu, Q.; Hennig, T.; Wang, W.; Wang, X.; Liu, S.; Nepal, S.; Samyn, D.; Hewitt, K.; et al. Glacial change and hydrological implications in the Himalaya and Karakoram. *Nat. Rev. Earth Environ.* **2021**, *2*, 91–106. [CrossRef]
92. Maharana, P.; Agnihotri, R.; Dimri, A.P. Changing Indian monsoon rainfall patterns under the recent warming period 2001–2018. *Clim. Dyn.* **2021**, *57*, 2581–2593. [CrossRef]
93. Seneviratne, S.I.; Zhang, X.; Adnan, M.; Badi, W.; Dereczynski, C.; Luca, A.D.; Ghosh, S.; Iskandar, I.; Kossin, J.; Lewis, S.; et al. Weather and climate extreme events in a changing climate. In *Climate Change 2021: The Physical Science Basis. Working Group I contribution to the Sixth Assessment Report of the Intergovernmental Panel on Climate Change*; Cambridge University Press: Cambridge, UK, 2021.
94. Garcia, C.; Schumann, D.A.; Howell, J.; Graeb, B.D.; Bertrand, K.N.; Klumb, R.A. Seasonality, floods and droughts structure larval fish assemblages in prairie rivers. *Ecol. Freshw. Fish* **2018**, *27*, 389–397. [CrossRef]
95. Jean de la Paix, M.; Lanhai, L.; Xi, C.; Varenyam, A.; Anming, B. Study of impacts of floods on the water quality in an arid zone: The case of the Tarim River in Northwest China. *Water Sci. Technol.* **2011**, *64*, 1973–1979. [CrossRef]
96. Qin, G.; Liu, J.; Xu, S.; Sun, Y. Pollution source apportionment and water quality risk evaluation of a drinking water reservoir during flood seasons. *Int. J. Environ. Res. Public Health* **2021**, *18*, 1873. [CrossRef]
97. Aldardasawi, A.M.; Eren, B. Floods and their impact on the environment. *Acad. Perspect. Procedia* **2021**, *4*, 42–49. [CrossRef]
98. Mosley, L.M. Drought impacts on the water quality of freshwater systems; review and integration. *Earth-Sci. Rev.* **2015**, *140*, 203–214. [CrossRef]
99. Obedzinski, M.; Nossaman Pierce, S.; Horton, G.E.; Deitch, M.J. Effects of flow-related variables on oversummer survival of juvenile Coho salmon in intermittent streams. *Trans. Am. Fish. Soc.* **2018**, *147*, 588–605. [CrossRef]
100. Damseth, S.; Thakur, K.; Kumar, R.; Kumar, S.; Mahajan, D.; Kumari, H.; Sharma, D.; Sharma, A.K. Assessing the impacts of river bed mining on aquatic ecosystems: A critical review of effects on water quality and biodiversity. *HydroResearch* **2024**, *7*, 122–130. [CrossRef]
101. Fahrig, L. Effects of habitat fragmentation on biodiversity. *Annu. Rev. Ecol. Evol. Syst.* **2003**, *34*, 487–515. [CrossRef]
102. Khani, S.; Rajaee, T. Modeling of dissolved oxygen concentration and its hysteresis behavior in rivers using wavelet transform-based hybrid models. *CLEAN–Soil Air Water* **2017**, *45*. [CrossRef]
103. Rajesh, M.; Rehana, S. Impact of climate change on river water temperature and dissolved oxygen: Indian riverine thermal regimes. *Sci. Rep.* **2022**, *12*, 9222. [CrossRef] [PubMed]

104. Sun, J.L.; Zhao, L.L.; Liao, L.; Tang, X.H.; Cui, C.; Liu, Q.; He, K.; Ma, J.D.; Jin, L.; Yan, T.; et al. Interactive effect of thermal and hypoxia on largemouth bass (*Micropterus salmoides*) gill and liver: Aggravation of oxidative stress, inhibition of immunity and promotion of cell apoptosis. *Fish Shellfish. Immunol.* **2020**, *98*, 923–936. [CrossRef]
105. Alfonso, S.; Gesto, M.; Sadoul, B. Temperature increase and its effects on fish stress physiology in the context of global warming. *J. Fish Biol.* **2021**, *98*, 1496–1508. [CrossRef] [PubMed]
106. Ilarri, M.; Souza, A.T.; Dias, E.; Antunes, C. Influence of climate change and extreme weather events on an estuarine fish community. *Sci. Total Environ.* **2022**, *827*, 154190. [CrossRef] [PubMed]
107. Chen, Q.; Li, Q.; Lin, Y.; Zhang, J.; Xia, J.; Ni, J.; Cooke, S.J.; Best, J.; He, S.; Feng, T.; et al. River damming impacts on fish habitat and associated conservation measures. *Rev. Geophys.* **2023**, *61*, e2023RG000819. [CrossRef]
108. Adnan, M.; Xiao, B.; Bibi, S.; Xiao, P.; Zhao, P.; Wang, H. Addressing current climate issues in Pakistan: An opportunity for a sustainable future. *Environ. Chall.* **2024**, *15*, 100887. [CrossRef]
109. National Climate Change Policy. 2021. Available online: https://mocc.gov.pk/SiteImage/Policy/NCCP%20Report.pdf (accessed on 20 December 2023).
110. Hlohowskyj, I.; Brody, M.S.; Lackey, R.T. Methods for assessing the vulnerability of African fisheries resources to climate change. *Clim. Res.* **1996**, *6*, 97–106. [CrossRef]
111. McDowall, R.M. Global climate change and fish and fisheries: What might happen in a temperate oceanic archipelago like New Zealand. *GeoJournal* **1992**, *28*, 29–37. [CrossRef]
112. Meisner, J.D.; Shuter, B.J. Assessing potential effects of global climate change on tropical freshwater fishes. *GeoJournal 28* **1992**, 21–27. [CrossRef]
113. Shuter, B.J.; Meisner, J.D. Tools for assessing the impact of climate change on freshwater fish populations. *GeoJournal 28* **1992**, 7–20. [CrossRef]
114. Maitland, P.S. Climatic change and fish in northern Europe: Some possible scenarios. *Proc. Inst. Fish. Manag. Annu. Study Course* **1991**, *22*, 97–110.
115. Tonn, W.M. Climate change and fish communities: A conceptual framework. *Trans. Am. Fish. Soc.* **1990**, *119*, 337–352. [CrossRef]
116. Magnuson, J.J.; Meisner, J.D.; Hill, D.K. Potential changes in the thermal habitat of Great Lakes fish after global climate warming. *Trans. Am. Fish. Soc.* **1990**, *119*, 254–264. [CrossRef]
117. Meisner, J.D. Effect of climatic warming on the southern margins of the native range of brook trout, Salvelinus fontinalis. *Can. J. Fish. Aquat. Sci.* **1990**, *47*, 1065–1070. [CrossRef]
118. Deangelis, D.L.; Cushman, R.M. Potential application of models in forecasting the effects of climate changes on fisheries. *Trans. Am. Fish. Soc.* **1990**, *119*, 224–239. [CrossRef]
119. Shuter, B.J.; Post, J.R. Climate, population viability, and the zoogeography of temperate fishes. *Trans. Am. Fish. Soc.* **1990**, *119*, 314–336. [CrossRef]
120. Hassan, B.; Qadri, H.; Ali, M.N.; Khan, N.A.; Yatoo, A.M. Impact of climate change on freshwater ecosystem and its sustainable management. In *Fresh Water Pollution Dynamics and Remediation*; Springer: Singapore, 2020; pp. 105–121.
121. Mckenzie, D.; Geffroy, B.; Farrell, A. Effects of global warming on fishes and fisheries. *J. Fish Biol.* **2021**, *98*, 1489–1492. [CrossRef]
122. Rogers, L.A.; Dougherty, A.B. Effects of climate and demography on reproductive phenology of a harvested marine fish population. *Glob. Chang. Biol.* **2019**, *25*, 708–720. [CrossRef] [PubMed]
123. Myers, B.J.; Lynch, A.J.; Bunnell, D.B.; Chu, C.; Falke, J.A.; Kovach, R.P.; Krabbenhoft, T.J.; Kwak, T.J.; Paukert, C.P. Global synthesis of the documented and projected effects of climate change on inland fishes. *Rev. Fish Biol. Fish.* **2017**, *27*, 339–361. [CrossRef]
124. Audzijonyte, A.; Richards, S.A.; Stuart-Smith, R.D.; Pecl, G.; Edgar, G.J.; Barrett, N.S.; Payne, N.; Blanchard, J.L. Fish body sizes change with temperature but not all species shrink with warming. *Nat. Ecol. Evol.* **2020**, *4*, 809–814. [CrossRef]
125. Goikoetxea, A.; Sadoul, B.; Blondeau-Bidet, E.; Aerts, J.; Blanc, M.O.; Parrinello, H.; Barrachina, C.; Pratlong, M.; Geffroy, B. Genetic pathways underpinning hormonal stress responses in fish exposed to short-and long-term warm ocean temperatures. *Ecol. Indic.* **2021**, *120*, 106937. [CrossRef]
126. Stillman, J.H. Heat waves, the new normal: Summertime temperature extremes will impact animals, ecosystems, and human communities. *Physiology* **2019**, *34*, 86–100. [CrossRef]
127. Cochrane, K.; De Young, C.; Soto, D.; Bahri, T. (Eds.) *Climate Change Implications for Fisheries and Aquaculture: Overview of Current Scientific Knowledge*; FAO Fisheries and Aquaculture Technical Paper: Rome, Italy, 2009; p. 212.
128. Parry, M.L.; Canziani, O.; Palutikof, J.; Van der Linden, P.; Hanson, C. (Eds.) *Climate Change 2007-Impacts, Adaptation and Vulnerability: Working Group II Contribution to the Fourth Assessment Report of the IPCC*; Cambridge University Press: Cambridge, UK, 2007; Volume 4.
129. Nyingi, D.W.; Gichuki, N.; Ogada, M.O. Freshwater ecology of Kenyan highlands and lowlands. In *Developments in Earth Surface Processes*; Elsevier: Amsterdam, The Netherlands, 2013; Volume 16, pp. 199–218.

130. Cooke, S.J.; Bergman, J.N.; Nyboer, E.A.; Reid, A.J.; Gallagher, A.J.; Hammerschlag, N.; Van de Riet, K.; Vermaire, J.C. Overcoming the concrete conquest of aquatic ecosystems. *Biol. Conserv.* **2020**, *247*, 108589. [CrossRef]
131. Liu, J.; Kattel, G.; Arp, H.P.H.; Yang, H. Towards threshold-based management of freshwater ecosystems in the context of climate change. *Ecol. Model.* **2015**, *318*, 265–274. [CrossRef]
132. Azevedo-Santos, V.M.; Frederico, R.G.; Fagundes, C.K.; Pompeu, P.S.; Pelicice, F.M.; Padial, A.A.; Nogueira, M.G.; Fearnside, P.M.; Lima, L.B.; Daga, V.S.; et al. Protected areas: A focus on Brazilian freshwater biodiversity. *Divers. Distrib.* **2019**, *25*, 442–448. [CrossRef]
133. Leidy, R.A.; Moyle, P.B. Keeping up with the status of freshwater fishes: A California (USA) perspective. *Conserv. Sci. Pract.* **2021**, *3*, e474. [CrossRef]
134. Usman, K.; Pervaiz, K.; Khan, H.; Rehman, H.U.; Aslam, S. Current status of ichthyofauna at Khaki site, River Siren Mansehra Khyber Pakhtunkhwa, Pakistan. *J. Entomol. Zool. Stud.* **2017**, *5*, 361–363.
135. Junk, W.J.; Soares, M.G.M.; Bayley, P.B. Freshwater fishes of the Amazon River basin: Their biodiversity, fisheries, and habitats. *Aquat. Ecosyst. Health Manag.* **2007**, *10*, 153–173. [CrossRef]
136. FAO Fisheries and Aquaculture Department. *The State of World Fisheries and Aquaculture 2012*; Food and Agriculture Organization of the United Nations: Rome, Italy, 2012.
137. Nelson, J.S. *Fishes of the World*; John Wiley & Sons, Inc.: Hoboken, NJ, USA, 2006.
138. Rafique, M.; Khan, N.U.H. Distribution and status of significant freshwater fishes of Pakistan. *Rec. Zool. Surv. Pak.* **2012**, *21*, 90–95.
139. IUCN. IUCN Freshwater Fish Specialist Group. 2015. Available online: http://www.iucnffsg.org/freshwaterfishes/freshwater-fish-diversity/ (accessed on 4 July 2017).
140. Ahmad, M.; Shah, A.H.; Maqbool, Z.; Khalid, A.; Khan, K.R.; Farooq, M. Ichthyofaunal diversity and conservation status in rivers of Khyber Pakhtunkhwa, Pakistan. *Proc. Int. Acad. Ecol. Environ. Sci.* **2020**, *10*, 131–143.
141. Mirza, M.R.; Mirza, Z.S. Longitudinal zonation in the fish fauna of the Indus River in Pakistan. *Biologia* **2014**, *60*, 149–152.
142. Allison, E.H.; Ellis, F.; Mvula, P.M.; Mathieu, L.F. Fisheries management and uncertainty: The causes and consequences of variability in inland fisheries in Africa, with special reference to Malawi. In Proceedings of the Lake Malawi Fisheries Management Symposium, Lilongwe, Malawi, 4–9 June 2001; Weyl, O.L.F., Weyl, M.V., Eds.; National Aquatic Resource Management Programme: Lilongwe, Malawi, 2001; pp. 66–79.
143. Daw, T.; Adger, W.N.; Brown, K.; Badjeck, M.-C. Climate change and capture fisheries: Potential impacts, adaptation and mitigation. Chapter 2. In *Climate Change Implications for Fisheries and Aquaculture: Overview of Current Scientific Knowledge*; Cochrane, K., De Young, C., Soto, D., Bahri, T., Eds.; FAO: Rome, Italy, 2009; pp. 107–150.
144. Welcomme, R.L. An overview of global catch statistics for inland fish. *ICES J. Mar. Sci.* **2011**, *68*, 1751–1756. [CrossRef]
145. Neiland, A.E. *Artisanal Fisheries of the Chad Basin in Africa, An Overview of Research and Annotated Bibliography 1920–1990*; FAO: Rome, Italy, 1992; pp. 1–89.
146. Food and Agriculture Organization of the United Nations. 2024. Available online: https://www.fao.org/asiapacific/news/news-detail/FAO-Report-Global-fisheries-and-aquaculture-production-reaches-a-new-record-high-/en (accessed on 12 September 2024).
147. Rountrey, A.N.; Coulson, P.G.; Meeuwig, J.J.; Meekan, M. Water temperature and fish growth: Otoliths predict growth patterns of a marine fish in a changing climate. *Glob. Chang. Biol.* **2014**, *20*, 2450–2458. [CrossRef]
148. Gamperl, A.K.; Ajiboye, O.O.; Zanuzzo, F.S.; Sandrelli, R.M.; Ellen de Fátima, C.P.; Beemelmanns, A. The impacts of increasing temperature and moderate hypoxia on the production characteristics, cardiac morphology and haematology of Atlantic Salmon (*Salmo salar*). *Aquaculture* **2020**, *519*, 734874. [CrossRef]
149. Heather, F.J.; Childs, D.Z.; Darnaude, A.M.; Blanchard, J.L. Using an integral projection model to assess the effect of temperature on the growth of gilthead seabream Sparus aurata. *PLoS ONE* **2018**, *13*, e0196092. [CrossRef]
150. Tao, J.; Che, R.; He, D.; Yan, Y.; Sui, X.; Chen, Y. Trends and potential cautions in food web research from a bibliometric analysis. *Scientometrics* **2015**, *105*, 435–447. [CrossRef]
151. Thackeray, S.J.; Henrys, P.A.; Feuchtmayr, H.; Jones, I.D.; Maberly, S.C.; Winfield, I.J. Food web de-synchronization in E ngland's largest lake: An assessment based on multiple phenological metrics. *Glob. Chang. Biol.* **2013**, *19*, 3568–3580. [CrossRef]
152. Queiros, Q.; McKenzie, D.J.; Dutto, G.; Killen, S.; Saraux, C.; Schull, Q. Fish shrinking, energy balance and climate change. *Sci. Total Environ.* **2024**, *906*, 167310. [CrossRef] [PubMed]
153. Carozza, D.A.; Bianchi, D.; Galbraith, E.D. Metabolic impacts of climate change on marine ecosystems: Implications for fish communities and fisheries. *Glob. Ecol. Biogeogr.* **2019**, *28*, 158–169. [CrossRef]
154. Murdoch, A.; Power, M. The effect of lake morphometry on thermal habitat use and growth in Arctic charr populations: Implications for understanding climate-change impacts. *Ecol. Freshw. Fish* **2013**, *22*, 453–466. [CrossRef]

155. Pilière, A.; Schipper, A.M.; Breure, A.M.; Posthuma, L.; de Zwart, D.; Dyer, S.D.; Huijbregts, M.A. Comparing responses of freshwater fish and invertebrate community integrity along multiple environmental gradients. *Ecol. Indic.* **2014**, *43*, 215–226. [CrossRef]
156. Tao, J.; Kennard, M.J.; Jia, Y.; Chen, Y. Climate-driven synchrony in growth-increment chronologies of fish from the world's largest high-elevation river. *Sci. Total Environ.* **2018**, *645*, 339–346. [CrossRef]
157. Ding, C.Z.; Jiang, X.M.; Chen, L.Q.; Juan, T.; Chen, Z.M. Growth variation of Schizothorax dulongensis Huang, 1985 along altitudinal gradients: Implications for the Tibetan Plateau fishes under climate change. *J. Appl. Ichthyol.* **2016**, *32*, 729–733. [CrossRef]
158. Conti, L.; Comte, L.; Hugueny, B.; Grenouillet, G. Drivers of freshwater fish colonisations and extirpations under climate change. *Ecography* **2015**, *38*, 510–519. [CrossRef]
159. Manjarrés-Hernández, A.; Guisande, C.; García-Roselló, E.; Heine, J.; Pelayo-Villamil, P.; Pérez-Costas, E.; González-Vilas, L.; González-Dacosta, J.; Duque, S.R.; Granado-Lorencio, C.; et al. Predicting the effects of climate change on future freshwater fish diversity at global scale. *Nat. Conserv.* **2021**, *43*, 1–24. [CrossRef]
160. Seastedt, T.R.; Oldfather, M.F. Climate change, ecosystem processes and biological diversity responses in high elevation communities. *Climate* **2021**, *9*, 87. [CrossRef]
161. Khan, A.M.; Ali, Z.; Shelly, S.Y.; Ahmad, Z.; Mirza, M.R. Aliens; a catastrophe for native fresh water fish diversity in Pakistan. *J. Anim. Plant Sci.* **2011**, *21*, 435–440.
162. Cambray, J.A. Impact on indigenous species biodiversity caused by the globalisation of alien recreational freshwater fisheries. *Hydrobiologia* **2003**, *500*, 217–230. [CrossRef]
163. Chen, S.; Billings, S.A. Representations of non-linear systems: The NARMAX model. *Int. J. Control* **1989**, *49*, 1013–1032. [CrossRef]
164. Gallardo, B.; Clavero, M.; Sánchez, M.I.; Vilà, M. Global ecological impacts of invasive species in aquatic ecosystems. *Glob. Chang. Biol.* **2016**, *22*, 151–163. [CrossRef] [PubMed]
165. Val, A.L. Biodiversity–the hidden risks. *An. Acad. Bras. Ciênc.* **2020**, *92*, 20200699. [CrossRef] [PubMed]
166. IUCN. *IUCN Red List of Threatened Species*; Version 2011.2; IUCN: Gland, Switzerland, 2011.
167. Morrison, W.E.; Nelson, M.W.; Howard, J.F.; Teeters, E.J.; Hare, J.A.; Griffis, R.B.; Alexander, M.A. *Methodology for Assessing the Vulnerability of Marine Fish and Shellfish Species to a Changing Climate*; NOAA: Washington, DC, USA, 2015.
168. Olusanya, H.O.; van Zyll de Jong, M. Assessing the vulnerability of freshwater fishes to climate change in Newfoundland and Labrador. *PLoS ONE* **2018**, *13*, e0208182. [CrossRef]
169. Hannah, L. Protected areas and climate change. *Ann. N. Y. Acad. Sci.* **2008**, *1134*, 201–212. [CrossRef]
170. Chessman, B.C. Identifying species at risk from climate change: Traits predict the drought vulnerability of freshwater fishes. *Biol. Conserv.* **2013**, *160*, 40–49. [CrossRef]
171. Johnson, S.L.; Penaluna, B.E. Climate Change and Interactions With Multiple Stressors in Rivers. In *Multiple Stressors in River Ecosystems*; Elsevier: Amsterdam, The Netherlands, 2019; pp. 23–44.
172. Shakir, A.S.; Ehsan, S. Climate change impact on river flows in Chitral watershed. *Pak. J. Eng. Appl. Sci.* **2010**, *7*, 12–23.
173. Salma, S.; Rehman, S.; Shah, M.A. Rainfall trends in different climate zones of Pakistan. *Pak. J. Meteorol.* **2012**, *9*, 37–47.
174. Altaf, M.; Javid, A.; Umair, M. *Biodiversity of Ramsar sites in Pakistan*; LAP: Saarbrucken, Germany, 2014; pp. 1–88.
175. Camara, M.; Jamil, N.R.; Abdullah, A.F.B. Impact of land uses on water quality in Malaysia: A review. *Ecol. Process.* **2019**, *8*, 1–10. [CrossRef]
176. Aziz, K.H.H.; Mustafa, F.S.; Omer, K.M.; Hama, S.; Hamarawf, R.F.; Rahman, K.O. Heavy metal pollution in the aquatic environment: Efficient and low-cost removal approaches to eliminate their toxicity: A review. *RSC Adv.* **2023**, *13*, 17595–17610. [CrossRef] [PubMed]
177. Setia, R.; Dhaliwal, S.S.; Kumar, V.; Singh, R.; Kukal, S.S.; Pateriya, B. Impact assessment of metal contamination in surface water of Sutlej River (India) on human health risks. *Environ. Poll.* **2020**, *265*, 114907. [CrossRef] [PubMed]
178. Zamora-Ledezma, C.; Negrete-Bolagay, D.; Figueroa, F.; Zamora-Ledezma, E.; Ni, M.; Alexis, F.; Guerrero, V.H. Heavy metal water pollution: A fresh look about hazards, novel and conventional remediation methods. *Environ. Tech. Innov.* **2021**, *22*, 101504. [CrossRef]
179. Pimm, S.L.; Jenkins, C.N.; Abell, R.; Brooks, T.M.; Gittleman, J.L.; Joppa, L.N.; Raven, P.H.; Roberts, C.M.; Sexton, J.O. The biodiversity of species and their rates of extinction, distribution, and protection. *Science* **2014**, *344*, 1246752. [CrossRef] [PubMed]
180. Li, W.P.; Liu, Z.F.; Guo, T.; Chen, H.; Xie, X. Using Optimal Environmental DNA Method to Improve the Fish Diversity Survey—From Laboratory to Aquatic Life Reserve. *Water* **2021**, *13*, 1468. [CrossRef]
181. Nørgaard, L.; Olesen, C.R.; Trøjelsgaard, K.; Pertoldi, C.; Nielsen, J.L.; Taberlet, P.; Ruiz-González, A.; De Barba, M.; Iacolina, L. EDNA metabarcoding for biodiversity assessment, generalist predators as sampling assistants. *Sci. Rep.* **2021**, *11*, 6820. [CrossRef]
182. Barnosky, A.D.; Matzke, N.; Tomiya, S.; Wogan, G.O.; Swartz, B.; Quental, T.B.; Marshall, C.; McGuire, J.L.; Lindsey, E.L.; Maguire, K.C.; et al. Has the Earth's sixth mass extinction already arrived? *Nature* **2011**, *471*, 51–57. [CrossRef]

183. Yasuhara, M.; Wei, C.L.; Kucera, M.; Costello, M.J.; Tittensor, D.P.; Kiessling, W.; Bonebrake, T.C.; Tabor, C.R.; Feng, R.; Baselga, A.; et al. Past and future decline of tropical pelagic biodiversity. *Proc. Natl. Acad. Sci. USA* **2020**, *117*, 12891–12896. [CrossRef]
184. Stafford, R.; Jones, P.J. Viewpoint–Ocean plastic pollution: A convenient but distracting truth? *Mar. Policy* **2019**, *103*, 187–191. [CrossRef]
185. Trathan, P.N.; García-Borboroglu, P.; Boersma, D.; Bost, C.A.; Crawford, R.J.; Crossin, G.T.; Cuthbert, R.J.; Dann, P.; Davis, L.S.; De La Puente, S.; et al. Pollution, habitat loss, fishing, and climate change as critical threats to penguins. *Conserv. Biol.* **2015**, *29*, 31–41. [CrossRef]
186. Kaimuddin, A.H.; Laë, R.; Tito De Morais, L. Fish species in a changing world: The route and timing of species migration between tropical and temperate ecosystems in Eastern Atlantic. *Front. Mar. Sci.* **2016**, *3*, 162. [CrossRef]
187. Desiderà, E.; Guidetti, P.; Panzalis, P.; Navone, A.; Valentini-Poirrier, C.A.; Boissery, P.; Gervaise, C.; Di Iorio, L. Acoustic fish communities: Sound diversity of rocky habitats reflects fish species diversity. *Mar. Ecol. Prog. Ser.* **2019**, *608*, 183–197. [CrossRef]
188. Staehr, P.A.; Dahl, K.; Buur, H.; Göke, C.; Sapkota, R.; Winding, A.; Panova, M.; Obst, M.; Sundberg, P. Environmental DNA Monitoring of Biodiversity Hotspots in Danish Marine Waters. *Front. Mar. Sci.* **2022**, *8*, 800474. [CrossRef]
189. Carvalho, C.O.; Gromstad, W.; Dunthorn, M.; Karlsen, H.E.; Schrøder-Nielsen, A.; Ready, J.S.; Haugaasen, T.; Sørnes, G.; de Boer, H.; Mauvisseau, Q. Harnessing eDNA metabarcoding to investigate fish community composition and its seasonal changes in the Oslo fjord. *Sci. Rep.* **2024**, *14*, 10154. [CrossRef]
190. Liu, Q.; Zhang, Y.; Wu, H.; Liu, F.; Peng, W.; Zhang, X.; Chang, F.; Xie, P.; Zhang, H. A review and perspective of EDNA application to eutrophication and HAB control in freshwater and marine ecosystems. *Microorganisms* **2020**, *8*, 417. [CrossRef]
191. Ficetola, G.F.; Miaud, C.; Pompanon, F.; Taberlet, P. Species detection using environmental DNA from water samples. *Biol. Lett.* **2008**, *4*, 423–425. [CrossRef]
192. Nagai, S.; Hida, K.; Urushizaki, S.; Onitsuka, G.; Yasuike, M.; Nakamura, Y.; Fujiwara, A.; Tajimi, S.; Kimoto, K.; Kobayashi, T.; et al. Influences of diurnal sampling bias on fixed-point monitoring of plankton biodiversity determined using a massively parallel sequencing-based technique. *Gene* **2016**, *576*, 667–675. [CrossRef]
193. Sildever, S.; Kawakami, Y.; Kanno, N.; Kasai, H.; Shiomoto, A.; Katakura, S.; Nagai, S. Toxic HAB species from the Sea of Okhotsk detected by a metagenetic approach, seasonality and environmental drivers. *Harmful Algae* **2019**, *87*, 101631. [CrossRef]
194. Iacolina, L.; Lukassen, M.B.; Fløjgaard, C.; Buttenschøn, R.; Nielsen, J.L.; Pertoldi, C. EDNA and metabarcoding for rewilding projects monitoring, a dietary approach. *Mamm. Biol.* **2020**, *100*, 411–418. [CrossRef]
195. Taberlet, P.; Bonin, A.; Zinger, L.; Coissac, E. *Environmental DNA for Biodiversity Research and Monitoring*; Oxford University Press: Oxford, UK, 2018.
196. Rees, H.C.; Maddison, B.C.; Middleditch, D.J.; Patmore, J.R.; Gough, K.C. The detection of aquatic animal species using environmental DNA–a review of EDNA as a survey tool in ecology. *J. Appl. Ecol.* **2014**, *51*, 1450–1459. [CrossRef]
197. Xu, N.; Zhu, B.; Shi, F.; Shao, K.; Que, Y.; Li, W.; Li, W.; Jiao, W.; Tian, H.; Xu, D.; et al. Monitoring seasonal distribution of an endangered anadromous sturgeon in a large river using environmental DNA. *Sci. Nat.* **2018**, *105*, 62. [CrossRef] [PubMed]
198. Laramie, M.B.; Pilliod, D.S.; Goldberg, C.S. Characterizing the distribution of an endangered salmonid using environmental DNA analysis. *Biol. Conserv.* **2015**, *183*, 29–37. [CrossRef]
199. Xing, Y.; Gao, W.; Shen, Z.; Zhang, Y.; Bai, J.; Cai, X.; Ouyang, J.; Zhao, Y. Review of Environmental DNA Field and Laboratory Protocols Applied in Fish Ecology and Environmental Health. *Front. Environ. Sci.* **2022**, *10*, 725360. [CrossRef]
200. Qu, C.; Stewart, K.A.; Clemente-Carvalho, R.; Zheng, J.; Wang, Y.; Gong, C.; Ma, L.; Zhao, J.; Lougheed, S.C. Comparing fish prey diversity for a critically endangered aquatic mammal in a reserve and the wild using EDNA metabarcoding. *Sci. Rep.* **2020**, *10*, 16715. [CrossRef] [PubMed]
201. Muha, T.P.; Rodriguez-Barreto, D.; O'Rorke, R.; Garcia de Leaniz, C.; Consuegra, S. Using EDNA metabarcoding to monitor changes in fish community composition after barrier removal. *Front. Ecol. Evol.* **2021**, *9*, 629217. [CrossRef]
202. Valdivia-Carrillo, T.; Rocha-Olivares, A.; Reyes-Bonilla, H.; Domínguez-Contreras, J.F.; Munguia-Vega, A. Integrating EDNA metabarcoding and simultaneous underwater visual surveys to describe complex fish communities in a marine biodiversity hotspot. *Mol. Ecol. Resour.* **2021**, *21*, 1558–1574. [CrossRef]
203. Sales, N.G.; Wangensteen, O.S.; Carvalho, D.C.; Deiner, K.; Præbel, K.; Coscia, I.; McDevitt, A.D.; Mariani, S. Space-time dynamics in monitoring neotropical fish communities using EDNA metabarcoding. *Sci. Total Environ.* **2021**, *754*, 142096. [CrossRef]
204. Rohtehrá, S.; Tournayre, O.; Windle, M.J.; Cormier, E.; Schwartz, K.; Wylie-Arbic, M.; Rundle, E.; Perron, M.A.; Francis, A.; Lougheed, S.C. Nearshore fish diversity changes with sampling method and human disturbance: Comparing eDNA metabarcoding and seine netting along the Upper St. Lawrence River. *J. Great Lakes Res.* **2024**, *50*, 102317.
205. Valdez-Moreno, M.; Ivanova, N.V.; Elías-Gutiérrez, M.; Pedersen, S.L.; Bessonov, K.; Hebert, P.D. Using EDNA to biomonitor the fish community in a tropical oligotrophic lake. *PLoS ONE* **2019**, *14*, e0215505. [CrossRef]
206. Beng, K.C.; Corlett, R.T. Applications of environmental DNA (eDNA) in ecology and conservation: Opportunities, challenges and prospects. *Biodivers. Conserv.* **2020**, *29*, 2089–2121.

207. Wee, A.K.; Salmo Iii, S.G.; Sivakumar, K.; Then, A.Y.; Basyuni, M.; Fall, J.; Habib, K.A.; Isowa, Y.; Leopardas, V.; Peer, N.; et al. Prospects and challenges of environmental DNA (eDNA) metabarcoding in mangrove restoration in Southeast Asia. *Front. Mar Sci.* **2023**, *10*, 1033258. [CrossRef]
208. Ahmed, N.; Lü, H.; Ahmed, S.; Adeyeri, O.E.; Ali, S.; Hussain, R.; Shah, S. Transboundary River Water Availability to Ravi Riverfront under Changing Climate: A Step towards Sustainable Development. *Sustainability* **2023**, *15*, 3526. [CrossRef]
209. Qazlbash, S.K.; Zubair, M.; Manzoor, S.A.; ul Haq, A.; Baloch, M.S. Socioeconomic determinants of climate change adaptations in the flood-prone rural community of Indus Basin, Pakistan. *Environ. Dev.* **2021**, *37*, 100603. [CrossRef]
210. Thomsen, P.F.; Willerslev, E. Environmental DNA–An emerging tool in conservation for monitoring past and present biodiversity. *Biol. Conserv.* **2015**, *183*, 4–18. [CrossRef]
211. Deiner, K.; Fronhofer, E.A.; Mächler, E.; Walser, J.C.; Altermatt, F. Environmental DNA reveals that rivers are conveyer belts of biodiversity information. *Nat. Commun.* **2016**, *7*, 12544. [CrossRef]
212. Addy, S.; Cooksley, S.; Dodd, N.; Waylen, K.; Stockan, J.; Byg, A.; Holstead, K. *River Restoration and Biodiversity*; IUCN: Gland, Switzerland, 2016; pp. 1–52.
213. Nakamichi, T.; Ono, M.; Hayashi, M.; Okamura, T.; Wada, T.; Saitoh, K. Environmental DNA Analysis in a River Detected a Possible Distribution of Fish Species Difficult to Capture. *Fishes* **2023**, *8*, 496. [CrossRef]
214. Boyd, S.H.; Niemiller, K.D.K.; Dooley, K.E.; Nix, J.; Niemiller, M.L. Using environmental DNA methods to survey for rare groundwater fauna: Detection of an endangered endemic cave crayfish in northern Alabama. *PLoS ONE* **2020**, *15*, e0242741. [CrossRef]
215. Yamanaka, H.; Minamoto, T. The use of environmental DNA of fishes as an efficient method of determining habitat connectivity. *Ecol. Indic.* **2016**, *62*, 147–153. [CrossRef]
216. Coster, S.S.; Dillon, M.N.; Moore, W.; Merovich, G.T., Jr. The update and optimization of an eDNA assay to detect the invasive rusty crayfish (Faxonius rusticus). *PLoS ONE* **2021**, *16*, e0259084. [CrossRef]
217. Wiggins, J.J.; Tobias, V.D.; Holcombe, E.F.; Karpenko, K.; Huber, E.R.; Goodman, A.C. Leveraging environmental DNA (eDNA) to optimize targeted removal of invasive fishes. *J. Freshw. Ecol.* **2024**, *39*, 2378841. [CrossRef]
218. Miller, A.D.; Roxburgh, S.H.; Shea, K. How frequency and intensity shape diversity–disturbance relationships. *Proc. Natl. Acad. Sci. USA* **2011**, *108*, 5643–5648. [CrossRef] [PubMed]
219. Gui, H.; Hou, L.; Wang, J.; Dong, X.; Han, S. Flood changed the community composition and increased the importance of stochastic process of vegetation and seed bank in a riparian ecosystem of the Yellow River. *Ecol. Indic.* **2023**, *154*, 110505. [CrossRef]
220. Lacoursière-Roussel, A.; Côté, G.; Leclerc, V.; Bernatchez, L. Quantifying relative fish abundance with eDNA: A promising tool for fisheries management. *J. Appl. Ecol.* **2016**, *53*, 1148–1157. [CrossRef]
221. Rourke, M.L.; Fowler, A.M.; Hughes, J.M.; Broadhurst, M.K.; DiBattista, J.D.; Fielder, S.; Wilkes Walburn, J.; Furlan, E.M. Environmental DNA (eDNA) as a tool for assessing fish biomass: A review of approaches and future considerations for resource surveys. *Environ. DNA* **2022**, *4*, 9–33. [CrossRef]
222. Takahara, T.; Minamoto, T.; Yamanaka, H.; Doi, H.; Kawabata, Z.I. Estimation of fish biomass using environmental DNA. *PLoS ONE* **2012**, *7*, e35868. [CrossRef]
223. Cristescu, M.E.; Hebert, P.D. Uses and misuses of environmental DNA in biodiversity science and conservation. *Annu. Rev. Ecol. Evol. Syst.* **2018**, *49*, 209–230. [CrossRef]
224. Yang, N.; Jin, D.; Govindarajan, A.F. Applying environmental DNA approaches to inform marine biodiversity conservation: The case of the Ocean Twilight Zone. *Mar. Policy* **2024**, *165*, 106151. [CrossRef]
225. Papaioannou, C.; Geladakis, G.; Kommata, V.; Batargias, C.; Lagoumintzis, G. Insights in Pharmaceutical Pollution: The Prospective Role of eDNA Metabarcoding. *Toxics* **2023**, *11*, 903. [CrossRef]
226. Rishan, S.T.; Kline, R.J.; Rahman, M.S. Exploitation of environmental DNA (eDNA) for ecotoxicological research: A critical review on eDNA metabarcoding in assessing marine pollution. *Chemosphere* **2024**, *351*, 141238. [CrossRef]

animals

MDPI

Article

Revealing Phenotypic Differentiation in *Ochetobius elongatus* from the Middle Yangtze River Through Geometric Morphometrics

Fangtao Cai [1,†], Zhiyuan Qi [1,†], Ziheng Hu [1], Dongdong Zhai [1,2], Yuanyuan Chen [1], Fei Xiong [1,2] and Hongyan Liu [1,2,*]

1 School of Life Sciences, Hubei Engineering Research Center for Protection and Utilization of Special Biological Resources in the Hanjiang River Basin, Jianghan University, Wuhan 430056, China; 18311558834@163.com (F.C.); q15055974063@163.com (Z.Q.); hzh01f@163.com (Z.H.); zhaidongdong@jhun.edu.cn (D.Z.); yychen@jhun.edu.cn (Y.C.); xiongfei@jhun.edu.cn (F.X.)

2 Hubei Key Laboratory of Environmental and Health Effects of Persistent Toxic Substances, Jianghan University, Wuhan 430056, China

* Correspondence: lhy9603@126.com

† These authors contributed equally to this work.

Simple Summary: Phenotypic differentiation reflected species adaptation to different habitats. *Ochetobius elongatus* was once an economically important species in the south of the Yangtze River, China. However, due to environmental degradation and human activities, its wild populations have declined drastically, and it is now classified as a critically endangered (CR) species. Following the implementation of the Ten-Year Fishing Ban in the Yangtze River, *O. elongatus* has reappeared in certain sections of the mainstem, the tributary and the river-connected lakes of the middle Yangtze River, which could provide valuable samples for studying its morphological variation and habitat adaptation. Using geometric morphometrics based on 14 landmarks, this study revealed significant phenotypic differentiation among *O. elongatus* populations from distinct habitats. These results indicated that phenotypic plasticity might underlie their adaptation to diverse aquatic environments. The findings provided important insights into the survival strategies of this species and could support its conservation and restoration efforts.

Abstract: *Ochetobius elongatus*, a critically endangered (CR) fish species of the Yangtze River Basin in China, has experienced a severe decline in its wild population. Understanding its mechanisms of phenotypic variation is essential for developing effective conservation and restoration strategies. Using geometric morphometrics based on 14 landmarks, we examined the phenotypic difference among five populations from the mainstem, the tributary, and the river-connected lakes of the middle Yangtze River. The results showed that significant phenotypic divergence was detected between river and lake populations. River individuals exhibited a more elongated body, smaller head, inferior mouth position, larger operculum, and narrower caudal peduncle, whereas lake individuals showed a deeper body, and anterior shift in the origin of pelvic fin. The first canonical variable effectively distinguished river and lake populations, with the accuracy of both original and cross-validation classification exceeding 90%, indicating that habitat heterogeneity was the primary driver of phenotypic differentiation. No significant correlation was found between morphological distance and geographical distance. Water temperature, flow velocity, water depth, and food abundance significantly influenced phenotypic variation, but their individual effects were limited, which suggested that environmental shaping of morphology depended more on synergistic effects. Our findings provide important insights into the adaptive evolution of this critically endangered species and offer a scientific basis for conservation efforts.

Keywords: *Ochetobius elongatus*; geometric morphometrics; phenotypic plasticity; phenotypic differentiation; the endangered species

1. Introduction

Intraspecific phenotypic variation, often shaped by selective pressures from different habitats, is a key strategy for species to adapt to heterogeneous environment [1–3]. As the most morphologically diverse group of vertebrates, fish are widely distributed across diverse aquatic habitats, making them ideal models for understanding the mechanisms by which habitat variation causes phenotypic differences [4,5]. Fish were highly sensitive to environmental changes, and their morphology could be influenced by factors such as water temperature, flow velocity, and food abundance [6,7]. Phenotypic plasticity enabled fish to respond rapidly to different habitats, resulting in obviously adaptive morphological changes [8,9]. For instance, flow gradients could directly induce phenotypic plasticity in species such as *Curimata mivartii* and *Pimelodus grosskopfii*, which modified their body shape to cope with varying water velocity [10]. Variations in food resources could drive adaptive phenotypic divergence in *Salvelinus alpinus*, with planktivorous populations exhibiting a deeper body shape, while benthivorous populations showed a slender body and larger head [11]. Habitat pressure influenced body structure, as seen in *Caquetaia kraussii*, where populations in river sections displayed a slenderer body and a larger operculum compared to those in shallow swamp areas [12].

The middle Yangtze River forms a complex network comprising the mainstem, the tributary, and the river-connected lakes (e.g., Dongting Lake and Poyang Lake), which together create a distinct habitat gradient [13,14]. The mainstem are characterized by high flow velocity and deep channels with limited food resources, the tributary exhibit moderate flow and depth, supporting richer plankton and benthic communities, and the river-connected lakes experience large water-level fluctuations and contain extensive wetland areas with high biodiversity and substantial biomass of food organisms [15,16]. These systematic differences in hydrology and resource availability may exert selective pressures on fish survival, feeding, and growth, potentially driving significant phenotypic adaptive divergence across these habitats.

O. elongatus was once an economically important species in the south of the Yangtze River, China [17,18]. However, due to habitat fragmentation, fishing pressure, and environmental pollution, its wild populations have declined drastically over the past four decades [19–21]. By the late 20th century, the species has nearly disappeared from field surveys. Now it is classified as a critically endangered (CR) species by both the China Red List of Vertebrates and the China Biodiversity Red List [22,23]. Following the implementation of the Ten-Year Fishing Ban in the Yangtze River in 2021, *O. elongatus* has reappeared in certain sections of the mainstem, the tributary, and the river-connected lakes of the middle Yangtze River, which offers a potential opportunity for population recovery. To develop effective conservation and restoration strategies, it is essential to clarify the population structure and ecological divergence within the species. Zhai et al. [24] found no significant genetic differentiation among populations in the middle Yangtze River using molecular markers, but genetic homogeneity was not synonymous with phenotypic homogeneity [7]. Therefore, examining phenotypic divergence through morphometric approaches can provide a scientific basis for precisely targeted conservation measures.

Geometric morphometrics is an analytical approach that quantifies biological morphological features through mathematical modeling and statistical methods to decipher patterns of phenotypic variation, their underlying causes, and ecological adaptations [5,25]. The methodology encompasses linear measurement, truss network, and geometric morphometrics [26,27]. In our previous study, we used the methods linear measurement and truss network for a preliminary morphological comparison of the populations of *O. elongatus*. However, as the two methods rely on linear variables, the results were highly correlated with individual size of the fish, thus making it difficult to effectively disentangle size-related changes from genuine shape variation. Although allometric adjustments (e.g., isometric standardization) can partially mitigate size effects, the traditional morphometrics remain limited in resolving subtle shape differences [26,28]. Geometric morphometrics use digitally annotated landmarks to accurately capture spatial configurations of biological structures. Generalized Procrustes Analysis (GPA) have effectively eliminated non-shape variations related to size, position, and orientation [29,30]. This enables more direct and intuitive quantification and comparison of morphological variation. Consequently, geometric morphometrics has become a predominant tool for studies of population discrimination and phenotypic divergence in fishes [27,31,32].

This study used geometric morphometrics to analyze phenotypic differences of the populations of *O. elongatus* in the mainstem, the tributary, and the river-connected lakes in the middle Yangtze River. The aims were to clarify the following questions: (1) whether significant phenotypic divergence existed among the populations from different habitats, (2) what were the key anatomical regions of phenotypic differentiation and their potential adaptive implications, and (3) whether the phenotypic variation correlated with geographical distance and environmental factors.

2. Materials and Methods

2.1. Study Area and Sampling

The study area encompassed three typical habitats in the middle Yangtze River: the mainstem river (Yichang and Jiayu section), the tributary river (Qianjiang section), and the river-connected lakes (Dongting and Poyang Lake) (Figure 1).

Sampling was conducted from 2022 to 2024 across these regions, covering two periods each year: May to July and September to November. In each region, a fish survey was performed continuously for 10 days per period. Over the three-year study, each region accumulated a total of 60 sampling days. All sampling activities were carried out under a special scientific fishing permit and under the supervision of local fisheries authorities, using stationary gillnets of multiple mesh sizes (2, 6, 10, and 14 cm), each of which measured 50 m in length and 2 m in height. Species identification followed the criteria in Fauna Sinica: Osteichthyes, Cypriniformes (Volume II) [17]. Biological measurements were taken immediately after capture. A total of 95 valid specimens were obtained (Table S1), which were frozen and stored for subsequent geometric morphometric analysis.

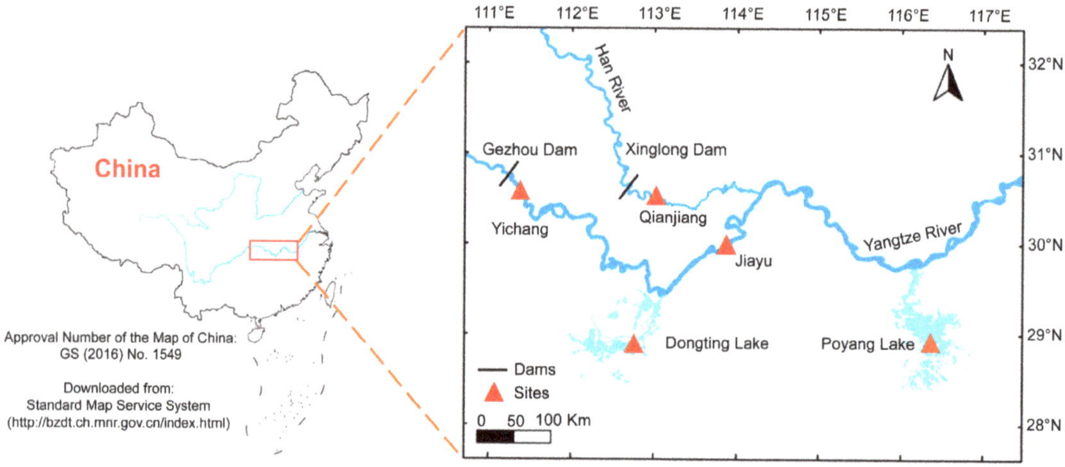

Figure 1. Sampling sites for *O. elongatus* in the middle Yangtze River. Map source: http://bzdt.ch.mnr.gov.cn/index.html, accessed on 20 May 2025.

2.2. Data Collecting and Processing

Lateral morphological images of *O. elongatus* were captured vertically using a digital camera, with a fixed height and focal length. Specimens were positioned to ensure natural body extension, full fin expansion, and alignment of the body axis parallel to the image plane. A standardized imaging with a light-colored background was used to minimize postural variation. Specimen placement, camera angle, and focal point were kept consistent across sessions. A scale bar (accuracy: 1 mm) was included for spatial calibration to ensure reproducibility and coordinate consistency. Images were converted to TPS format using tpsUtil (v1.83) software. A total of 14 landmarks were then digitized on each image using tpsDig2 (v2.32) software (Figure 2).

Figure 2. Position of the landmarks used in the geometric morphometric analysis: 1: oral position; 2: distal tip of occiput; 3: origin of dorsal fin; 4: posterior end of dorsal fin base; 5: upper insertion of caudal fin base; 6: distal tip of caudal peduncle; 7: lower insertion of caudal fin base; 8: posterior end of anal fin base; 9: origin of anal fin; 10: origin of pelvic fin; 11: origin of pectoral fin; 12: origin of the preopercle on the ventral; 13: the posterior margin of the eye; and 14: the anterior margin of the eye.

To assess data reliability, 20% of the samples were randomly selected and re-digitized by the same operator after an interval of at least two weeks. Procrustes ANOVA was performed to evaluate sources of variation, and the significance of operator-induced error (p-value) was assessed along with the intraclass correlation coefficient (ICC) to ensure measurement repeatability [33,34].

2.3. Geometric Morphometric Analysis

Generalized Procrustes Analysis (GPA) was performed in MorphoJ (v1.08) to align all specimens by removing differences in position, scale, and orientation, yielding a consensus configuration representing the overall morphology and group-specific mean shapes. Thin-plate spline (TPS) deformation grids were generated to visualize shape changes in key anatomical regions between populations. Principal component analysis (PCA) was conducted based on the Procrustes coordinates. Principal components (PCs) accounting for over 95% of cumulative variance were retained. Canonical variate analysis (CVA) was utilized to quantify phenotypic differences among the populations, with Mahalanobis and Procrustes distances used as metrics. The Mahalanobis distance, derived from the covariance matrix, measured the degree of morphological overlap among groups by evaluating the deviation of an individual relative to others. The Procrustes distance was used to assess morphological divergence between group mean shapes. Statistical significance was assessed by 10,000 permutation tests. To examine the effect of habitat type on morphology, CVA was performed with specimens grouped into river and lake habitats, and the accuracy of both original and cross-validated classification were computed.

All statistical analyses were carried out in R (v4.4.1) and MorphoJ (v1.08) software. The geomorph R package (v4.0.1) was used for landmark data analysis and visualization, while the vegan (v2.7-1) and pls (v2.8-5) R packages facilitated multivariate statistical testing. GPA, PCA, CVA, mean shape calculation, and TPS deformation grid generation were conducted in MorphoJ (v1.08) [25,35].

2.4. Correlation of Morphology with Geographical Distance and Environmental Factors

Mantel test was applied to assess the correlation between morphological distance and geographical distance [36]. The morphological distance matrix was computed as the Euclidean distance between specimens based on GPA-aligned Procrustes coordinates. The geographical distance matrix was calculated from the longitude and latitude coordinates of the sampling sites using the Haversine formula. The correlation was evaluated using the Pearson method, with significance determined by 9999 permutation tests.

Environmental variables included mean annual water temperature, mean annual flow velocity, water depth, and food abundance (Table S2). Data for temperature, flow velocity, and depth were sourced from the National Earth Data Science Center (https://www.geodata.cn, accessed on 1 May 2025). Food abundance was scored on a 0–5 scale according to the Evaluation System for Aquatic Biological Integrity in the Yangtze River Basin (Table S3). Partial least squares regression (PLS) model was implemented to predict phenotypic variation from environmental factors [12]. Geometric morphometric data were first reduced by PCA, and PCs explaining $\geq$ 95% of cumulative variance were retained as response variables. Environmental factors were standardized and one-hot encoded where necessary, then used in PLS modeling against morphological PCs. Model performance was evaluated through cross-validation, and the key environmental variables were identified based on variable importance in projection (VIP).

3. Results

3.1. Data Reliability Validation

The data quality control indicated high consistency in operator re-landmarking (Table S4), with an intraclass correlation coefficient (ICC) of 0.9268, exceeding the acceptable threshold of 0.75 [33]. Procrustes ANOVA (Table 1) revealed no significant measurement error (residual MS = 0.000038 < individual MS = 0.000129). Strong agreement between Procrustes sum of squares (0.0935) and tangent space sum of squares (0.0933) further confirmed the reliability of landmark data for geometric morphometric analyses.

Table 1. Procrustes-based analysis of variance in geometric morphometrics. Values include sum of squares (SS), mean square (MS), degrees of freedom (df), and *F*-statistic.

Centroid Size:							
Effect	**SS**	**MS**	**df**	***F***	***p* (param.)**		
Individual	866,562.758	216,640.690	4	27.43	<0.01		
Residual	710,864.676	7898.496	90				
Shape, Procrustes ANOVA:							
Effect	**SS**	**MS**	**df**	***F***	***p* (param.)**	**Pillai tr.**	***p* (param.)**
Individual	0.012400	0.000129	96	3.45	<0.01	2.36	<0.01
Residual	0.080980	0.000038	2160				

3.2. Geometric Morphometric Differentiation Among Populations

GPA analysis indicated deviations from the consensus configuration in all populations. River populations (Yichang, Jiayu, and Qianjiang) exhibited a more elongated body shape, while lake populations (Dongting and Poyang Lake) showed a relatively deeper body (Figure 3). TPS results further revealed phenotypic variation among populations, with regions of high phenotypic disparity localized in the head (oral position and operculum), the abdomen (origin of pelvic fin) and the tail (caudal peduncle). Specifically, river populations exhibited a smaller head, lower oral position, larger operculum, and narrower caudal peduncle, whereas lake populations were characterized by a more anterior origin of pelvic fin (Figure 4).

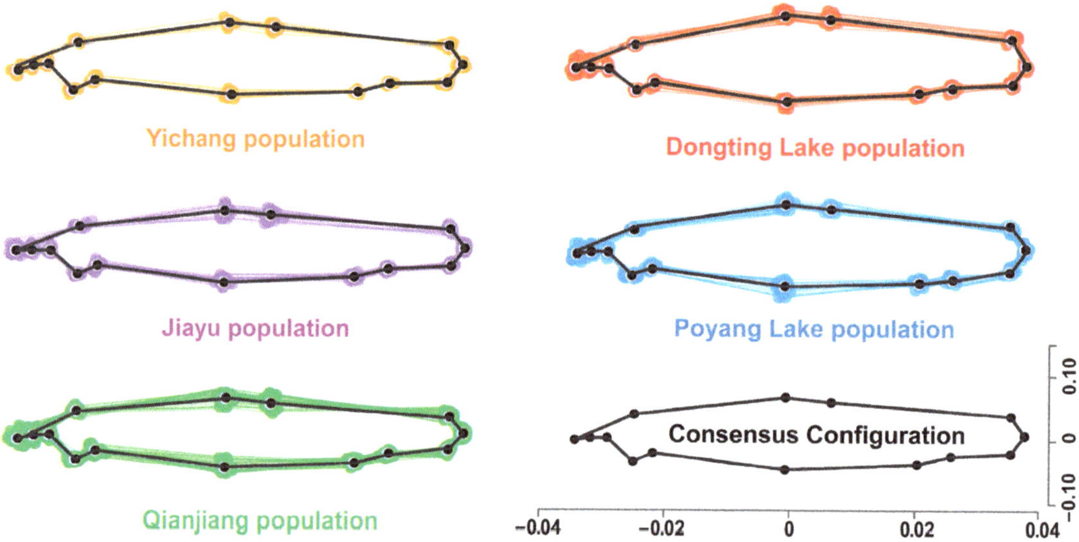

Figure 3. Generalized Procrustes Analysis (GPA) of the *O. elongatus* populations.

The first two canonical variates collectively accounted for 82.53% of the phenotypic variation (CV1: 65.16%; CV2: 17.37%). Populations exhibited clear phenotypic segregation by different habitat types. The mainstem populations (Yichang and Jiayu) clustered toward the negative axis of CV1, and lake populations (Dongting and Poyang Lake) occupied the positive axis of CV1, while the tributary population (Qianjiang) aligned toward the positive axis of CV2 (Figure 5).

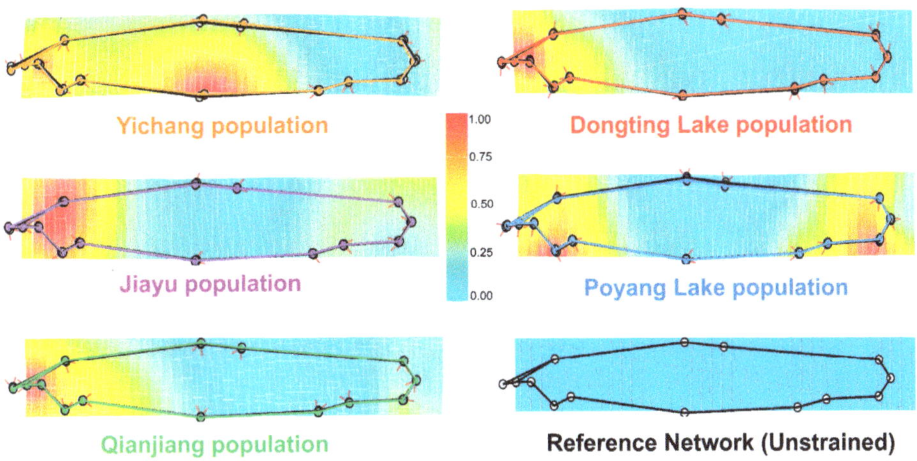

Figure 4. Thin-plate spline analysis (TPS) of the *O. elongatus* populations.

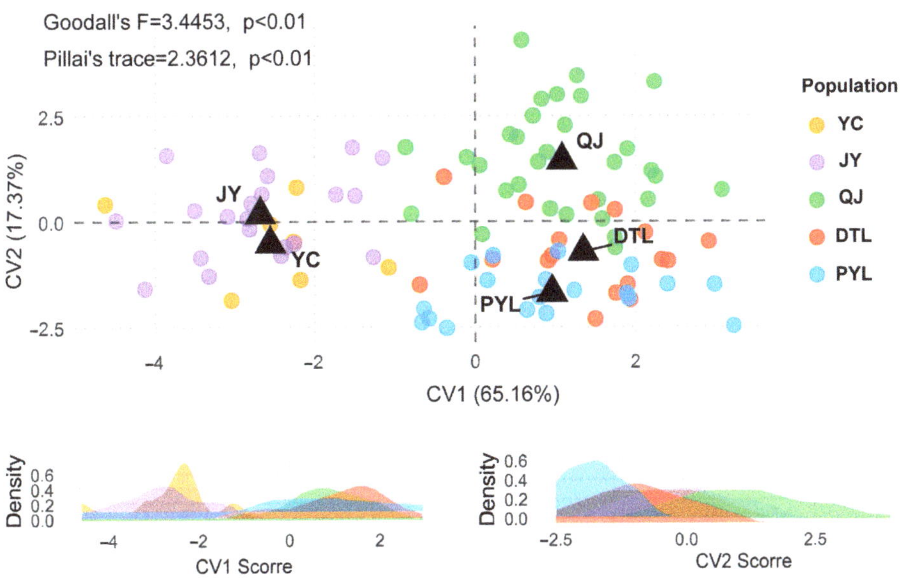

Figure 5. Canonical variate analysis (CVA) of phenotypic variation among *O. elongatus* populations from different habitats. Population abbreviations: YC: Yichang, JY: Jiayu, QJ: Qianjiang, DTL: Dongting Lake, PYL: Poyang Lake.

To clarify phenotypic differences between river and lake habitats of *O. elongatus*, populations from Yichang, Jiayu, and Qianjiang were combined into a river group, while those from Dongting and Poyang Lake were combined into a lake group for CVA analysis. The CV1 alone could explain all variations between the river and lake group (eigenvalue = 3.5325), and significant differentiation was detected between the two groups (Pillai's trace = 0.779, $p < 0.01$; Goodall's $F = 3.743$, $p < 0.05$) (Figure 6). Both the Mahalanobis distance (3.833, $p < 0.01$) and Procrustes distance (0.013, $p < 0.05$) were significant, further supporting pronounced phenotypic differentiation between river and lake populations.

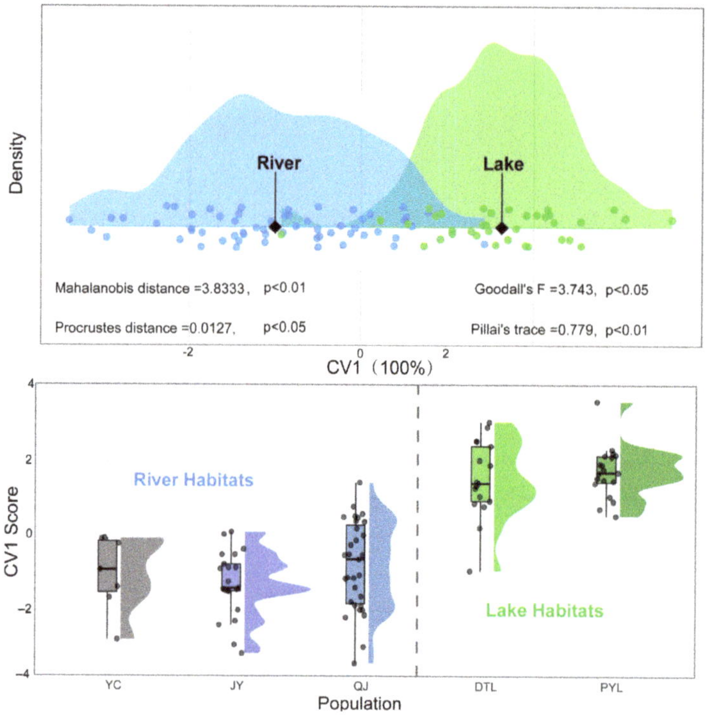

Figure 6. Canonical variate analysis (CVA) of morphometric differentiation in *O. elongatus* between river and lake habitats. Population abbreviations: YC: Yichang, JY: Jiayu, QJ: Qianjiang, DTL: Dongting Lake, PYL: Poyang Lake.

Discriminant analysis based on river and lake habitat type showed high classification accuracy, with original and cross-validated accuracy rates of 95.8% and 91.6% (Table 2), respectively, indicating a strong association between phenotypic variation and river versus lake habitat types.

Table 2. Classification accuracy for *O. elongatus* from river versus lake habitats.

Classification Method	Group	Correct Size	Misclassified Size	Accuracy	Overall Accuracy
Original Classification	River	57	2	96.6%	95.8%
	Lake	34	2	94.4%	
Cross-Validation	River	54	5	91.5%	91.6%
	Lake	33	3	91.7%	

3.3. *Influence of Geographic and Environmental Variables on Morphological Traits*

The Mantel test revealed no significant correlation between morphological distance and geographical distance in populations of *O. elongatus* ($r = 0.07$, $p = 0.08$) (Figure 7).

VIP analysis indicated that temperature, flow velocity, water depth, and food abundance all exerted significant influences on phenotypic variation (VIP $\geq$ 1). However, all VIP values were only marginally above the significance threshold, with limited individual contributions (Figure 8a). The comparison between observed and predicted values demonstrated a weak overall model explanation ($R^2 = 0.043$), indicating that environmental factors accounted for only 4.3% of the total phenotypic variation (Figure 8b).

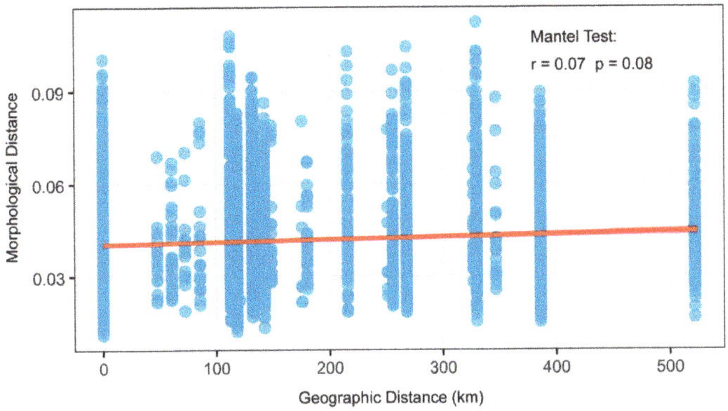

Figure 7. Correlation between morphological distance and geographical distance.

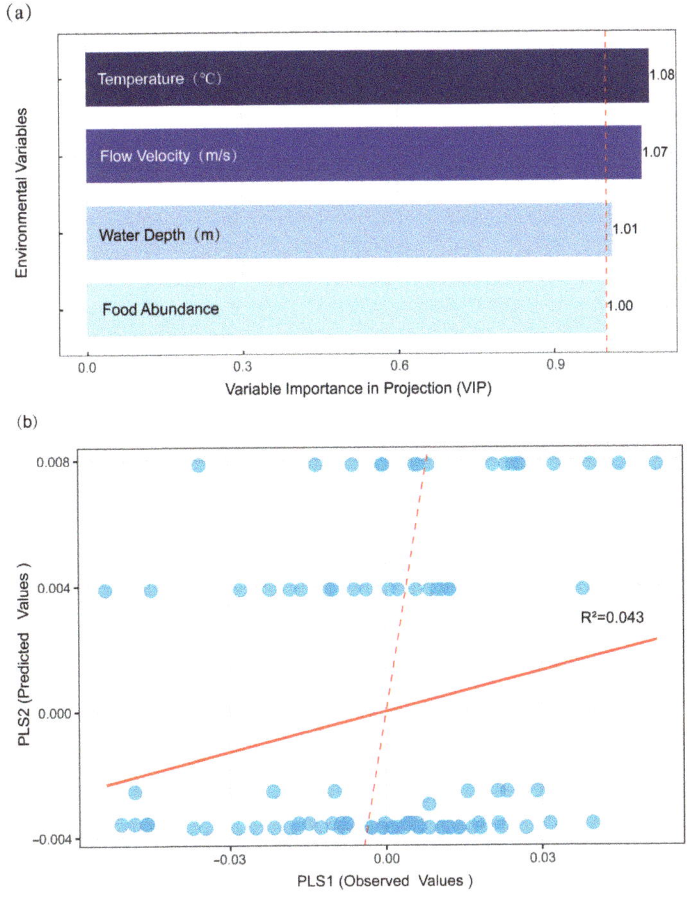

Figure 8. Partial least squares (PLS) regression analysis of phenotypic variation and environmental factors. (**a**) Variable importance in projection (VIP) scores of environmental factors. Factors with VIP ≥ 1 (dashed line) were considered influential. (**b**) Scatter plot of predicted vs. observed morphological values. The solid red line indicated the fitted regression, and the dashed red line represented the line of ideal prediction.

4. Discussion

Phenotypic variation in fish often reflects functional and ecological adaptations. This study revealed significant differentiation across multiple morphological regions among populations of *O. elongatus* in the middle Yangtze River, with the highest divergence observed in the head, the abdomen, and the caudal peduncle. As a key structure for feeding and sensory functions, the variation in head was associated with dietary divergence and foraging strategies. Differential food resources could induce corresponding adaptive head morphologies; for instance, Andersson [37] demonstrated in controlled experiments that *Salvelinus alpinus* feeding on zooplankton developed smaller heads, whereas those consuming benthic chironomid larvae exhibited larger heads. Hernandez et al. [12] suggested that fish in river environments developed larger operculum to cope with oxygen availability. Abdominal morphology, particularly the position of the pelvic fins, was linked to swimming stability and habitat complexity. Andersson et al. [38] reported that fish inhabiting structurally complex environments tended to have more anteriorly positioned pelvic fins, which facilitated rapid stopping, turning and hovering, thereby enhancing maneuverability. Caudal peduncle morphology was closely tied to swimming and escape behavior, and varied with flow velocity. Studies on species such as *Culter alburnus* and *Salmo salar* had identified the caudal peduncle as a region of high variability [39,40].

In this study, river populations (Yichang, Jiayu, and Qianjiang) exhibited a more elongated body shape, whereas lake populations (Dongting and Poyang Lake) had a deeper body and more anterior origin of pelvic fins. The river sections in the middle Yangtze were characterized largely by unidirectional flow and have become structurally simplified due to channel dredging and dam construction. In contrast, Dongting and Poyang Lakes exhibited complex hydrology and diverse habitats. These findings aligned with the documented patterns wherein fish in simple habitats tended to be more elongated, while those in complex environments displayed deeper bodies and anteriorly shifted pelvic fins [41]. Our stomach content analysis indicated that the Jiayu and Qianjiang populations primarily consumed zooplankton, whereas the Poyang Lake group incorporated fish into their diet (unpublished data), suggesting a trophic divergence between river and lake populations. This may partly explain the observed differences in head and oral morphology. Furthermore, the larger operculum and narrower caudal peduncle in river populations were consistent with previously reported adaptations to flowing water. Therefore, the phenotypic divergence observed in the head, abdominal, and caudal regions of *O. elongatus* reflected functional adaptations related to locomotion and feeding, illustrating phenotypic plasticity in response to local habitat conditions and resource availability.

The selective role of divergent habitats in driving phenotypic differentiation has been documented across various fish species. For instance, Li et al. [42] identified lake, marine, and river ecotypes of *Coilia nasus*, while Wang et al. [43] reported distinct lake and river ecotypes in *Schizopygopsis stoliczkai*. Similarly, studies by Langerhans et al. [8] and Hernandez et al. [12] on neotropical fishes (*Bryconops caudomaculatus* and *Biotodoma wavrini*) and *Caquetaia kraussii*, revealed phenotypic differentiation between channel–lake and river–swamp habitats, respectively. These findings suggested that such ecological divergence might be widespread among fishes. Our study revealed a consistent pattern in the *O. elongatus* populations, in that phenotypic structures were significantly influenced by habitat type, allowing classification into mainstem, tributary, and lake groups. The phenotypic variation followed a gradient along habitat types, while minimal divergence between the mainstem and tributary populations was shown, so we merged the mainstem and the tributary samples into a river group and the two lake samples into a lake group to enhance clarity in distinguishing river and lake phenotypic differentiation. Discriminant analysis based on river and lake habitat types showed that the CV1 effectively distinguished river and lake groups, with the accuracy of both original and cross-validation classification exceeding 90%. This strongly supported the finding that the phenotypic differentiation between river and lake populations of *O. elongatus* in the middle Yangtze River were not random but reflected significant adaptive divergence.

The Mantel test revealed no significant correlation between morphological distance and geographical distance in the *O. elongatus* populations, indicating that geographical distance was not the primary driver of phenotypic differentiation. This finding contrasted with the common isolation by distance pattern in many freshwater fish species. Zhai et al. [24] indicated extensive gene flow among populations of *O. elongatus* in the middle Yangtze River, with no significant genetic differentiation observed. Previous studies had shown that gene flow between populations could constrain genetic differentiation, but it did not necessarily limit phenotypic divergence resulting from phenotypic plasticity [44]. In this study, significant phenotypic differentiation was detected between river and lake populations, suggesting that habitat heterogeneity was a key factor promoting phenotypic divergence. Analysis of environmental variables indicated that temperature, flow velocity, water depth, and food abundance all significantly influenced phenotypic variation, but each factor exhibited limited independent explanatory power. This implied that environmental shaping of morphology likely depended on synergistic effects of multiple factors, which was consistent with the findings of Hetzel and Forsythe [45] in *Semotilus atromaculatus*, where phenotypic plasticity was often regulated by a combination of environmental cues rather than a single dominant factor. The multi-factor synergy might reflect an adaptive strategy to replicated environment, as these factors were often interdependent (e.g., water temperature affects food distribution; flow velocity is correlated with depth) and might collectively shape phenotype through indirect pathways such as behavioral adjustments and energy allocation. Moreover, the overall explanation of environmental factors was low, indicating that some unquantified factors might contribute to the unexplained phenotypic variation.

5. Conclusions

Our study revealed significant phenotypic differentiation between river and lake populations of *O. elongatus* in the middle Yangtze River, reflecting local adaptation and phenotypic plasticity in response to distinct hydrological environments. The pattern of phenotypic divergence was closely associated with habitat type. The river populations exhibited a more elongated body shape, whereas lake populations displayed a deeper body, with notable variations in functional regions such as the head, abdomen, and caudal peduncle. These variations suggested that *O. elongatus* could modulate its body structure to cope with different flow regimes and adjust head morphology to exploit varied foraging resources, demonstrating phenotypic plasticity that enabled broad adaptation to diverse habitats. These findings provide important insights into the adaptive evolution of this critically endangered species and offer a scientific basis for conservation strategies. Based on our findings, we propose that the *O. elongatus* population in the middle Yangtze River should be divided into two management units: a riverine unit and a lacustrine unit. For the riverine unit, we recommend maintaining natural hydrological regimes and river connectivity, protecting critical habitats such as spawning sites, and mitigating impacts from dams and embankments. For the lacustrine unit, efforts should emphasize integrated aquatic ecosystem conservation, including water quality management and regulation of water levels, to sustain its adapted habitats.

Supplementary Materials: The following supporting information can be downloaded at: https://www.mdpi.com/article/10.3390/ani15192870/s1, Table S1: Sampling information of *O. elongatus*; Table S2: Geographic coordinates and environmental factors of sampling sites for *O. elongatus* populations in the middle Yangtze River; Table S3: Scoring criteria for the food abundance based on biological indicators; Table S4: Operator consistency analysis summary.

Author Contributions: F.C. and Z.Q. processed the data and prepared the manuscript. Z.H., D.Z. and Y.C. collected the samples and analyzed data. F.X. designed the experiments. H.L. designed the experiments and revised the manuscript. All authors have read and agreed to the published version of the manuscript.

Funding: This research was supported by the Natural Science Foundation Program of Hubei Province (2024AFB461), the Monitoring of Fishery Resources in Hubei Province-Monitoring of Aquatic Living Resources in the Lower Reaches of the Hanjiang River.

Institutional Review Board Statement: All animal protocols have been reviewed and approved by the experimental animal welfare and ethics review committee of Jianghan University (approval no. JHDXKYLL2025-181).

Informed Consent Statement: Not applicable.

Data Availability Statement: Data are contained within the article and Supplementary Materials.

Conflicts of Interest: The authors declare no conflicts of interest.

References

1. Langerhans, R.B. Predictability of phenotypic differentiation across flow regimes in fishes. *Integr. Comp. Biol.* **2008**, *48*, 750–768. [CrossRef]
2. Gómez, J.M.; González-Megías, A.; Armas, C.; Narbona, E.; Navarro, L.; Perfectti, F. The role of phenotypic plasticity in shaping ecological networks. *Ecol. Lett.* **2023**, *26*, S47–S61. [CrossRef]
3. Shuai, F.; Yu, S.; Lek, S.; Li, X. Habitat effects on intra-species variation in functional morphology: Evidence from freshwater fish. *Ecol. Evol.* **2018**, *8*, 10902–10913. [CrossRef]
4. Pigliucci, M. *Phenotypic Plasticity: Beyond Nature and Nurture*; Johns Hopkins University Press: Baltimore, MD, USA, 2001.
5. Schlichting, C.D.; Pigliucci, M. *Phenotypic Evolution: A Reaction Norm Perspective*; Sinauer Associates: Sunderland, MA, USA, 1998.
6. Imre, I.; McLaughlin, R.L.; Noakes, D.L.G. Phenotypic plasticity in brook charr: Changes in caudal fin induced by water flow. *J. Fish Biol.* **2005**, *61*, 1171–1181. [CrossRef]
7. Rovito, S.M. Lineage Divergence and Species Formation in Plethodontid Salamanders. Ph.D. Thesis, University of California, Berkeley, CA, USA, 2009.
8. Langerhans, R.B.; Layman, C.A.; Langerhans, A.K.; DeWitt, T.J. Habitat-associated morphological divergence in two Neotropical fish species. *Biol. J. Linn. Soc.* **2003**, *80*, 689–698. [CrossRef]
9. Franssen, N.R.; Stewart, L.K.; Schaefer, J.F. Morphological divergence and flow-induced phenotypic plasticity in a native fish from anthropogenically altered stream habitats. *Ecol. Evol.* **2013**, *3*, 4648–4657. [CrossRef] [PubMed]
10. Hincapié-Cruz, J.; Márquez, E. Variación fenotípica de los peces *Curimata mivartii* (Characiformes: Curimatidae) y *Pimelodus grosskopfii* (Siluriformes: Pimelodidae) en hábitats lóticos y lénticos. *Rev. Biol. Trop.* **2021**, *69*, 434–444. [CrossRef]
11. Kristjánsson, B.K.; Leblanc, C.A.-L.; Skúlason, S.; Snorrason, S.S.; Noakes, D.L.G. Phenotypic plasticity in the morphology of small benthic Icelandic Arctic charr (*Salvelinus alpinus*). *Ecol. Freshw. Fish* **2018**, *27*, 636–645. [CrossRef]
12. Hernandez, J.; Villalobos-Leiva, A.; Bermúdez, A.; Ahumada, C.D.; Suazo, M.J.; Correa, M.; Díaz, A.; Benítez, H.A. Ecomorphology and Morphological Disparity of *Caquetaia kraussii* (Perciformes: Cichlidae) in Colombia. *Animals* **2022**, *12*, 3438. [CrossRef]
13. Wang, Z.T.; Duan, P.X.; Akamatsu, T.; Chen, Y.W.; An, X.; Yuan, J.; Lei, P.Y.; Li, J.; Zhou, L.; Liu, M.C.; et al. Riverside underwater noise pollution threaten porpoises and fish along the middle and lower reaches of the Yangtze River, China. *Ecotoxicol. Environ. Saf.* **2021**, *226*, 112860. [CrossRef]
14. Du, J.; Tian, H.W.; Xiang, Z.Y.; Zhao, K.S.; Yu, L.Y.; Duan, X.B.; Chen, D.Q.; Xu, J.; Liu, M.D. Impact of the fishing ban on fish diversity and population structure in the middle reaches of the Yangtze River, China. *Front. Environ. Sci.* **2025**, *12*, 1530716. [CrossRef]
15. Cui, L.J.; Gao, C.J.; Zhao, X.S.; Ma, Q.F.; Zhang, M.Y.; Li, W.; Song, H.T.; Wang, Y.F.; Li, S.G.; Zhang, Y. Dynamics of the lakes in the middle and lower reaches of the Yangtze River basin, China, since late nineteenth century. *Environ. Monit. Assess.* **2013**, *185*, 4005–4018. [CrossRef] [PubMed]
16. Chen, J. Hydrological Characteristics of the Yangtze River. In *Evolution and Water Resources Utilization of the Yangtze River*; Springer: Singapore, 2020. [CrossRef]
17. Chen, Y.Y. *Fauna Sinica, Osteichthyes Cypriniformes II*; Science Press: Beijing, China, 1998; pp. 106–107.

18. Chen, W.T.; Duan, X.B.; Gao, L.; Li, X.H.; Yang, J.P.; Wang, D.Q. Genetic structure analysis of *Ochetobius elongatus* between Yangtze River and Pearl River using multiple loci. *South China Fish. Sci.* **2022**, *18*, 19–25. [CrossRef]
19. Yang, J.P.; Li, C.; Chen, W.T.; Li, Y.F.; Li, X.H. Genetic diversity and population demographic history of *Ochetobius elongatus* in the middle and lower reaches of the Xijiang River. *Biodiv. Sci.* **2018**, *26*, 1289–1295. [CrossRef]
20. Yang, J.P.; Li, Y.F.; Zhu, S.L.; Chen, W.T.; Li, J.; Xue, H.M.; Li, X.H. Development and characterization of 26 SNP markers in *Ochetobius elongatus* based on restriction site-associated DNA sequencing (RAD-seq). *Conserv. Genet. Resour.* **2020**, *12*, 53–55. [CrossRef]
21. Li, L.K.; Liang, X.G.; Zhang, J.T.; Xu, Q.; Wang, L.; Gao, X.P.; Song, X.C.; Zhang, B.; Huang, D.; Wang, H.; et al. A chromosome-level genome assembly of critically endangered *Ochetobius elongatus*. *Sci. Data.* **2024**, *11*, 1399. [CrossRef]
22. Jiang, Z.G.; Jiang, J.P.; Wang, Y.Z.; Zhang, E.; Zhang, Y.Y.; Li, L.L.; Xie, F.; Cai, B.; Cao, L.; Zheng, G.M.; et al. Red List of China's vertebrates. *Biodiv. Sci.* **2016**, *24*, 500–551. [CrossRef]
23. Ministry of Ecology and Environment; Chinese Academy of Sciences. *Announcement on Releasing the China Biodiversity Red List: Volume of Vertebrates (2020) and China Biodiversity Red List: Volume of Higher Plants (2020)*; Ministry of Ecology and Environment; Chinese Academy of Sciences: Beijing, China, 2023. Available online: https://www.mee.gov.cn/xxgk2018/xxgk/xxgk01/202305/t20230522_1030745.html (accessed on 20 June 2025).
24. Zhai, D.D.; Wang, D.; He, T.Q.; Cai, F.T.; Duan, X.B.; Hu, X.K.; Zhu, B.; Xie, Z.G.; Xia, M.; Liu, H.Y.; et al. Genetic diversity and population historical dynamics of *Ochetobius elongatus* in the middle reaches of the Yangtze River. *Acta Hydrobiol. Sin.* **2024**, *48*, 1736–1744. [CrossRef]
25. Baken, E.K.; Collyer, M.L.; Kaliontzopoulou, A.; Adams, D.C. geomorph v4.0 and gmShiny: Enhanced analytics and a new graphical interface for a comprehensive morphometric experience. *Methods Ecol. Evol.* **2021**, *12*, 2355–2363. [CrossRef]
26. Zelditch, M.L.; Swiderski, D.L.; Sheets, H.D. *Geometric Morphometrics for Biologists: A Primer*, 2nd ed.; Academic Press: San Diego, CA, USA, 2012. [CrossRef]
27. Adams, D.C.; Otárola-Castillo, E. Geomorph: An R package for the collection and analysis of geometric morphometric shape data. *Methods Ecol. Evol.* **2013**, *4*, 393–399. [CrossRef]
28. Adams, D.C.; Rohlf, F.J. Ecological character displacement in Plethodon: Biomechanical differences found from a geometric morphometric study. *Proc. Natl. Acad. Sci. USA* **2000**, *97*, 4106–4111. [CrossRef] [PubMed]
29. Webb, P.W. Body form, locomotion and foraging in aquatic vertebrates. *Amer. Zool.* **1984**, *24*, 107–120. [CrossRef]
30. Mitteroecker, P.; Schaefer, K. Thirty years of geometric morphometrics: Achievements, challenges, and the ongoing quest for biological meaningfulness. *Am. J. Biol. Anthropol.* **2022**, *178*, 181–210. [CrossRef]
31. Bookstein, F.L. Thin-plate splines and the atlas problem for biomedical images. In *Information Processing in Medical Imaging*; Colchester, A.C.F., Hawkes, D.J., Eds.; Lecture Notes in Computer Science; Springer: Berlin/Heidelberg, Germany, 1991; Volume 511, pp. 326–342. [CrossRef]
32. Zhu, G.P.; Liu, F.Q. Geometric morphometrics and its application in fish ecology: A review. *J. Shanghai Ocean Univ.* **2022**, *31*, 1180–1189. [CrossRef]
33. Koo, T.K.; Li, M.Y.A. Guideline of Selecting and Reporting Intraclass Correlation Coefficients for Reliability Research. *J. Chiropr. Med.* **2016**, *15*, 155–163. [CrossRef]
34. Luo, D. Quantitative Analysis of Fish Morphology Through Landmark and Outline-based Geometric Morphometrics with Free Software. *Bio. Protoc.* **2024**, *14*, e5087. [CrossRef]
35. Klingenberg, C.P. MorphoJ: An integrated software package for geometric morphometrics. *Mol. Ecol. Resour.* **2011**, *11*, 353–357. [CrossRef]
36. Amoutchi, A.I.; Ugbor, O.N.; Kouamelan, E.P.; Mehner, T. Morphological variation of African snakehead (*Parachanna obscura*) populations along climate and habitat gradients in Côte d'Ivoire, West Africa. *Environ. Biol. Fishes.* **2023**, *106*, 1233–1246. [CrossRef]
37. Andersson, J. Effects of diet- induced resource polymorphism on performance in arctic charr (*Salvelinus alpinus*). *Evol. Ecol. Res.* **2003**, *5*, 213–228.
38. Andersson, J.; Bystrom, P.; Persson, L.; De Roos, A.M. Plastic resource polymorphism: Effects of resource availability on Arctic char (*Salvelinus alpinus*) morphology. *Biol. J. Linn. Soc.* **2005**, *85*, 341–351. [CrossRef]
39. Wang, W.; Chen, L.Q.; Gu, Z.M.; Peng, S.M.; Li, Y.K. Analysis of Morphological Variations among Seven Populations of *Erythroculter ilishaeformis*. *Fresh. Fish.* **2007**, *37*, 40–44.
40. Winans, G.A. Multivariate Morphometric Variability in Pacific Salmon: Technical Demonstration. *Can. J. Fish. Aquat. Sci.* **1984**, *41*, 1150–1159. [CrossRef]
41. Hjelm, J.; Svanbäck, R.; Byström, P.; Persson, L.; Wahlström, E. Diet-dependent body morphology and ontogenetic reaction norms in Eurasian perch. *Oikos* **2001**, *95*, 311–323. [CrossRef]
42. Li, Y.; Feng, G.P.; Zhang, T.; Zhuang, P.; Song, C.; Wang, H.H.; Zhang, Y.P.; Chen, J.H. Morphological Characteristics and Cluster Analysis of *Coilia nasus* Populations from Six Different Waters. *J. Hydroecol.* **2022**, *43*, 116–123. [CrossRef]

43. Wang, H.K.Y.; Tang, Y.T.; Li, G.G.; Zhang, R.Y.; Feng, C.G.; Tong, C.; Liu, S.J.; Zhang, C.F.; Tian, F.; Zhao, K. Geometric Morphometrics of the Cephalic Contour and Its Morphological Variation in *Schizopygopsis stoliczkanii* (Teleostei: Cyprinidae). *Acta Hydrobiol. Sin.* **2017**, *41*, 182–193. [CrossRef]
44. Sultan, S.E.; Spencer, H.G. Metapopulation structure favors plasticity over local adaptation. *Am. Nat.* **2002**, *160*, 271–283. [CrossRef]
45. Hetzel, C.; Forsythe, P. Phenotypic plasticity of a generalist fish species resident to lotic environments: Insights from the Great Lakes region. *Ecol. Evol.* **2023**, *13*, e10715. [CrossRef]

MDPI AG
Grosspeteranlage 5
4052 Basel
Switzerland
Tel.: +41 61 683 77 34

MDPI Books Editorial Office
E-mail: books@mdpi.com
www.mdpi.com/books

www.ingramcontent.com/pod-product-compliance
Lightning Source LLC
Chambersburg PA
CBHW061449210926
51850CB00008B/1661
9783725856732